Finite Mathematics, Models, and Structure

Finite Mathematics, Models, and Structure

Revised Printing

WILLIAM J. ADAMS
Pace University

with illustrations by
Ramunė B. Adams

KENDALL/HUNT PUBLISHING COMPANY
4050 Westmark Drive   Dubuque, Iowa 52002

The front cover is reproduced from TDS 188, a painting in The Third Dimensional Series of Kazimieras Zoromskis (b. 1913). It is in the private collection of Ramunė B. Adams.

The author expresses appreciation to Prentice-Hall for permission to reprint selected material from Chapters 4–12 of his book *Fundamentals of Mathematics for Business, Social, and Life Sciences,* published by Prentice-Hall.

Revised Printing 1999

Copyright © 1995, 1999 by William J. Adams

ISBN 0-7872-6444-X

Kendall/Hunt Publishing Company has the exclusive rights to reproduce this work,
to prepare derivative works from this work, to publicly distribute this work,
to publicly perform this work and to publicly display this work.

All rights reserved. No part of this publication may be reproduced,
stored in a retrieval system, or transmitted, in any form or by any
means, electronic, mechanical, photocopying, recording, or otherwise,
without the prior written permission of Kendall/Hunt Publishing Company.

Printed in the United States of America
10 9 8 7 6 5 4 3

*To
Andrius Adams*

Contents

Preface *xi*

Chapter 1
Linear Structures *1*

1.1 Linear Equations and Lines *1*
1.2 Systems of Linear Equations: A First Look *6*
1.3 Application to Finding a Regression Line *12*
1.4 The Tableau Method *20*
1.5 Noteworthy Applications *35*
1.6 Systems with Infinitely Many Solutions *41*
1.7 Inequalities, Linear Inequalities, and Graphs *48*
1.8 Preface to Linear Programming *62*

Chapter 2
Introduction to Mathematical Modeling *65*

2.1 Math Models for a Vacation Trip *65*
2.2 The Mathematical Model Building Process *69*

Chapter 3

Linear Program Models 77

3.1 The Birth and Rebirth of Linear Programming 77
3.2 A Tale of Two Linear Programs 78
3.3 The Corner Point Solution Method 81
3.4 Which Solution Should Be Implemented? 89
3.5 How Could It Be Wrong? I Used a Computer 93
3.6 The Scope of Linear Programming Applications 94
3.7 A Linear Programming Shortfall 112
3.8 Sometimes You Can't Have It All 114
3.9 Problems Requiring Solutions in Integers 116
3.10 Mathematical Duality 130

Chapter 4

The Simplex Method 139

4.1 Solutions of Maximum Type Linear Programs 139
4.2 An Interpretation of the Simplex Method 153
4.3 Solutions of Minimum Type Linear Programs 155

Chapter 5

Life in the Land of Matrices 161

5.1 Algebraic Operations on Matrices 161
5.2 Properties of Matrices 169
5.3 Matrix Inversion 173
5.4 Matrix Solutions to Systems of Linear Equations 177
5.5 Return to Service Charge Allocation and Income Consolidation Problems 180
5.6 Leontief Input-Output Models 183

SELF-TESTS FOR CHAPTERS 1–5 188

Chapter 6

Basic Concepts of Probability 195

- 6.1 Set Notation and Language *195*
- 6.2 Preface to Probability *198*
- 6.3 Structure of a Probability Model *204*
- 6.4 A Tale of Three Probability Models *206*
- 6.5 Probability Models for Random Processes *209*
- 6.6 Derived Events and Their Probabilities *216*
- 6.7 Interpretations of Probability *221*
- 6.8 Probabilities and Odds *226*

Chapter 7

Equally Likely Outcome Models 229

- 7.1 Tools for Counting *229*
- 7.2 Return to Equally Likely Outcomes *239*

Chapter 8

Conditional Probability Models 253

- 8.1 Return to a Production Process Situation *253*
- 8.2 Conditional Probability *255*
- 8.3 Bayes's Theorem *257*
- 8.4 Markov Chains *266*

Chapter 9

Bernoulli Trial Models 279

- 9.1 Independent Events *279*
- 9.2 Bernoulli Trials *282*
- 9.3 Normal Curves *287*
- 9.4 Normal Curve Approximation of Bernoulli Trial Probabilities *295*

SELF-TESTS FOR CHAPTERS 6–9 *301*

Chapter 10

Mathematical Models for Conflict Situations 311

10.1 Introduction to Game Theory *311*
10.2 A Point of Contact with Linear Programming *325*
10.3 Self-Tests for Chapter 10 *334*

Chapter 11

Structure: Validity Versus Truth 337

11.1 Deductive Reasoning, Validity, and Truth *337*
11.2 Dubious Deductions *343*
11.3 More on Proof and Postulate Systems *347*
11.4 Consistency and Independence *355*
11.5 A Classical Problem of Independence and Its Aftermath *358*
11.6 Mathematical Models for Real World Phenomena *365*
11.7 Which Geometry is Right for Space? *368*
11.8 Self-Tests for Chapter 11 *369*

Appendix on Tables 377

1. Table A: Standard Normal Curve Areas *378*

Answers to Odd-Numbered Exercises and Self-Tests 379

Index *433*

Preface

My objective in this book is to present a lucid exposition of topics in mathematics along with a spectrum of applications. These are intended to show off the power of mathematics as a tool for studying the "real" world. The topics undertaken for the most part fall into the category generally called finite mathematics, hence the appearance of this term in the book's title.

Much attention is given to developing the mathematical modeling perspective to make clear what mathematics can and cannot do for us. A point continually stressed on every appropriate occasion is that reaching a mathematical answer to a problem is not the end of the story. It is in a sense the end of a chapter, but the next chapter is concerned with questions of whether and how the mathematical answer should be implemented. Also addressed is the question of what to consider when more than one mathematical answer is obtained for a situation. These issues are most thoroughly explored in the context of Chapters 2 (Introduction to Mathematical Modeling), 3 (Linear Program Models), 6 (Basic Concepts of Probability), and some of the subsequent sections on probability applications as opportunities arise.

Structural issues involving mathematical proof and the all-important distinction between a conclusion's validity and its "real-world" truth are addressed in detail in Chapter 11.

The intended audience is mainly composed of students who wish to pursue studies outside of mathematics itself, but make use of mathematical methods. This includes, but is not limited to, students of business in its various dimensions, economics, education, and the life and social sciences.

Self-tests have been included to help students see the unity of the topics and get blocks of material under better control.

As to which topics are taken up and the order pursued, this will depend on course requirements and the instructor's judgment. Basic chapter dependencies are indicated by the tree diagram that follows.

I am greatly indebted to my daughter Ramunė for preparing the illustrations. I should also like to express my appreciation to my colleague Dr. Joshua Yarmish for his helpful comments and suggestions.

W.J.A.

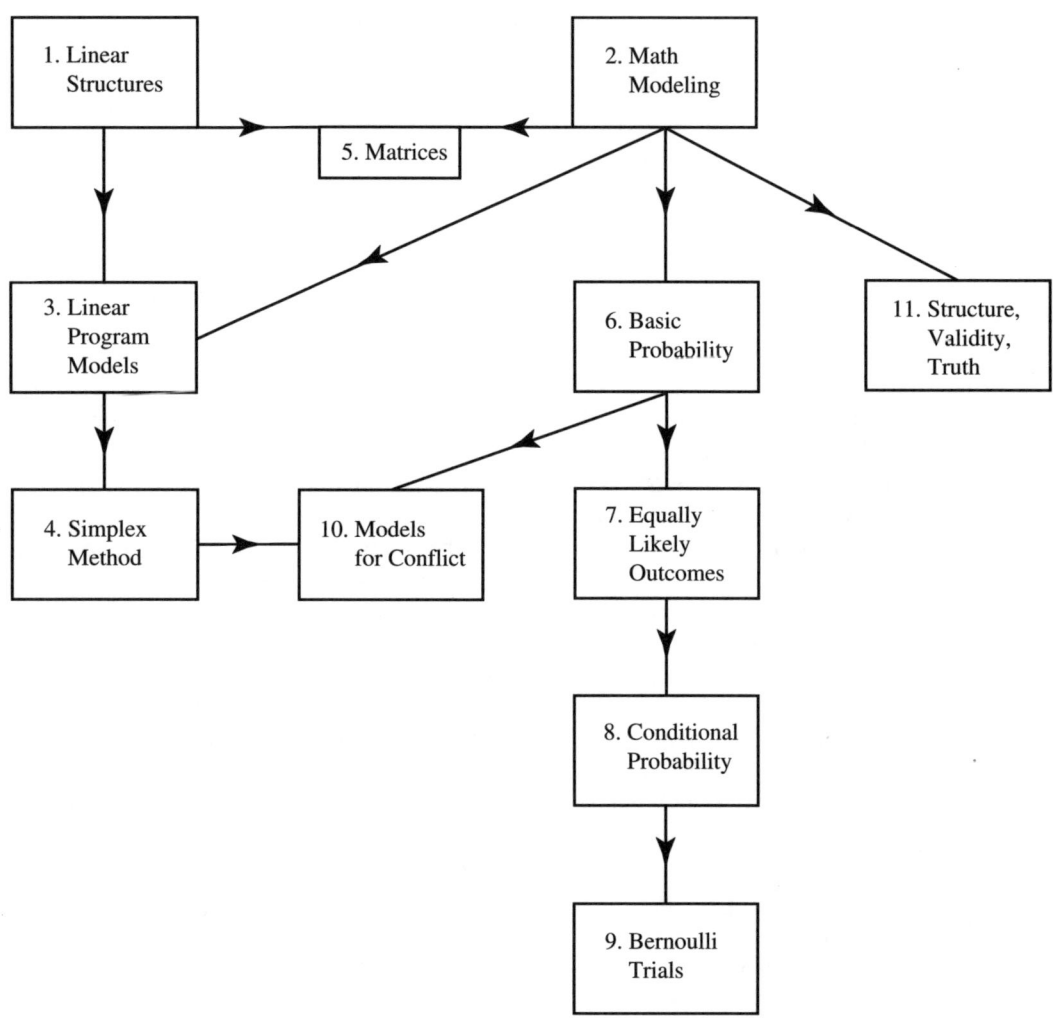

CHAPTER 1

Linear Structures

1.1. LINEAR EQUATIONS AND LINES

The equation $2x + y = 2$ illustrates a **linear equation** in two variables, whose general structure is

$$Ax + By = C$$

where A, B, and C are constants, with not both A and B equal to zero. It is not coincidental that the term "linear" is used here and that $2x + y = 2$ represents a line. The term linear suggests a connection with the line, and this is indeed the case. The graph of every linear equation in two variables

$$Ax + By = C$$

in a line; moreover, every line is the graph of a linear equation in two variables. Because of this intimate connection between two-variable linear equations and lines, it is acceptable practice to simplify language and say, for example, "consider the line $2x + y = 2$," rather than "consider the line that is the graph of $2x + y = 2$."

The problem we turn to now is that of determining the equation of a given line. Let L denote a nonvertical line (see Figure 1.1). L is described by some linear equation that, upon solving for y in terms of x, can be written as

$$y = mx + b$$

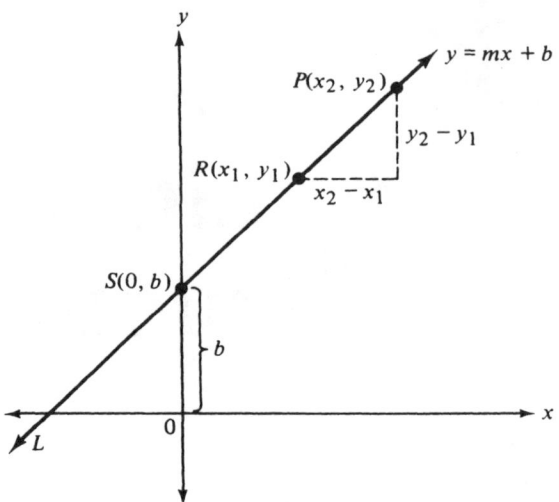

Figure 1.1

where m and b are constants. Let $R(x_1, y_1)$ and $P(x_2, y_2)$ denote two points on L. Then we have
$$y_2 = mx_2 + b$$
and
$$y_1 = mx_1 + b$$
Subtracting and simplifying yields
$$y_2 - y_1 = mx_2 + b - (mx_1 + b)$$
$$y_2 - y_1 = mx_2 + b - mx_1 - b$$
$$y_2 - y_1 = mx_2 - mx_1$$
$$y_2 - y_1 = m(x_2 - x_1)$$
By dividing both sides by $x_2 - x_1$, we obtain
$$\frac{y_2 - y_1}{x_2 - x_1} = m$$

The constant m is called the **slope** of line L. Geometrically, m expresses the ratio of the vertical climb to the horizontal run in going from any one point $R(x_1, y_1)$ on L to any other point $P(x_2, y_2)$ on L. The slope m is a measure of the inclination of line L to the x-axis and is a fundamental characteristic of the line. To determine the equation of L, it suffices to know the slope of L and a point on L, or two points on L.

Let us note that if we substitute 0 for x in the equation $y = mx + b$, $y = b$ is obtained. The constant b is called the **y-intercept** of line L, since it is the y-value of the point of intersection of L with the y-axis (see Figure 1.1).

EXAMPLE 1

Find the equation of the line L that passes through $R(1, 3)$ and has slope 2.

If $P(x, y)$ is any point on L, then
$$y = 2x + b$$

Since $R(1, 3)$ is on L, we have
$$3 = 2(1) + b$$
$$1 = b$$

Thus
$$y = 2x + 1$$

is an equation of line L (see Figure 1.2).

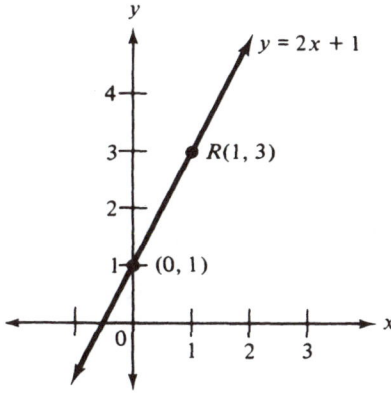

Figure 1.2

EXAMPLE 2

Find the equation of the line L that passes through $R(1, 1)$ and $S(3, 2)$.

We first determine the slope of L from the points $R(1, 1)$ and $S(3, 2)$.
$$m = \frac{2 - 1}{3 - 1} = \frac{1}{2}$$

Thus
$$y = \tfrac{1}{2}x + b$$

Since $R(1, 1)$ is on L, we have
$$1 = \tfrac{1}{2}(1) + b$$
$$\tfrac{1}{2} = b$$

Thus
$$y = \tfrac{1}{2}x + \tfrac{1}{2}$$
is an equation for L (see Figure 1.3).

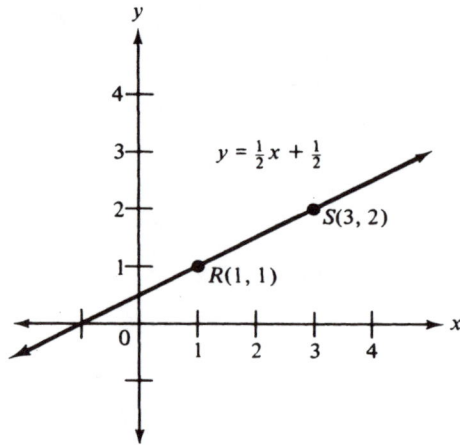

Figure 1.3

Horizontal and Vertical Lines

If L is a horizontal line that passes through the point $S(0, b)$ (see Figure 1.4),

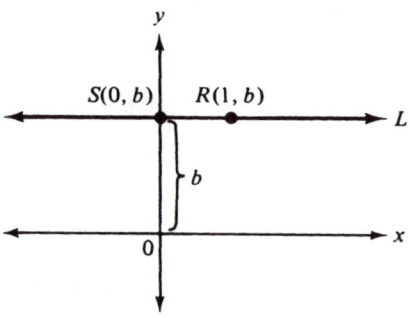

Figure 1.4

then all points on L have y-value b. Thus $R(1, b)$ is another point on L, and the slope of L is
$$m = \frac{b - b}{1 - 0} = \frac{0}{1} = 0$$
As the equation of L, we obtain
$$y = mx + b = 0 + b = b$$

Therefore, $y = 2$ is the equation of the horizontal line that passes through $(0, 2)$, $y = -3$ is the equation of the horizontal line that passes through $(0, -3)$, and $y = 0$ is the equation of the horizontal line that passes through the origin $(0, 0)$, the x-axis (see Figure 1.5).

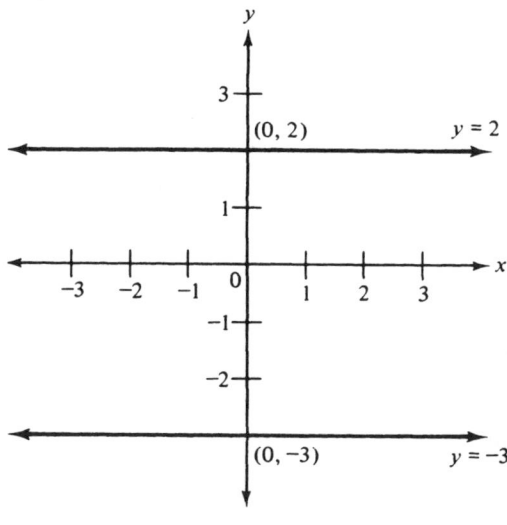

Figure 1.5

The slope of a vertical line, on the other hand, is undefined. To see why this is the case, consider the vertical line L that passes through the point $(1, 0)$ (see Figure 1.6).

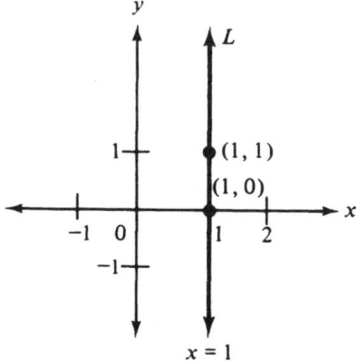

Figure 1.6

Since all points on L have the same x-value, 1, this line is described by the equation $x = 1$. The slope of L is undefined, since the difference between the x-values for any two points is $1 - 1 = 0$, and division by 0 is undefined. More generally, the vertical line L that passes through $(b, 0)$ (see Figure 1.7) is described by the equation $x = b$.

6 Finite Mathematics, Models, and Structure

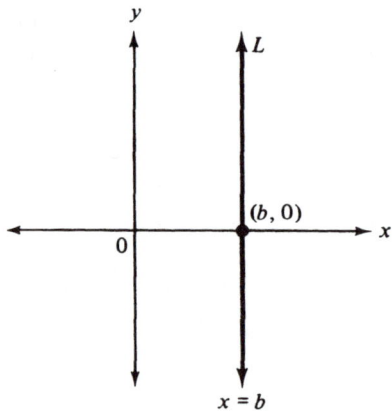

Figure 1.7

EXERCISES

Write an equation of the line satisfying the following conditions.

1. Passing through (1, 4) with slope 2.
2. Passing through (−1, 3) with slope 3.
3. Passing through (2, 1) with slope −1.
4. Passing through (4, − 1) with slope $-\frac{1}{2}$.
5. Passing through (1, − 3) with slope −2.
6. Passing through (−1, 2) and (3, 4).
7. Passing through (1, 1) and (4, 2).
8. Passing through (−2, 1) and (4, 1).
9. Passing through (1, 4) and (1, 5).
10. Passing through (−1, −1) and (2, 5).

1.2. SYSTEMS OF LINEAR EQUATIONS: A FIRST LOOK

In many problems of interest a number of linear equations arise, and it is required to find values of the variables that satisfy all the equations. For example, consider the equations

$$2x + 3y = 14 \quad (1.1)$$
$$x - 2y = -7 \quad (1.2)$$

These equations, considered together as a unit, are said to form a system of two linear equations. A solution of such a system is any ordered pair of numbers, one for x and the other for y, that satisfies both equations in the system. Thus $x = 1$, $y = 4$, also denoted by (1, 4), is a solution of this system since substitution of 1 for x and 3 for y in Equations (1.1) and (1.2) yields

$$2(1) + 3(4) = 14$$
$$1 - 2(4) = -7$$

both of which are correct statements. Although (7, 0) satisfies Equation (1.1), it does not satisfy Equation (1.2), and thus is not a solution of the system consisting of (1.1) and (1.2).

To find solutions of systems of linear equations, we shall employ, in a systematic way, techniques that leave the solutions of a system unchanged. Consider the following principles.

> **1.** If both sides of an equation are multiplied by a nonzero constant, then the resulting equation has the same solutions as the original equation.

For example, the equations $x - 2y = -7$ and $-2(x - 2y = -7)$, that is, $-2x + 4y = 14$, have the same solutions.

> **2.** If two equations have a common solution, then this common solution is also a solution of the sum of the two equations.

For example, the equations $2x + 3y = 14$ and $x - 2y = -7$ have (1, 4) as a solution in common. Their sum, $3x + y = 7$, also has (1, 4) as a solution.

Principles 1 and 2 taken together yield the following:

> **3.** If two equations have a common solution, then this common solution is also a solution of the sum of one of the equations and a nonzero constant multiple of the other.

For example, the equations $2x + 3y = 14$ and $x - 2y = -7$ have (1, 4) as a solution in common. The sum of $2x + 3y = 14$ and $-2(x - 2y = -7)$ is $0x + 7y = 28$, which also has (1, 4) as a solution.

These principles lead to the following rules of operation, which leave the solutions of a system unchanged.

> **Rule 1.** An equation in a system of linear equations may be multiplied by a nonzero constant.
>
> **Rule 2.** Any equation in a system of linear equations may be added to any other equation in the system.
>
> **Rule 3.** A nonzero constant multiple of any equation in a system of linear equations may be added to any other equation in the system.

To illustrate one way of employing these rules to determine solutions, consider again the system

$$2x + 3y = 14 \tag{1.1}$$
$$x - 2y = -7 \tag{1.2}$$

The difficulty is that there are two variables to contend with. Thus we shall direct our efforts toward taking one of the variables out of the scene and reducing the problem to solving one linear equation in one unknown. The easiest way to proceed is to take x out of the picture by multiplying Equation (1.2) by -2, yielding

$$-2x + 4y = 14 \tag{1.3}$$

and adding (1.3) to (1.1), thereby obtaining

$$7y = 28$$
$$y = 4$$

To determine x, we substitute 4 for y in one of the equations, $x - 2y = -7$, for example, and solve for x. This yields

$$x - 2(4) = -7$$
$$x = 1$$

Thus, if our system has a solution, then (1, 4) is the solution. We saw earlier that (1, 4) is indeed a solution since it satisfies both (1.1) and (1.2).

EXAMPLE 1

Solve the system

$$x + 3y = 90 \tag{1.4}$$
$$2x + y = 80 \tag{1.5}$$

Whether x or y is eliminated from the scene is up to us. For the sake of argument, let's eliminate y. To do so we multiply (1.5) by -3, thus yielding

$$x + 3y = 90 \tag{1.6}$$
$$-6x - 3y = -240 \tag{1.7}$$

By adding (1.6) and (1.7) we obtain

$$-5x = -150$$
$$x = 30$$

To determine y we substitute 30 for x in one of our equations, $2x + y = 80$, let us say, and solve for y. We have

$$2(30) + y = 80$$
$$y = 20$$

Thus, if our system has a solution, (30, 20) is the solution. To check if (30, 20) is a solution, we substitute 30 for x and 20 for y in (1.4) and (1.5) and see whether they are satisfied. This yields

$$30 + 3(20) = 30 + 60 = 90$$
$$2(30) + 20 = 60 + 20 = 80$$

Thus (30, 20) is a solution of our system.

EXAMPLE 2

Solve the system

$$3x - 4y = 10 \tag{1.8}$$
$$2x - 5y = 9 \tag{1.9}$$

Let us begin by eliminating x. To do so we must multiply both equations by constants so that x drops out when the resulting equations are added. The simplest way to achieve this is to multiply (1.8) by -2 and multiply (1.9) by 3. We obtain

$$-6x + 8y = -20 \tag{1.10}$$
$$6x - 15y = 27 \tag{1.11}$$

Adding (1.10) and (1.11) yields

$$-7y = 7$$
$$y = -1$$

Substituting $y = -1$ into (1.8) yields

$$3x - 4(-1) = 10$$
$$x = 2$$

By substituting 2 for x and -1 for y in (1.8) and (1.9), it is easily verified that $(2, -1)$ is a solution of this system.

EXAMPLE 3

Solve the system

$$x + y = 2 \tag{1.12}$$

$$x + y = 1 \tag{1.13}$$

It is clear by inspection that this system has no solution, since the sum of two numbers cannot be equal to 2 [required by (1.12)] and at the same time be equal to 1 [required by (1.13)]. But let's see what happens when we eliminate a variable, x, for example. Multiplying (1.13) by -1 and adding the result to (1.12) yields

$$0x + 0y = 1 \tag{1.14}$$

which has no solution. (For all values of x and y, $0x + 0y = 0$.) The appearance of such an equation with no solution is interpreted in the following way. We have shown that, if (1.12) and (1.13) have a solution in common, then this solution must also satisfy (1.14) (principle 3). Since (1.14) has no solution, there is no common solution to (1.12) and (1.13), and thus this system has no solution.

EXAMPLE 4

Solve the system

$$2x + 4y = 12 \tag{1.15}$$

$$-x - 2y = -6 \tag{1.16}$$

To eliminate x we multiply (1.16) by 2 and add the result to (1.15). This yields

$$0x + 0y = 0 \tag{1.17}$$

which has all ordered pairs of numbers as solutions. (For all values of x and y, $0x + 0y = 0$ is satisfied.) Let us observe that (1.15) is a multiple of (1.16); by multiplying (1.16) by -2 we obtain (1.15). Thus our system has infinitely many solutions—all ordered pairs of numbers that satisfy $2x + 4y = 12$. By infinitely many solutions we mean that there are more solutions than can be described by any positive integer.

The solution structures that emerged in these examples illustrate the possibilities in general. A system of two linear equations in two variables has either one solution, no solution, or infinitely many solutions. From a geometric point of view, the one-solution system corresponds to a pair of lines that intersect in one point (whose coordinates describe the solution), the no-solution system corresponds to a pair of parallel lines (no points of intersection), and the infinitely many solution system corresponds to two lines that coincide. In connection with Examples 2 through 4, these possibilities are shown in Figures 1.8 through 1.10.

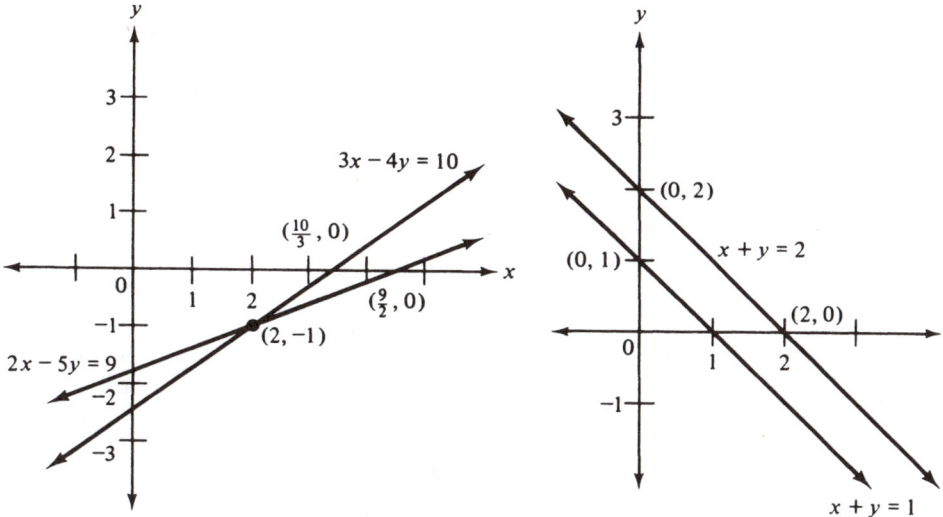

Figure 1.8

Figure 1.9

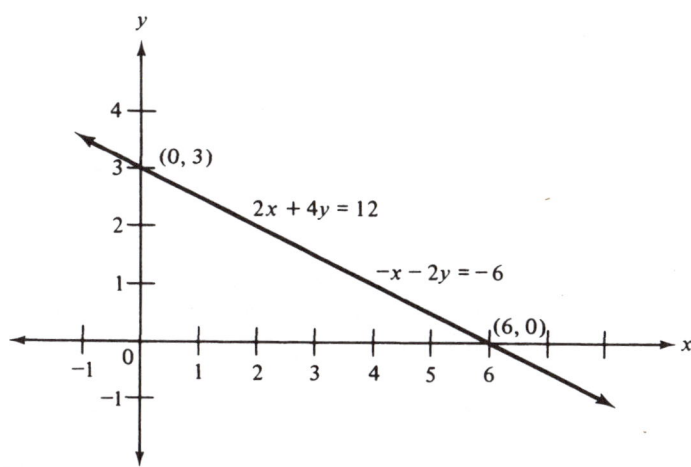

Figure 1.10

EXERCISES

Solve the following systems of linear equations.

1. $x + y = 9$
   $x - y = 1$

2. $x + y = 25$
   $-x + 4y = 0$

3. $x + 3y = 13$
   $2x - y = -2$

4. $3x - y = 1$
   $2x - 2y = 6$

5. $2x + 3y = 13$
   $5x - 2y = 4$

6. $-3x + 6y = 12$
   $2x - 4y = -8$

7. $2x + y = 23$
   $4x + 5y = 75$

8. $s + 2t = 10$
   $3s + t = 15$

9. $4x + 3y = 320$
   $5x + 2y = 330$

10. $2x + 3y = 120$
    $3x + 2y = 150$

11. $2x + 3y = 120$
    $x + y = 55$

12. $x + 1.2y = 12$
    $x + 2y = 16$

13. $5x + 3y = 95$
    $3x + 5y = 88$

14. $3.25x + 2y = 225$
    $4x + 3y = 320$

1.3. APPLICATION TO FINDING A REGRESSION LINE*

The marketing department of the Rasa Company wants to predict sales volume on the basis of advertising expenditures. To establish a quantitative relationship between these variables, data (shown in Table 1.1) were obtained from reports for eight monthly peri-

Table 1.1

Advertising Expenditures (x) (thousands of dollars)	Sales Volume (y) (thousands of dollars)
2	120
4	150
7	180
9	195
12	210
17	240
22	275
27	300

ods chosen at random from the company's files. Letting x denote advertising expenditures, y denote sales volume, and plotting the eight pairs of values obtained yields the scatter diagram shown in Figure 1.11. Although not all points lie on a line, it does seem that a linear function offers a reasonably good description of the relationship between sales volume and advertising expenditures. Assuming this to be the case, the problem of determining the line of best fit to the eight sample points arises. The most widely employed criterion for best fit is provided by the principle of least squares.

To describe the idea that underlies the least-squares principle, consider the three points (X_1, Y_1), (X_2, Y_2), (X_3, Y_3) and line $y = mx + b$ shown in Figure 1.12. The criterion defining line of best fit should, in some sense, minimize the deviation of the points from the line. The least-squares principle does this by defining the **least-squares line of best fit** as that line for which the sum of the squares of the vertical distances from the data points to the line, that is,

$$d_1^2 + d_2^2 + d_3^2$$

is smallest.

---

*May be omitted without loss of continuity.

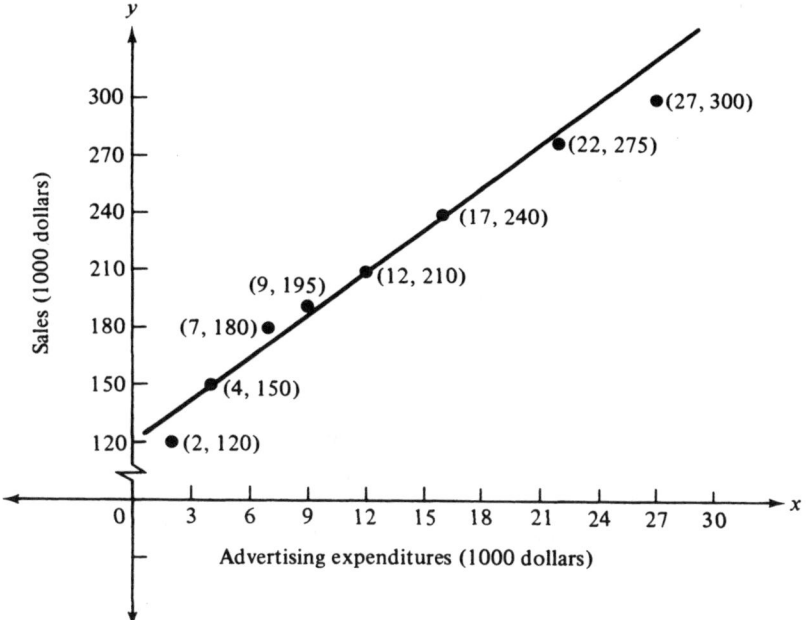

Figure 1.11

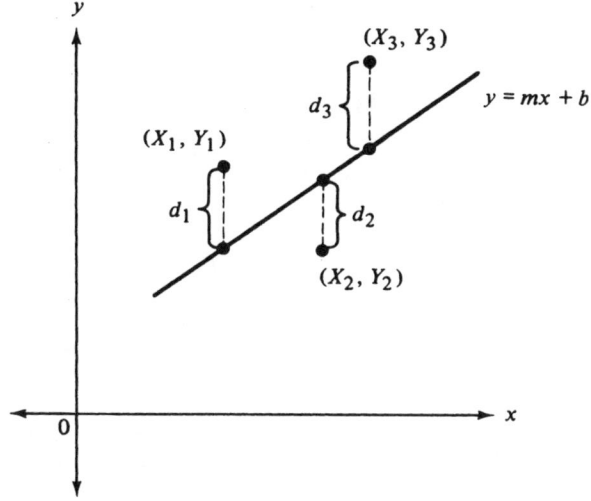

Figure 1.12

It can be shown that for the least-squares line of best fit

$$y = mx + b$$

m and b satisfy the system of equations

$$3b + Am = B$$
$$Ab + Cm = D$$

where 3 is the number of data points,

$$A = X_1 + X_2 + X_3$$

is the sum of the x-values of the data points,

$$B = Y_1 + Y_2 + Y_3$$

is the sum of the y-values of the data points,

$$C = X_1^2 + X_2^2 + X_3^2$$

is the sum of the squared x-values of the data points, and

$$D = X_1Y_1 + X_2Y_2 + X_3Y_3$$

is the sum of the mixed products of corresponding x and y data values.

More generally, if there are n data points, the same system of equations, with 3 replaced by n, is obtained. That is, we have

$$nb + Am = B$$
$$Ab + Cm = D$$

where A, B, C, and D are the same above defined sums, only with n, instead of 3, terms. This system looks more unfriendly than it actually is; the constants A, B, C, and D for a given situation are easily obtained when the data are organized in column form. To illustrate, we present the data for the Rasa Company's problem in the first two columns of Table 1.2.

Table 1.2

X_i	Y_i	X_i^2	X_iY_i
2	120	4	240
4	150	16	600
7	180	49	1260
9	195	81	1755
12	210	144	2520
17	240	289	4080
22	275	484	6050
27	300	729	8100
Sums: $A = 100$	$B = 1670$	$C = 1796$	$D = 24{,}605$

Column 3, containing X_i^2 values, is obtained by squaring each of the X_i values in column 1; column 4, containing X_iY_i values, is obtained by multiplying corresponding X_i and Y_i entries in columns 1 and 2. The sums A, B, C, and D are the sums of these four respective columns. Since there are 8 data points, $n = 8$, and we obtain the following system of equations to be solved for m and b:

$$8b + 100m = 1670 \quad (1.18)$$
$$100b + 1796m = 24{,}605 \quad (1.19)$$

To simplify matters we multiply Equation (1.18) by $\frac{1}{2}$, thus obtaining

$$4b + 50m = 835 \quad (1.20)$$
$$100b + 1796m = 24{,}605 \quad (1.19)$$

Since the easiest variable to eliminate is b, we multiply (1.20) by -25, thus obtaining

$$-100b - 1250m = -20{,}875 \quad (1.21)$$
$$100b + 1796m = 24{,}605 \quad (1.19)$$

Adding (1.21) and (1.19) gives us

$$546m = 3730$$
$$m = \frac{3730}{546}$$
$$= 6.832 \text{ (rounded off)}$$

Substituting $m = 6.832$ into (1.20) and solving for b yields

$$4b + 50(6.832) = 835$$
$$4b + 341.6 = 835$$
$$4b = 493.4$$
$$b = 123.35$$

If you have not dropped your calculator recently and are on otherwise friendly terms with it, it is easily verified as a check that $m = 6.832$ and $b = 123.35$ satisfy the system of Equations (1.18) and (1.19). Thus the (least squares) line of best fit to our data is

$$y = 6.832x + 123.35 \quad (1.22)$$

Since the eight pairs of data values are a sample of the totality, or population, of such pairs of values, Equation (1.22) is termed an **estimated regression line**. It serves as an estimate for the **population regression line**

$$y = Mx + E$$

which specifies the average value of y in terms of x. Other randomly chosen samples yield other estimates for the population regression line.

To illustrate the use of Equation (1.22) as a tool for prediction, let us suppose that the Rasa Company is interested in predicting the sales volume corresponding to advertising expenditures of $15 thousand per month. Substituting 15 for x in Equation (1.22) yields

$$y = 6.832(15) + 123.35$$
$$= 225.83$$

For advertising expenditures of $15 thousand per month, we predict an average monthly sales volume of $225.83 thousand. Note that the predicted y-value is interpreted as an average. We are not predicting $225.83 thousand in sales for each month that $15 thousand is spent on advertising.

Since the system of equations

$$nb + Am = B \qquad (1.23)$$
$$Ab + Cm = D \qquad (1.24)$$

often referred to as the system of **normal equations**, occurs in connection with all linear regression problems, it would be advantageous to obtain a general solution for m and b in terms of n, A, B, C, and D that could be applied to any problem. We begin by eliminating b; multiplying (1.23) by $-A$ and (1.24) by n yields

$$-nAb - A^2m = -AB \qquad (1.25)$$
$$nAb + nCm = nD \qquad (1.26)$$

Adding (1.25) and (1.26) gives us

$$nCm - A^2m = nD - AB$$

By factoring m we obtain

$$(nC - A^2)m = nD - AB$$

Dividing both sides by $nC - A^2$ yields

$$m = \frac{nD - AB}{nC - A^2}$$

By solving Equation (1.23) for b, we obtain

$$nb + Am = B$$
$$nb = B - Am$$
$$b = \frac{B - Am}{n}$$

In summary, then, we have

$$m = \frac{nD - AB}{nC - A^2} \tag{1.27}$$

$$b = \frac{B - Am}{n} \tag{1.28}$$

That is,

$$m = \frac{n(\text{sum of } X_i Y_i) - (\text{sum of } X_i)(\text{sum of } Y_i)}{n(\text{sum of } X_i^2) - (\text{sum of } X_i)^2} \tag{1.27}$$

$$b = \frac{(\text{sum of } Y_i) - (\text{sum of } X_i) m}{n} \tag{1.28}$$

To use these expressions for the computation of m and b, m must be determined first from (1.27) and then substituted into (1.28) for the computation of b.

EXAMPLE 1

The institutional research department of a large university wishes to predict the income of graduates six years after graduation on the basis of the graduate's grade-point average (GPA). For this purpose the department conducted a random sample of 10 alumni who had graduated six years earlier. The results of this sample are shown in Table 1.3.

Table 1.3

GPA (x)	2.0	2.3	2.5	2.5	3.0	3.2	3.3	3.5	3.5	3.8
Income (y) (1000 dollars)	19	25	25	30	32	35	37	42	45	40

Corresponding to these data, find the estimated linear regression equation that would enable us to predict income on the basis of grade-point average. Find the predicted income for a grade—point average of 3.4 and interpret the value obtained.

We begin by arranging the data in column form, determining the X_i^2 and $X_i Y_i$ columns, and summing these columns to obtain A (the sum of the X_i values), B (the sum of the Y_i values), C (the sum of X_i^2 values), and D (the sum of $X_i Y_i$ values). From the data obtained in Table 1.4, we obtain m and b from Equations (1.27) and (1.28).

Table 1.4

GPA (X_i)	Income (Y_i)	X_i^2	X_iY_i
2.0	19	4.00	38.0
2.3	25	5.29	57.5
2.5	25	6.25	62.5
2.5	30	6.25	75.0
3.0	32	9.00	96.0
3.2	35	10.24	112.0
3.3	37	10.89	122.1
3.5	42	12.25	147.0
3.5	45	12.25	157.5
3.8	40	14.44	152.0

Sums: $A = 29.6$ $B = 330$ $C = 90.86$ $D = 1019.6$

$$m = \frac{n(\text{sum of } X_iY_i) - (\text{sum of } X_i)(\text{sum of } Y_i)}{n(\text{sum of } X_i^2) - (\text{sum of } X_i)^2}$$

$$= \frac{10(1019.6) - (29.6)(330)}{10(90.86) - (29.6)^2}$$

$$= \frac{10{,}196 - 9768}{908.6 - 876.16}$$

$$= \frac{428}{32.44}$$

$$= 13.194$$

$$b = \frac{(\text{sum of } Y_i) - (\text{sum of } X_i)m}{n}$$

$$= \frac{330 - (29.6)(13.194)}{10}$$

$$= \frac{330 - 390.542}{10}$$

$$= -6.054$$

Thus the estimated linear regression equation is

$$y = 13.194x - 6.054$$

Substitution of 3.4 for x in this regression equation yields $y = 38.81$. For students with a 3.4 grade-point average, the predicted average income six years after graduation is $38.81 thousand. Here, too, we should be careful to note that we are not predicting the income of a specific student six years after graduation, but rather the average income of all students six years after graduating from the university. Specific incomes making up that average may differ considerably from each other.

The nature of the problems considered makes it clear that discretion must be exercised in choosing values of x to be substituted into the regression equation. In general, the further removed x is from the data values, the less reliable we can expect the predicted average value of y to be. It is well known, for example, that wheat yields improve as rainfall increases, and one could determine a regression line for given data. But it is also well known that wheat yields improve up to a certain point, and that too much rain has a destructive rather than beneficial effect. If the regression line were used to predict average wheat yields for suitably large rainfall values, the predicted average yield would differ considerably from the average yield actually obtained.

Let us also note that in establishing a linear regression equation relating x and y for prediction purposes, we are not claiming that the behavior of one variable causes the behavior of the other variable. Cause and effect is an entirely separate issue altogether. In connection with Example 1, it is most clear that, although grade-point average might be used to predict income, income level is not caused by the grade-point average.*

EXERCISES

1. The following data show the annual salary increment (y) in hundreds of dollars at a certain company and the number of times (x) the president of the company was complimented per week for his intellectual keenness.

    (a) Find the estimated linear regression equation specifying the average value of y in terms of x.

x	1	2	3
y	3	7	12

    (b) Find the predicted value of y for $x = 5$ and interpret the result obtained.

    (c) Plot the original data as well as the regression line on one diagram.

2. The following table shows equipment maintenance expenditures (x) and profit before taxes (y) for a random sample of 10 firms in the same industry. All figures are in thousands of dollars.

x	5	10	13	15	18	21	27	31	37	41
y	24	36	28	36	36	54	52	76	76	100

    (a) Find the estimated linear regression equation specifying the average value of y in terms of x.

---

*For further discussion of regression analysis see, for example, W. J. Adams, I. Kabus, M. P. Preiss, *Statistics: Basic Principles and Applications* (Dubuque: Kendall/Hunt Pub. Co., 1994), Ch. 12.

(b) Find the predicted value of y corresponding to an equipment maintenance expenditure of $50 thousand and interpret the result obtained.

(c) Plot the original data as well as the regression line on one diagram.

3. The following table shows scores (x) on an aptitude test for salesmen employed by the Algis Company and first-year sales (y) in thousands of dollars for a random sample of eight dossiers chosen from the company's files.

x	24	16	20	17	15	25	13	25
y	310	160	280	200	200	290	140	320

(a) Find the estimated linear regression equation specifying the average value of y in terms of x.

(b) Find the predicted value of y corresponding to a test score of 30 and interpret the result obtained.

(c) Plot the original data as well as the regression line on one diagram.

4. The following table shows scores (x) on a mathematics proficiency exam given to all entering freshmen at Ecap University and the final exam grade (y) in a calculus course for a random sample of 10 students.

x	20	25	30	32	40	50	60	60	65	70
y	50	50	60	64	70	75	80	85	90	90

(a) Find the estimated linear regression equation specifying the average value of y in terms of x.

(b) Find the predicted value of y for $x = 55$ and interpret the result obtained.

(c) Plot the original data as well as the regression line on one diagram.

5. The following table shows the grade point average (y) in the last two years of university studies and the student's age at the time of graduation (x) for a random sample of five students in a large university.

x	20	24	26	30	40
y	2.2	2.4	2.4	3.0	3.2

(a) Find the estimated regression equation specifying the average value of y in terms of x.

(b) Find the predicted general point average for age 35 and interpret the result obtained.

(c) Plot the original data as well as the regression line on one diagram.

1.4. INTRODUCTION TO THE TABLEAU METHOD

Although the elimination-of-a-variable technique works well for systems of two linear equations in two unknowns, its extension to larger systems of linear equations is awkward. We now turn our attention to developing a general solution method that is effective for a wide spectrum of systems of linear equations and whose procedures do not depend on the number of equations or variables in the system. First, some preliminary definitions.

Two or more linear equations involving the same variables, considered as a unit, are said to form a **system of linear equations**. If there are m equations and n unknowns, the system is called an m **by** n **system**. A **solution of an** m **by** n **system** is an ordered collection of n numbers that satisfies each of the m equations in the system. For example,

$$x - 2y + 2z = -1 \qquad (1.29)$$
$$3x + 2y + 2z = 9 \qquad (1.30)$$
$$2x - 3y - 3z = 6 \qquad (1.31)$$

is a 3 by 3 system. The ordered triple $(3, 1, -1)$ is a solution of this system since it satisfies all three equations of the system. The 2 by 3 system

$$x + 2y + z = 2 \qquad (1.32)$$
$$-x - y + 2z = 4 \qquad (1.33)$$

has infinitely many solutions, two of which are $(0, 0, 2)$ and $(-10, 6, 0)$. As is the case with 2 by 2 systems, a general m by n system has either one solution, no solution, or infinitely many solutions.

The tableau method for solving systems of linear equations, or Gauss-Jordan elimination procedure, as it is sometimes called, is based on the three earlier cited rules of operation, which, while changing the form of a system, leave its solutions undisturbed. We restate these rules here for convenience.

> **Rule 1.** Any equation in a system of linear equations may be multiplied by a nonzero constant.
>
> **Rule 2.** Any equation in a system of linear equations may be added to any other equation in the system.
>
> **Rule 3.** A nonzero constant multiple of any equation in a system of linear equations may be added to any other equation in the system.

In short, the application of rules 1 through 3 reduce a given system of linear equations to an **equivalent system**, that is, one with the same solutions. To illustrate the tableau method, consider the following system:

$$4x + 3y = 10 \qquad (1.34)$$
$$2x + 5y = 12 \qquad (1.35)$$

Finite Mathematics, Models, and Structure

If we could reduce this system to an equivalent system with the structure

$$0x + 1y = 2$$
$$1x + 0y = 1$$

then the solution common to both systems could easily be read off as $x = 1$, $y = 2$. The tableau method is a systematic procedure for applying rules 1 through 3 to replace the given system by an equivalent system, with a structure characterized by an arrangement of 0's and 1's such that the solution can be read off at a glance. There are two key features to this arrangement:

1. Each column involving a variable has one 1 with all other values being 0.
2. No row contains more than one 1 as a multiplier of a variable.

To obtain this structure, we begin by choosing a term in either (1.34) or (1.35), $2x$ in Equation (1.35), for example, and convert the coefficient 2 to 1. This is done by applying rule 1 and multiplying Equation (1.35) by $\frac{1}{2}$, the reciprocal of 2. The result is shown in Figure 1.13. The value to be converted to 1 is called the **pivot value**. It is useful to indicate the chosen pivot value by circling it, as shown in Figure 1.13. To **pivot** on the chosen pivot value is to convert it to 1 by multiplying the equation containing the pivot value by its reciprocal. It is desirable to maintain relative positions of equations, so that Equation (1.35) is replaced by Equation (1.37) as shown.

$$4x + 3y = 10 \qquad (1.34)$$
$$②x + 5y = 12 \qquad (1.35)$$
$$\qquad (1.36)$$
$$1x + \tfrac{5}{2}y = 6 \qquad (1.37)$$

Figure 1.13

The next step is to convert the term $4x$ in the column of our pivot value to $0x$ (see Figure 1.13); in short, convert 4 in the column of 2 to 0. To do this we work with Equation (1.37) by applying rule 3; multiply (1.37) by -4, thus obtaining $-4x - 10y = -24$, and add this result to (1.34), $4x + 3y = 10$, thereby obtaining $0x - 7y = -14$, which is recorded as (1.36) in Figure 1.14.

Our next step is to repeat the process. Choose a pivot value and convert it to 1, at the same time observing the guideline that no row should contain more than one 1

$$4x + 3y = 10 \tag{1.34}$$
$$\boxed{2}x + 5y = 12 \tag{1.35}$$
$$0x + \boxed{-7}y = -14 \tag{1.36}$$
$$1x + \tfrac{5}{2}y = 6 \tag{1.37}$$

Figure 1.14

as a multiplier of a variable. Thus we must stay out of Equation (1.37), and the only candidate available for the office of pivot value is -7 of $-7y$ in Equation (1.36). We indicate -7 as our choice of pivot value by circling it as shown in Figure 1.14. To pivot on -7, multiply Equation (1.36) by $-\tfrac{1}{7}$, the reciprocal of -7, and record the result as Equation (1.38), as shown in Figure 1.15.

$$4x + 3y = 10 \tag{1.34}$$
$$\boxed{2}x + 5y = 12 \tag{1.35}$$
$$0x + \boxed{-7}y = -14 \tag{1.36}$$
$$1x + \tfrac{5}{2}y = 6 \tag{1.37}$$
$$0x + 1y = 2 \tag{1.38}$$
$$\tag{1.39}$$

Figure 1.15

Next we convert the term $\tfrac{5}{2}y$ in the column of our pivot value of $0y$; in short, convert $\tfrac{5}{2}$ in the column of -7 to 0. To do this we work through Equation (1.38) by applying rule 3; multiply (1.38) by $-\tfrac{5}{2}$, thus obtaining $0x - \tfrac{5}{2}y = -5$, and add this result to (1.37), $1x + \tfrac{5}{2}y = 6$, thereby obtaining $1x + 0y = 1$, which is recorded as (1.39) in Figure 1.16. From Figure 1.16 we obtain $x = 1$, $y = 2$. As a check against error, it is easily verified that $(1, 2)$ is a solution of the original system (1.34), (1.35).

$$0x + 1y = 2, \quad y = 2 \tag{1.38}$$
$$1x + 0y = 1, \quad x = 1 \tag{1.39}$$

Figure 1.16

Let us observe that in carrying out these procedures the arithmetic operations are performed on the numerical values that define the equations. The variables x and y are just carried along for the ride, so to speak, and play no part in these procedures. Therefore, there is no need to continually write down the variables; it suffices to make note of which columns belong to which variables, identify the column of constants, and perform the indicated procedures on the resulting tableau of numbers. Rewriting (1.34) and (1.35), leaving out variables x and y and the equality sign, yields the tableau of numbers shown in Figure 1.17. The procedures described earlier in terms of equations would now be described in terms of the rows in the tableau that describe the equations.

(1.34) $4x + 3y = 10$
(1.35) $2x + 5y = 12$

Rows	x	y	Constant column
①	4	3	10
②	2	5	12

Figure 1.17

---

1. Choose a pivot value in the part of the tableau corresponding to the variables, ② in row 2, for example, and pivot on it.

This is done by multiplying row ② by $\tfrac{1}{2}$, thus yielding row ④, $[1 \;\; \tfrac{5}{2} \;\; 6]$, shown in Figure 1.18.

	Rows	x	y	Constant column
T_1	①	4	3	10
	②	②	5	12
T_2	③			
	④	1	$\tfrac{5}{2}$	6

Figure 1.18

---

2. Convert 4 in the column of pivot value 2 to 0.

To do this we work through row ④ (which we shall term the **one-row** since it is the result of pivoting). Multiply row ④ by -4, thus obtaining $[-4 \;\; -10 \;\; -24]$, and add this result to row ①, $[4 \;\; 3 \;\; 10]$, thereby obtaining $[0 \;\; -7 \;\; -14]$, which is recorded as row ③ in Figure 1.19.

Chapter 1 / Linear Structures   25

	Rows	x	y	Constant column
T_1	①	4	3	10
	②	②	5	12
T_2	③	0	−7	−14
	④	1	$\frac{5}{2}$	6

Figure 1.19

3. Repeat the process on tableau T_2. Choose a pivot value in a row that does not contain a 1 from a previous pivot operation, and pivot on it.

Row ④ is excluded by this guideline, and thus −7 in row ③ is the only available pivot value. We pivot on −7 by multiplying row ③ by $-\frac{1}{7}$, the reciprocal of −7, and record the result as row ⑤, [0 1 2], in Figure 1.20.

	Rows	x	y	Constant column
T_1	①	4	3	10
	②	②	5	12
T_2	③	0	⊖7	−14
	④	1	$\frac{5}{2}$	6
T_3	⑤	0	1	2
	⑥			

Figure 1.20

4. Convert $\frac{5}{2}$ in the column of the pivot value −7 to 0.

To do this we work through row ⑤, our one row. Multiply row ⑤ by $-\frac{5}{2}$, thus obtaining [0 $-\frac{5}{2}$ −5], and add to row ④, [1 $\frac{5}{2}$ 6], thus obtaining [1 0 1], which is recorded as row ⑥ in Figure 1.21. From rows ⑤ and ⑥ in tableau T_3 we read off the solution, $x = 1$, $y = 2$.

Finite Mathematics, Models, and Structure

	Rows	x	y	Constant column	
T_1	①	4	3	10	
	②	②	5	12	
T_2	③	0	-⑦	-14	
	④	1	$\frac{5}{2}$	6	
T_3	⑤	0	1	2	$y = 2$
	⑥	1	0	1	$x = 1$

Figure 1.21

In summary, the procedures of the tableau method consist of the following sequence of steps.

> 1. Express the given system of equations in tableau form.
>
> 2. Choose any column associated with a variable of the system, select a nonzero number as pivot value, and pivot on that value (that is, convert it to 1).

The resulting row, called the **one-row**, is recorded in the analogous position in the next tableau and serves as the tool for converting all other numbers in the column of the pivot value to 0.

> 3. Convert all nonzero numbers in the column of the pivot value to 0.

To convert a number c to 0, multiply the one-row by $-c$ and add the result to the row containing c.

> 4. Repeat steps 2 and 3, observing the guideline that pivot values should not be chosen from rows that contain a 1 from a previous pivot operation.
>
> 5. Steps 2 and 3 are repeated until every column (other than the column of constants) contains exactly one 1 and all other values in the column are 0. Such a column is said to be in **zero-one form**. When all columns are in zero-one form, the solution is read off.

For some systems it is not possible to convert all columns to zero-one form. Such situations will be discussed later.

In brief, the basis of the tableau method is the conversion of all possible columns connected with variables of the system to **zero-one form**. To further illustrate, we turn to the following examples.

EXAMPLE 1

Solve the system

$$x + 3y = 90 \qquad (1.40)$$
$$2x + y = 80 \qquad (1.41)$$

The tableau corresponding to this system is shown in Figure 1.22.

	Rows	x	y	
$x + 3y = 90$	①	1	3	90
$2x + y = 80$	②	2	1	80

Figure 1.22

We begin by choosing as our pivot value the number 1 in row ①. In choosing a pivot value, it's a good idea to look around for 1's and −1's since they are easiest to work with. Row ① is carried over to tableau T_2 as row ③ (see Figure 1.23).

	Rows	x	y	
T_1	①	①	3	90
	②	2	1	80
T_2	③	1	3	90
	④			

Figure 1.23

Row ③, our one-row, serves as our tool for converting 2 in the column of our pivot value to 0. To convert 2 to 0 we multiply row ③ by −2, thus obtaining [−2 −6 −180], and add to row ②, [2   1   80], thereby obtaining row ④, [0   −5   −100], shown in Figure 1.24. To help the reader keep track of the sequence of operations, the origin of each row is indicated on the right side of the tableaus. This, of course, is not necessary in the privacy of one's study.

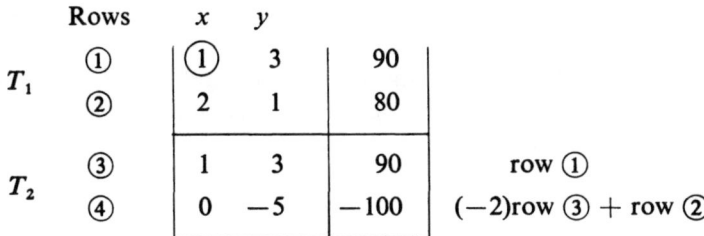

Figure 1.24

Since not all appropriate columns in tableau T_2 are in zero-one form, we repeat the conversion steps on T_2. The value -5 in row ④ is the only eligible pivot candidate. (3 in row ③ is ineligible since row ③ contains a 1 from a previous pivot operation.) We pivot on -5 by multiplying row ④ by $-\frac{1}{5}$, the reciprocal of -5, thereby obtaining row ⑥ in Figure 1.25.

Rows		x	y		
T_1	①	①	3	90	
	②	2	1	80	
T_2	③	1	3	90	row ①
	④	0	⊖	-100	(-2)row ③ + row ②
T_3	⑤				
	⑥	0	1	20	$(-\frac{1}{5})$row ④

Figure 1.25

Row ⑥, our one-row, now serves as our tool for converting 3 in the column of our pivot value to 0. To convert 3 to 0, we multiply row ⑥ by -3, thus obtaining [0 -3 -60], and add to row ③, [1 3 90], thereby obtaining row ⑤, [1 0 30], shown in Figure 1.26.

Since the x and y columns in tableau T_3 are in zero-one form, we can read off the solution $x = 30$, $y = 20$. It is easily verified that $(30, 20)$ satisfies the given system of Equations (1.40), (1.41).

Let us suppose that in applying the tableau method to a system of equations there arises, as illustrated by Figure 1.27, a row of zeros, except for a non-zero term in the constant column. In terms of the variables of this system, x and y, let us say, this row represents the equation $0x + 0y = 3$. Since this equation has no solution, the system to which it belongs has no solution. Thus:

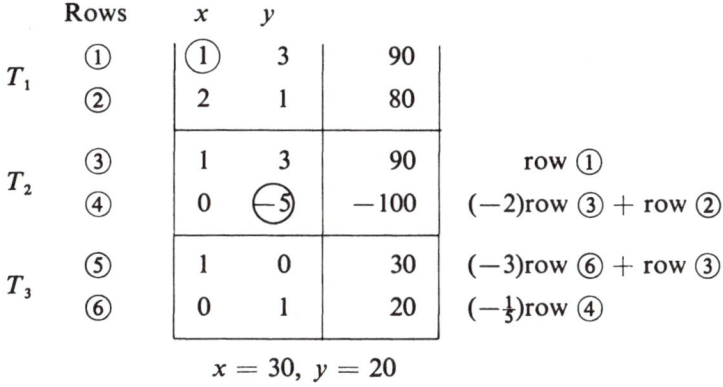

$x = 30, y = 20$

Figure 1.26

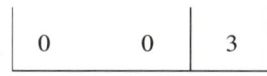

Figure 1.27

> The appearance of a row of zeros, except for a non-zero value in the constant column, implies that the given system of equations has no solution.

On the other hand, suppose that a row arises consisting entirely of zeros, as illustrated by Figure 1.28. This row represents the equation $0x + 0y = 0$, which is

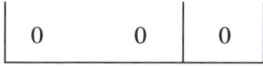

Figure 1.28

satisfied by all values of x and y. Such an equation, and its corresponding row may be omitted from the system, or tableau, since it has no further point of interest. Any common solution to the other equations in the system will satisfy this equation as well and there is no need to reproduce it.

> Discard from a tableau any row consisting entirely of zeros and continue the solution generation process.

Systems of two equations in two unknowns serve as a vehicle here for the introduction of the row and column procedures that underlie the tableau method. The real power of this technique is to be felt in dealing with larger systems, which we take up in the three sections that follow.

For a more complete perspective on the row and column operations introduced here, it is noteworthy that these procedures find application elsewhere. We develop one application to matrix inversion and matrix solutions to systems of linear equations in Sections 5.3 and 5.4.

EXERCISES

Solve the following systems of linear equations by the tableau method.

1. $x + y = 9$
   $x - y = 1$

2. $x + y = 25$
   $-x + 4y = 0$

3. $x + 3y = 13$
   $2x - y = -2$

4. $3x - y = 1$
   $2x - 2y = 6$

5. $2x + y = 23$
   $4x + 5y = 75$

6. $x - 2y = 6$
   $-3x + 6y = -18$

7. $3x + y = 9$
   $x + 2y = 8$

8. $5x + 3y = 1400$
   $4x + y = 756$

9. $2x - y = -5$
   $-6x + 3y = 15$

10. $2x + 3y = 15$
    $x + y = 6$

11. $x + y = 250{,}000$
    $2x + y = 400{,}000$

The Tableau Method for Larger Systems

In applying the tableau method to larger systems we proceed in the same way. It's just that there is more to be done.

EXAMPLE 2

Solve the system

$$x - 2y + 2z = -1 \qquad (1.42)$$
$$3x + 2y + 2z = 9 \qquad (1.43)$$
$$2x - 3y - 3z = 6 \qquad (1.44)$$

The tableau corresponding to this system is shown in Figure 1.29. We begin by choosing as our pivot value the number 1 in row ①. Row ① is carried over to tableau T_2 as row ④ (see Figure 1.30).

$$x - 2y + 2z = -1$$
$$3x + 2y + 2z = 9$$
$$2x - 3y - 3z = 6$$

⟷

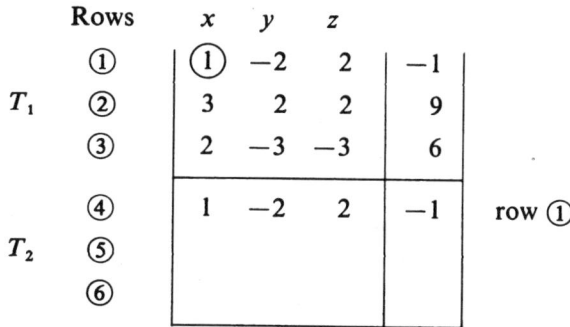

Figure 1.29

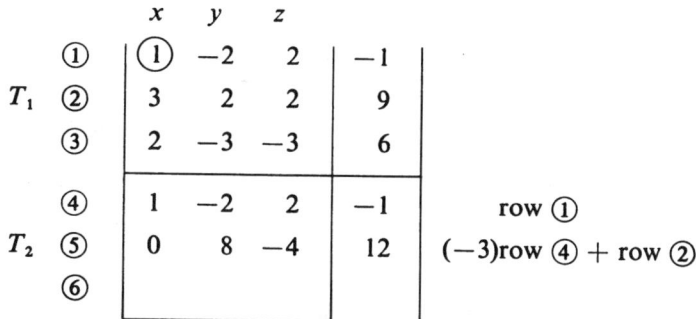

Figure 1.30

Row ④, our one-row, serves as our tool for converting 3 and 2 in the first column of T_1 to 0. To convert 3 to 0, we multiply row ④ by -3, thus obtaining $[-3 \ \ 6 \ \ -6 \ \ 3]$, and add to row ②, thereby obtaining row ⑤, $[0 \ \ 8 \ \ -4 \ \ 12]$, shown in Figure 1.31. To convert 2 to 0, we multiply row ④ by -2 and add to row ③, thereby obtaining row ⑥ in Figure 1.32.

		x	y	z		
T_1	①	①	-2	2	-1	
	②	3	2	2	9	
	③	2	-3	-3	6	
T_2	④	1	-2	2	-1	row ①
	⑤	0	8	-4	12	(-3)row ④ + row ②
	⑥					

Figure 1.31

Since not all appropriate columns in tableau T_2 are in zero-one form, we repeat the process on T_2. Staying out of row ④, which has a 1 from a previous pivot operation, we choose a pivot value, that is, 1, in row ⑥. (The other eligible pivot candi-

		x	y	z		
T_1	①	①	−2	2	−1	
	②	3	2	2	9	
	③	2	−3	−3	6	
T_2	④	1	−2	2	−1	row ①
	⑤	0	8	−4	12	(−3)row ④ + row ②
	⑥	0	①	−7	8	(−2)row ④ + row ③

Figure 1.32

dates and −7, −4, and 8, but 1 is easier to work with.) Row ⑥ is carried over to tableau T_3 as row ⑨ (see Figure 1.33). Row ⑨ serves as our tool for converting 8 and −2 in the column of our pivot value to 0. To convert 8 to 0, we multiply row ⑨ by −8 and add to row ⑤, thus obtaining row ⑧ in tableau T_3. To convert −2 to 0, we multiply row ⑨ by 2 and add to row ④, thus obtaining row ⑦ in tableau T_3.

		x	y	z		
T_1	①	①	−2	2	−1	
	②	3	2	2	9	
	③	2	−3	−3	6	
T_2	④	1	−2	2	−1	row ①
	⑤	0	8	−4	12	(−3)row ④ + row ②
	⑥	0	①	−7	8	(−2)row ④ + row ③
T_3	⑦	1	0	−12	15	(2)row ⑨ + row ④
	⑧	0	0	㊿	−52	(−8)row ⑨ + row ⑤
	⑨	0	1	−7	8	row ⑥

Figure 1.33

Since not all appropriate columns in T_3 are in zero-one form, we repeat the process on T_3. The number 52 in row ⑧ is the only eligible pivot candidate (why?); we pivot on 52 by multiplying row ⑧ by $\frac{1}{52}$, thereby obtaining row ⑪ in tableau T_4 (see Figure 1.34). Row ⑪ serves as our tool for converting −12 and −7 in the column of our pivot value to 0. To convert −12 to 0, we multiply row ⑪ by ⑫ and add to row ⑦, thus obtaining row ⑩ in tableau T_4. To convert −7 to 0, we multiply row ⑪ by 7 and add to row ⑨, thus obtaining row ⑫ in T_4.

Since the x, y, and z columns of T_4 are in zero-one form, we can read off the solution $x = 3$, $y = 1$, $z = -1$. It is easily verified that $(3, 1, -1)$ satisfies the given system of equations.

		x	y	z		
	①	①	-2	2	-1	
T_1	②	3	2	2	9	
	③	2	-3	-3	6	
	④	1	-2	2	-1	row ①
T_2	⑤	0	8	-4	12	(-3)row ④ + row ②
	⑥	0	①	-7	8	(-2)row ④ + row ③
	⑦	1	0	-12	15	(2)row ⑨ + row ④
T_3	⑧	0	0	㊾	-52	(-8)row ⑨ + row ⑤
	⑨	0	1	-7	8	row ⑥
	⑩	1	0	0	3	(12)row ⑪ + row ⑦
T_4	⑪	0	0	1	-1	$(\frac{1}{52})$row ⑧
	⑫	0	1	0	1	(7)row ⑪ + row ⑨

$$x = 3, y = 1, z = -1$$

Figure 1.34

EXAMPLE 3

Solve the system

$$x + 2y - 4z = -4$$
$$-x - 4y - 2z = -2$$
$$4x + 2y - 3z = 4$$

A sequence of tableaus leading to the solution is shown in Figure 1.35. Tableau T_2 is obtained by pivoting on 1 in row ① and converting -1 and 4 in the column of 1 to 0. Pivoting on -2 in tableau T_2 avoids the introduction of fractions. Thus -2 is the best choice among the available pivot candidates $(-2, -6, 13, -6)$ since messy arithmetic is avoided. In general, it's a good idea to choose a pivot value that divides into all the numbers in its row so as to avoid the introduction of fractions. Unfortunately, this is not always possible. Tableau T_3 is obtained by pivoting on -2 in T_2 and converting 2 and -6 in the column of -2 to 0. Since not all columns in tableau T_3 are in zero-one form, we must continue the conversion process. The value 31 in row ⑨ of T_3 is the only available pivot choice (why?), and thus we must pivot on 31. Tableau T_4

34 Finite Mathematics, Models, and Structure

		x	y	z		
	①	①	2	−4	−4	
T_1	②	−1	−4	−2	−2	
	③	4	2	−3	4	
	④	1	2	−4	−4	row ①
T_2	⑤	0	⟨−2⟩	−6	−6	row ④ + row ②
	⑥	0	−6	13	20	(−4)row ④ + row ③
	⑦	1	0	−10	−10	(−2)row ⑧ + row ④
T_3	⑧	0	1	3	3	$(-\tfrac{1}{2})$row ⑤
	⑨	0	0	㉛	38	(6)row ⑧ + row ⑥
	⑩	1	0	0	$\tfrac{70}{31}$	(10)row ⑫ + row ⑦
T_4	⑪	0	1	0	$-\tfrac{21}{31}$	(−3)row ⑫ + row ⑧
	⑫	0	0	1	$\tfrac{38}{31}$	$(\tfrac{1}{31})$row ⑨

$$x = \tfrac{70}{31},\ y = -\tfrac{21}{31},\ z = \tfrac{38}{31}$$

Figure 1.35

is obtained by pivoting on 31 and converting 3 and -10 in the column of 31 to 0. The x, y, and z columns in T_4 are in zero-one form, and thus the solution $x = \tfrac{70}{31}$, $y = -\tfrac{21}{31}$, $z = \tfrac{38}{31}$ can be read off.

EXAMPLE 4

Solve the system

$$x - 2y + z = 3$$
$$2x - 3y - z = 7$$
$$5x - 8y - z = 20$$

A sequence of solution tableaus is shown in Figure 1.36. Since row 9 consists of 0's, except for a nonzero constant term, our system has no solution.

EXERCISES

Solve the following systems by the tableau method.

12. $x + 3y + 2z = -1$
    $2x + 3y + 3z = -2$
    $-2x + 2y - 3z = 7$

13. $x + 2y - 2z = 1$
    $2x + 3y + 2z = -3$
    $2x - 3y - z = 6$

14. $2s + t - u = 2$
    $3s - 2t + u = 7$
    $s - 3t - 2u = -7$

		x	y	z		
	①	①	−2	1	3	
T_1	②	2	−3	−1	7	
	③	5	−8	−1	20	
	④	1	−2	1	3	row ①
T_2	⑤	0	①	−3	1	(−2)row ④ + row ②
	⑥	0	2	−6	5	(−5)row ④ + row ③
	⑦	1	0	−5	5	(2)row ⑧ + row ④
T_3	⑧	0	1	−3	1	row ⑤
	⑨	0	0	0	3	(−2)row ⑧ + row ⑥

Figure 1.36

15. $\begin{aligned} 4u + v - 2w &= -3 \\ 2u + 3v - 2w &= 7 \\ 7u + 2v + 5w &= -13 \end{aligned}$

16. $\begin{aligned} 2x + 3y + z &= 4 \\ x + y - 3z &= 11 \\ 4x + 5y - 5z &= 9 \end{aligned}$

17. $\begin{aligned} 8s + t - u &= -7 \\ 4s + 3t + v &= -3 \\ 2s + t + 4u + 4v &= 14 \\ 4s + 2t - 3u - v &= -12 \end{aligned}$

18. $\begin{aligned} 2s + t + 4u &= 8 \\ 3s - 2t - 7u &= 12 \\ -4s + 5t + 10u &= -16 \end{aligned}$

We next examine two applications from the world of accounting which lead to large systems.

1.5. NOTEWORTHY APPLICATIONS*

EXAMPLE 1. *SERVICE CHARGE ALLOCATION*

The Arkin Company, which makes television sets, has three production departments, which we shall denote by P_1, P_2, and P_3, and four service departments—accounting, maintenance, marketing, and purchasing. Each service department's cost must be distributed to the production departments and to other service departments based on their respective usages of the services provided. For each service department listed in the leftmost column of Table 1.5, the fraction of its total cost assigned to the service and production departments of the firm is given. Thus from row 1 we have that 2 percent of the total cost of accounting is assigned to maintenance, 10 percent of the total cost of accounting is assigned to marketing, and so on. Column 1 specifies the fraction of the service departments' costs that is assigned to accounting, column 2 specifies the

---

*The applications considered here are examined from a matrix algebra point of view in Section 5.5. Otherwise they may be omitted without loss of continuity.

Table 1.5

SERVICE DEPARTMENT	SERVICE DEPARTMENTS				PRODUCTION DEPARTMENTS			JANUARY OVERHEAD
	Acc.	Main.	Mar.	Pur.	P_1	P_2	P_3	
Accounting	0	0.02	0.10	0.10	0.24	0.26	0.28	$20,000
Maintenance	0.10	0	0.20	0.10	0.20	0.20	0.20	$18,000
Marketing	0	0	0	0	0.30	0.30	0.40	$80,000
Purchasing	0.10	0.10	0.10	0	0.20	0.20	0.30	$10,000

fraction of the service departments' costs that is assigned to maintenance, and so on. (Such determinations can be made in various ways; one way is based on the floor space occupied by each department.) The problem is to determine each service department's total costs (overhead plus charges for services provided by other departments) and to allocate these costs to the production departments.

Let x, y, z, and w denote the total costs of the accounting, maintenance, marketing, and purchasing departments, respectively. From column 1 we have that the costs of the maintenance, marketing, and purchasing departments that are assigned to accounting are $0.10y$, $0z$, and $0.10w$, respectively. The overhead of the accounting department (costs that are directly assigned to accounting such as salaries of employees in accounting, equipment, supplies, etc.) for January is $20,000. Thus x, the total cost of accounting, must satisfy the following condition:

$$x = 20{,}000 + 0.1y + 0.1w$$

A similar analysis for the maintenance, marketing, and purchasing departments yields the following conditions:

$$y = 18{,}000 + 0.02x + 0.1w$$
$$z = 80{,}000 + 0.1x + 0.2y + 0.1w$$
$$w = 10{,}000 + 0.1x + 0.1y$$

Rearranging terms gives us the following system:

$$
\begin{aligned}
x - 0.1y \quad\quad\quad - 0.1w &= 20{,}000 \\
-0.02x + \quad y \quad\quad - 0.1w &= 18{,}000 \\
-0.1x - 0.2y + z - 0.1w &= 80{,}000 \\
-0.1x - 0.1y \quad\quad + w &= 10{,}000
\end{aligned}
$$

A sequence of tableaus leading to the solution of this system is shown in Figure 1.37. From tableau T_4 we obtain the total cost values for the service departments.

Accounting: $23,423
Maintenance: $19,902

T_1	①	−0.1	0	−0.1	20,000
	−0.02	1	0	−0.1	18,000
	−0.1	−0.2	1	−0.1	80,000
	−0.1	−0.1	0	1	10,000
T_2	1	−0.1	0	−0.1	20,000
	0	0.998	0	−0.102	18,400
	0	−0.21	1	−0.11	82,000
	0	−0.11	0	0.99	12,000
T_3	1	0	0	−0.1102204	21,843.69
	0	1	0	−0.1022044	18,436.87
	0	0	1	−0.1314629	85,871.74
	0	0	0	0.9787576	14,028.06
T_4	1	0	0	0	23,423.43
	0	1	0	0	19,901.72
	0	0	1	0	87,755.93
	0	0	0	1	14,332.52

$$x = 23{,}423.43 \qquad z = 87{,}755.93$$
$$y = 19{,}901.72 \qquad w = 14{,}332.52$$

Figure 1.37

Marketing: $87,756
Purchasing: $14,333

From these cost values and the percentages given in columns 5, 6, and 7 of Table 1.5, we obtain the following allocation of service departments' costs to the production departments:

Department P_1: (0.24)(23,423) + (0.20)(19,902) + (0.30)(87,756) + (0.20)(14,333) = $38,795

Department P_2: (0.26)(23,423) + (0.20)(19,902) + (0.30)(87,756) + (0.20)(14,333) = $39,264

Department P_3: (0.28)(23,423) + (0.20)(19,902) + (0.40)(87,756) + (0.30)(14,333) = $49,941

If we were to replace the January overhead values of $20,000, $18,000, $80,000, and $10,000 by values for another month, we would have made a small change in the overall nature of the problem. Mathematically, the change is not small because we would be faced by a different system of equations.

The question is, can we develop a way that would allow us to change the overhead values with only a small change needed in the mathematical analysis needed to deal with the modified system of equations? The answer is yes, and the means for doing this is developed in Section 5.5.

EXAMPLE 2   AN INCOME CONSOLIDATION PROBLEM

The accompanying affiliation diagram shows the interdependency structure between the Russel, Ferrara, and Thomas Companies.

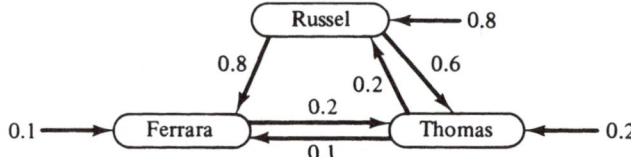

This diagram shows, for example, that the Russel Company owns 60 percent of the corporate stock of the Thomas Company, while the Thomas Company owns 20 percent of the corporate stock of the Russel Company. Eighty percent of the Russel Company's stock is not controlled by affiliates. Let us suppose that the net incomes of the Russel, Ferrara, and Thomas Companies from their own operations are $100,000, $80,000, and $60,000, respectively. The consolidated basis net income of a company is its net income determined in terms of its interdependency with other affiliates. The problem we consider here is that of determining the net incomes of the Russel, Ferrara, and Thomas Companies on a consolidated basis and the allocation of net incomes to these affiliate companies.

To do so we introduce the following notation:

1. Let x denote the net income of the Russel Company on a consolidated basis.
2. Let y denote the net income of the Ferrara Company on a consolidated basis.
3. Let z denote the net income of the Thomas Company on a consolidated basis.

Then we have the following relations:

$$x = 100{,}000 + 0.8y + 0.6z$$
$$y = 80{,}000 + 0.2z$$
$$z = 60{,}000 + 0.2x + 0.1y$$

Transposing and combining similar terms yields the following system:

$$x - 0.8y - 0.6z = 100{,}000$$
$$y - 0.2z = 80{,}000$$
$$-0.2x - 0.1y + z = 60{,}000$$

A sequence of tableaus leading to the solution of this system is shown in Figure 1.38. From tableau T_4 we obtain the following net income statements on a consolidated basis:

Net income of the Russel Company on a consolidated basis = \$256,521.74
Net income of the Ferrara Company on a consolidated basis = \$104,347.82
Net income of the Thomas Company on a consolidated basis = \$121,739.13

In determining the allocation of net incomes of affiliate companies, only the nonaffiliate shareholders in the parent company constitute the majority interest. For example, 20 percent of the Russel Company's stock is held by the Thomas Company; accordingly, the outside interest in the Russel Company of 80 percent is the equity multiplier in determining the consolidated net income of this company. In summary, we have the following allocation of net incomes of affiliate companies:

	x	y	z	
T_1	①	−0.8	−0.6	100,000
	0	1	−0.2	80,000
	−0.2	−0.1	1	60,000
T_2	1	−0.8	−0.6	100,000
	0	①	−0.2	80,000
	0	−0.26	0.88	80,000
T_3	1	0	−0.76	164,000
	0	1	−0.2	80,000
	0	0	⓪.828	100,800
T_4	1	0	0	256,521.74
	0	1	0	104,347.82
	0	0	1	121,739.13

$x = 256{,}521.74, \ y = 104{,}347.82, \ z = 121{,}739.13$

Figure 1.38

Consolidated Net Income

Russel Company: 80 percent of the Russel Company's consolidated basis net income (80 percent of \$256,521.74) = \$205,217.39.

Ferrara Company: 10 percent of the Ferrara Company's consolidated basis net income (10 percent of $104,347.82) = $10,434.78.

Thomas Company: 20 percent of the Thomas Company's consolidated basis net income (20 percent of $121,739.13) = $24,347.83.

EXERCISES

1. The Sonin Company, which makes typewriters, has two production departments, P_1 and P_2, and three service departments, S_1, S_2, and S_3. Each service department's total cost must be distributed to the production departments and to the other service departments based on their respective usages of the services provided. For each service department listed in the leftmost column of Table 1.6, the fraction of its total cost assigned to the service and production departments of the firm is given.

Table 1.6

SERVICE DEPARTMENT	SERVICE DEPARTMENT			PRODUCTION DEPARTMENT		MARCH OVERHEAD
	S_1	S_2	S_3	P_1	P_2	
S_1	0	0.10	0.05	0.40	0.45	$40,000
S_2	0.10	0	0.10	0.40	0.40	$30,000
S_3	0.20	0.05	0	0.35	0.40	$20,000

(a) Set up the system of equations that describes the conditions to be satisfied by the total costs of the service departments.

(b) Solve this system to determine the total costs of the service departments.

(c) Determine the allocation of the total costs of the service departments to the production departments.

2. The accompanying affiliation diagram shows the interdependency structure between the Ramunė, Algis, and Charles companies.

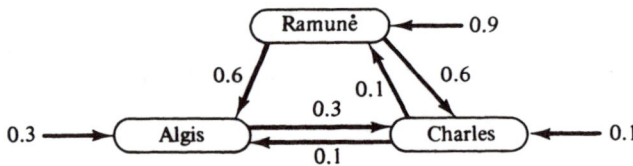

These companies have net incomes from their own operations of $80,000, $60,000 and $50,000, respectively.

(a) Set up and solve a system of linear equations to determine the net incomes of these affiliate companies on a consolidated basis.

(b) Determine the allocation of net incomes to these affiliate companies.

3. Letting x denote the state income tax owed, y denote the state capital stock tax owed, and z denote the federal income tax owed, a tax consultant developed the following relationships among these taxes for a client corporation.

$$x = 0.095(27{,}257 - y)$$
$$y = 0.01(152{,}970 - x - y - z)$$
$$z = 0.48(27{,}263 - x - y) - 833$$

Set up a system of linear equations from these relations and solve it to determine the taxes owed.

4. The federal income tax of the Sommers Company is computed after deducting payroll bonuses and the contribution to the profit-sharing plan. Payroll bonuses are based on income after deducting the profit-sharing contribution and federal income tax. The profit-sharing contribution is based on income after deducting bonuses and taxes. The tax rate is 50 percent, the bonus rate is 10 percent, the profit-sharing contribution is 5 percent, and profit, before consideration of taxes, bonuses, and profit sharing, is $2,000,000.

(a) Set up a system of linear equations to describe the interrelationships among taxes, bonuses, and the profit-sharing contribution.

(b) Solve this system to determine these amounts.

1.6. SYSTEMS WITH INFINITELY MANY SOLUTIONS*

EXAMPLE 1

Solve the system

$$x + 2y + z = 2 \quad (1.45)$$
$$-x - y + 2z = 4 \quad (1.46)$$
$$3x + 4y - 3z = -6 \quad (1.47)$$

A sequence of three solution tableaus for this system is shown in Figure 1.39. Observe that row ⑨ in tableau T_3 consists entirely of 0's and may thus be omitted. At the same time, another curious feature emerges; the third column of tableau T_3 cannot be converted to zero-one form without violating our standing rule that a pivot value is not to be chosen from a row that has a 1 from a previous pivot operation. This signals us to stop and write the equations described by the tableau. Doing this yields

---

*May be omitted without loss of continuity.

		x	y	z		
	①	①	2	1	2	
T_1	②	−1	−1	2	4	
	③	3	4	−3	−6	
	④	1	2	1	2	row ①
T_2	⑤	0	①	3	6	row ④ + row ②
	⑥	0	−2	−6	−12	(−3)row ④ + row ③
	⑦	1	0	−5	−10	(−2)row ⑧ + row ④
T_3	⑧	0	1	3	6	row ⑤
	⑨	0	0	0	0	(2)row ⑧ + row ⑥

Figure 1.39

$$x \quad - 5z = -10$$
$$y + 3z = 6$$

By transposing the z term in both equations, we obtain

$$x = -10 + 5z$$
$$y = 6 - 3z$$

If we give z the value 1, we obtain $x = -5$ and $y = 3$; if we give z the value 0, we obtain $x = -10$ and $y = 6$. Our system has infinitely many solutions, and a specific solution can be obtained by giving z any specific value of our choice and determining x and y from $x = -10 + 5z$ and $y = 6 - 3z$. We describe these solutions by writing

$$x = -10 + 5z$$
$$y = 6 - 3z$$
$$z \text{ is arbitrary}$$

As a check against error, we prove that these relations correctly describe the solutions of our system by substituting $-10 + 5z$ for x and $6 - 3z$ for y in (1.45) through (1.47), and verifying that 2, 4, and -6, respectively, are obtained.

$$(-10 + 5z) + 2(6 - 3z) + z = -10 + 5z + 12 - 6z + z = 2 \quad (1.45)$$
$$-(-10 + 5z) - (6 - 3z) + 2z = 10 - 5z - 6 + 3z + 2z = 4 \quad (1.46)$$
$$3(-10 + 5z) + 4(6 - 3z) - 3z = -30 + 15z + 24 - 12z - 3z = -6 \quad (1.47)$$

When a tableau is obtained in which not all columns connected with variables are in zero-one form, and it is not possible to convert the remaining columns to zero-one form, write the equations that correspond to the tableau. By transposing those

EXAMPLE 2

Solve the system

$$x - 3y + 4z + w = 8 \quad (1.48)$$
$$-x + 4y - z + 2w = 4 \quad (1.49)$$

A sequence of solution tableaus for this system is shown in Figure 1.40.

		x	y	z	w		
T_1	①	①	−3	4	1	8	
	②	−1	4	−1	2	4	
T_2	③	1	−3	4	1	8	row ①
	④	0	①	3	3	12	row ③ + row ②
T_3	⑤	1	0	13	10	44	(3)row ⑥ + row ③
	⑥	0	1	3	3	12	row ④

Figure 1.40

Writing the equations that correspond to tableau T_3 yields

$$x + 13z + 10w = 44$$
$$y + 3z + 3w = 12$$

By transposing the z and w terms in both equations, we obtain the following description of the solutions of our system:

$$x = 44 - 13z - 10w$$
$$y = 12 - 3z - 3w$$
$$z \text{ is arbitrary}$$
$$w \text{ is arbitrary}$$

Our system has infinitely many solutions, and a specific solution can be obtained by giving z and w specific values of our choice, and determining x and y from $x = 44 - 13z - 10w$ and $y = 12 - 3z - 3w$. If we give z and w the value 0, we obtain $x = 44$ and $y = 12$; for $z = 1$ and $w = 2$, we obtain $x = 11$ and $y = 3$. To establish in general that the cited relations describe the solutions of our system, substitute $44 - 13z - 10w$ for x and $12 - 3z - 3w$ for y in (1.48) and (1.49), and verify that 8 and 4, respectively, are obtained.

$$44 - 13z - 10w - 3(12 - 3z - 3w) + 4z + w$$
$$= 44 - 13z - 10w - 36 + 9z + 9w + 4z + w \tag{1.48}$$
$$= 8$$

$$-(44 - 13z - 10w) + 4(12 - 3z - 3w) - z + 2w$$
$$= -44 + 13z + 10w + 48 - 12z - 12w - z + 2w \tag{1.49}$$
$$= 4$$

EXAMPLE 3

Part of a traffic network being designed to service a component of the Johnson City Airport is shown in Figure 1.41. The roads are one way, as shown by the arrows, and the given values express the expected number of cars entering and leaving the network per hour during a heavy load period. The total number of cars entering the network, 800, equals the total number leaving the network, so that a basic traffic-flow equilibrium condition is satisfied. Another basic equilibrium condition is that the total number of cars entering each intersection point equal the total number of cars leaving the intersection point. Assuming this to be the case, let us explore conclusions that can be drawn about traffic flow in the interior branches of the network.

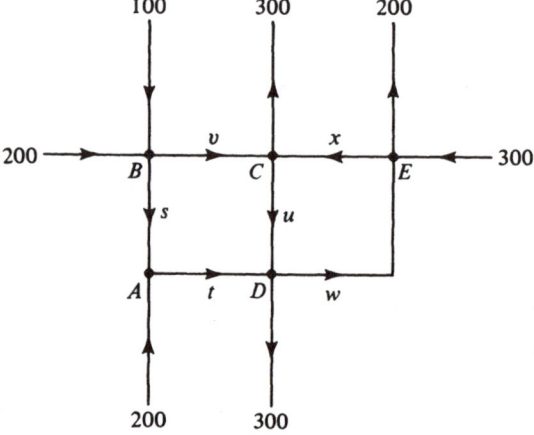

Figure 1.41

To do so, we introduce variables s, t, u, v, w, and x as shown in Figure 1.41. Variable x denotes the number of cars passing between intersection points E and C per hour, t denotes the number of cars passing between intersection points A and D per hour, and so on. Since $200 + s$ cars enter A while t cars leave A, for equilibrium we have

$$A: \quad s + 200 = t$$

Similarly, for B, C, D, and E we obtain

B: $\quad 200 + 100 = s + v$
C: $\quad v + x = u + 300$
D: $\quad t + u = w + 300$
E: $\quad w + 300 = x + 200$

Rewriting these equations so that the variables appear on one side of an equation and the constant appears on the other side yields the system

$$s - t \qquad\qquad\qquad\quad = -200 \qquad (1.50)$$
$$s \qquad + v \qquad\qquad\quad = 300 \qquad (1.51)$$
$$\qquad -u + v \qquad + x = 300 \qquad (1.52)$$
$$t + u \qquad - w \qquad = 300 \qquad (1.53)$$
$$\qquad\qquad\qquad -w + x = 100 \qquad (1.54)$$

A sequence of solution tableaus is shown in Figure 1.42. Writing the equations that correspond to tableau T_5 yields

$$s \qquad + v \qquad\qquad = 300$$
$$t \qquad + v \qquad\qquad = 500$$
$$\qquad u - v \qquad - x = -300$$
$$\qquad\qquad\qquad w - x = -100$$

By transposing the terms arising from columns in T_5 that are not in zero-one form (v and x columns), we obtain the following description of the solutions of our system.

$$s = 300 - v$$
$$t = 500 - v$$
$$u = -300 + v + x$$
$$w = -100 + x$$

v is arbitrary

x is arbitrary

We leave it as an exercise to verify that these relations describe the solutions of (1.50) through (1.54). Although this system has infinitely many solutions, only a small number of them make sense in terms of the network. Since the variables express the number of cars passing per hour between branches connecting intersection points of the network, the values given to these variables must, at the very least, be restricted to nonnegative integers (0, 1, 2, etc.). The requirement $s \geq 0$ yields

$$s = 300 - v \geq 0, \qquad v \leq 300$$

		s	t	u	v	w	x		
T_1	①	①	−1	0	0	0	0	−200	
	②	1	0	0	1	0	0	300	
	③	0	0	−1	1	0	1	300	
	④	0	1	1	0	−1	0	300	
	⑤	0	0	0	0	−1	1	100	
T_2	⑥	1	−1	0	0	0	0	−200	row ①
	⑦	0	①	0	1	0	0	500	(−1)row ⑥ + row ②
	⑧	0	0	−1	1	0	1	300	row ③
	⑨	0	1	1	0	−1	0	300	row ④
	⑩	0	0	0	0	−1	1	100	row ⑤
T_3	⑪	1	0	0	1	0	0	300	row ⑫ + row ⑥
	⑫	0	1	0	1	0	0	500	row ⑦
	⑬	0	0	−1	1	0	1	300	row ⑧
	⑭	0	0	①	−1	−1	0	−200	(−1)row ⑫ + row ⑨
	⑮	0	0	0	0	−1	1	100	row ⑩
T_4	⑯	1	0	0	1	0	0	300	row ⑪
	⑰	0	1	0	1	0	0	500	row ⑫
	⑱	0	0	0	0	−1	1	100	row ⑲ + row ⑬
	⑲	0	0	1	−1	−1	0	−200	row ⑭
	⑳	0	0	0	0	⊖1	1	100	row ⑮
T_5	㉑	1	0	0	1	0	0	300	row ⑯
	㉒	0	1	0	1	0	0	500	row ⑰
	㉓	0	0	0	0	0	0	0	row ㉕ + row ⑱
	㉔	0	0	1	−1	0	−1	−300	row ㉕ + row ⑲
	㉕	0	0	0	0	1	−1	−100	(−1)row ⑳

Figure 1.42

From $w \geq 0$ we have

$$w = -100 + x \geq 0, \qquad x \geq 100$$

From $u \geq 0$ we obtain

$$u = -300 + v + x \geq 0, \qquad v + x \geq 300$$

Thus traffic equilibrium cannot be maintained if more than 300 cars per hour pass between B and C, or fewer than 100 cars per hour pass between E and C, or the sum of the number of cars passing per hour between B and C and E and C is less than 300. This tells us something about the practical feasibility of the network.

EXERCISES

Solve the systems of equations stated in 1-6. Set up and solve the traffic network problem described in 7.

1. $x + 3y - 2z = 5$
    $2x - y + 3z = 10$
    $-3x + 5y - 8z = -15$

2. $3s + t - 2u = 8$
    $2s - 3t + 5u = 15$
    $-s + 7t - 12u = -22$

3. $2x + 2y - 2z = 8$
    $x - y + 3z = 16$

4. $x + 2y + 3z = 6$
    $-2x + y - z = 3$
    $3x + y + 4z = 3$
    $4x + 3y + 7z = 9$

5. $x - 2y + 3z - w = 2$
    $-2x + 5y - z + 2w = 4$
    $-x + y - 2z + w = 2$

6. $2s + t + 2u + w = 3$
    $5s + 3t - u + 2w = 2$

7. Part of a traffic network being designed to service an envisioned shopping center is shown in Figure 1.43. The total number of cars entering the network per hour during a peak period, 900, equals the total number leaving the network per hour, so that a basic equilibrium condition is satisfied. On the basis of the equilibrium condition that the total number of cars entering each intersection point equals the total number leaving the intersection point, set up a system of linear equations to describe traffic flow in the interior of the network. Solve this system of equations and interpret the results obtained in terms of the network.

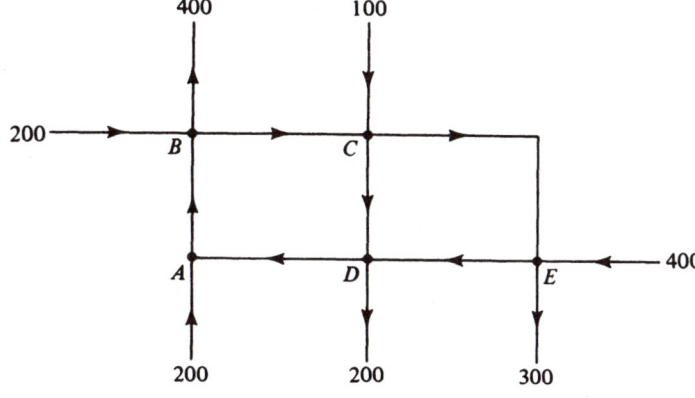

Figure 1.43

1.7. INEQUALITIES, LINEAR INEQUALITIES, AND GRAPHS

Inequalities

Inequalities come in a variety of forms. Any statement involving one of the symbols $>$ (greater than), $\geq$ (greater than or equal to), $<$ (less than), or $\leq$ (less than or equal to) is termed an **inequality**. Inequalities of the $<$ type and $>$ type are sometimes called **strict inequalities**. Two kinds of inequalities, absolute inequalities and conditional inequalities, are to be distinguished. An inequality that involves numbers only or holds for all permissible values of the variables involved is called an **absolute inequality**.

$$-3 < -2, \quad 8 > \tfrac{1}{2}, \quad \text{and} \quad x^2 \geq 0$$

are absolute inequalities. An inequality that does not hold for all permissible values of the variables involved is called a **conditional inequality**, or more simply, an **inequality**.

$$2x + 3 < 7, \quad x^2 \leq 16, \quad \text{and} \quad 2x + 3y \geq 6$$

are conditional inequalities.

When 1 is substituted for x in the inequality

$$2x + 3 < 7$$

we obtain $2(1) + 3 < 7$, or $5 < 7$, which is a correct statement. The number 1 or any other number that yields a correct statement when substituted for x in $2x + 3 < 7$ (such as $\tfrac{3}{2}, \tfrac{1}{2}, 0, -4$) is said to be a solution of $2x + 3 < 7$. The number 1 and the other solutions of $2x + 3 < 7$ are said to satisfy this inequality. The number 2 is not a solution of (does not satisfy) $2x + 3 < 7$ since substitution of 2 in $2x + 3 < 7$ yields $2(2) + 3 < 7$, or $7 < 7$, which is an incorrect statement. More generally, a number is said to be a **solution** of a one-variable inequality if a correct statement is obtained when the number is substituted for the variable of the inequality.

A **solution** of an inequality in two variables, x and y, let us say, is any ordered pair of numbers that yields a correct statement when substituted for x and y, respectively. For example, $(3, 0)$ is a solution of

$$2x + 3y \geq 6$$

since substitution of 3 for x and 0 for y yields $2(3) + 3(0) \geq 6$, or $6 \geq 6$, which is a correct statement. $(3, 0)$ is said to satisfy $2x + 3y \geq 6$. $(0, 1)$ is not a solution of (does not satisfy) $2x + 3y \geq 6$, since substitution of 0 for x and 1 for y yields $2(0) + 3(1) \geq 6$, or $3 \geq 6$, which is an incorrect statement.

A **solution** of an inequality in three variables, x, y, and z, is any ordered triple of numbers that yields a correct statement when substituted for x, y, and z, respectively. Thus $(-1, 3, 4)$ is a solution of

$$x^2 y + z < 10$$

since substitution of -1 for x, 3 for y, and 4 for z yields $(-1)^2(3) + 4 < 10$, or $7 < 10$, which is a correct statement. Here too we say that $(-1, 3, 4)$ satisfies $x^2y + z < 10$. The ordered triple of numbers $(2, 4, -1)$ does not satisfy (is not a solution of) $x^2y + z < 10$, since substitution of 2 for x, 4 for y, and -1 for z yields $(2)^2(4) - 1 < 10$, or $15 < 10$, which is an incorrect statement. The notion of solution of an inequality in any number of variables is defined in an analogous way.

EXERCISES

Determine which of $(0, 2)$, $(3, -2)$, $(-4, 0)$, $(\frac{1}{2}, 5)$, $(2, 3)$, *and* $(6, 2)$ *satisfy the following inequalities.*

1. $3x + 2y \leq 6$
2. $2x - y^2 < 4$
3. $2x + xy \geq 10$
4. $2x - 3y < 4$
5. $5x^2 - y > 6$
6. $2x^2 - y^2 \leq 4$

Determine which of $(2, 3, 1)$, $(1, 3, -1)$, $(4, 2, 1)$, $(3, -1, 4)$, $(\frac{1}{2}, \frac{1}{3}, -3)$, *and* $(-2, 3, 1)$ *satisfy the following inequalities.*

7. $3x + 2y - z \leq 8$
8. $2x + 5y - 3z \leq 10$
9. $3x^2y - 2z > 6$
10. $x^2 + 2y^2 - z \leq 12$
11. $\dfrac{2x^2 - y^2}{z} \leq 6$
12. $4x - 2y + 3z \leq 8$

Operations of Equivalence

Two inequalities are said to be **equivalent** if they have the same solutions. The following operations of equivalence lead to equivalent inequalities:

> **1.** The same constant may be added to, or subtracted from, both sides of an inequality; the direction of the inequality is maintained. That is, if $a \leq b$ and c is any number, then $a + c \leq b + c$ (and $a - c \leq b - c$).

For example, adding 5 to both sides of $2x - 5 \leq 3$ yields the equivalent inequality $2x \leq 8$.

> **2.** A term may be transferred from one side of an inequality to the other side, provided that its sign is changed; the direction of the inequality is maintained. This operation is called **transposition**.

For example, transposing $2x$ from the left side to the right side of $2x + y \leq 8$ yields the equivalent inequality $y \leq 8 - 2x$.

> **3.** Both sides of an inequality may be multiplied or divided by the same nonzero constant; multiplying or dividing by a positive constant maintains the direction of the inequality; multiplying or dividing by a negative constant reverses the direction of the inequality. That is, if $a \leq b$ and c is positive, then $ac \leq bc$ and $a/c \leq b/c$. If $a \leq b$ and c is negative, then $ac \geq bc$ and $a/c \geq b/c$.

For example, multiplying both sides of $2 < 3$ by -1 yields $-2 > -3$. The direction of the inequality is reversed since -1 is negative. Dividing both sides of $2x \leq 8$ by 2 yields the equivalent inequality $x \leq 4$. The direction of the original inequality is maintained since 2 is positive.

These operations of equivalence for inequalities are analogous to the ones cited for equations, with one difference. In multiplying or dividing both sides of an inequality by a value one must pay particular attention to the sign of the value. Positive values maintain the direction of an inequality; negative values reverse the direction of an inequality.

EXAMPLE 1

Determine the solutions of $10 - 3x \geq 7$.

Subtracting 10 from both sides of $10 - 3x \geq 7$ yields

$$-3x \geq -3$$

By dividing both sides of $-3x \geq -3$ by -3 (which is negative), we obtain

$$x \leq 1$$

Since the operations employed lead to equivalent inequalities, $10 - 3x \geq 7$ is equivalent to $x \leq 1$. Every number that is less than or equal to 1 is a solution of $10 - 3x \geq 7$, and every solution of $10 - 3x \geq 7$ is less than or equal to 1.

EXAMPLE 2

Show that $y \geq \frac{1}{5}(x + y)$ is equivalent to $-x + 4y \geq 0$.

Multiplying both sides of $y \geq \frac{1}{5}(x + y)$ by 5 yields

$$5y \geq x + y$$

By transposing $x + y$ we obtain

$$-x + 4y \geq 0$$

EXAMPLE 3

Show that $200 - x - y \geq 0$ is equivalent to $x + y \leq 200$.

Subtracting 200 from both sides of $200 - x - y \geq 0$ yields

$$-x - y \geq -200$$

Multiplying both sides of $-x - y \geq -200$ by -1 (which is negative) gives us

$$x + y \leq 200$$

EXAMPLE 4

Show that $2(x - 4) > 3x + 2$ is equivalent to $x < -10$.

$$2(x - 4) > 3x + 2$$

Since $2(x - 4) = 2x - 8$, we have

$$2x - 8 > 3x + 2$$

Transposing $3x$ yields

$$-x - 8 > 2$$

Adding 8 to both sides gives us

$$-x > 10$$

Multiplying both sides by -1 yields

$$x < -10$$

EXERCISES

13. Show that $100 - x \geq 0$ is equivalent to $x \leq 100$.
14. Show that $300 - y \geq 0$ is equivalent to $y \leq 300$.
15. Show that $1000 - x - y \geq 0$ is equivalent to $x + y \leq 1000$.
16. Show that $2x + y < 6$ is equivalent to $y < 6 - 2x$.

17. Show that $x \leq 2$ is equivalent to $3x - 4 \leq 2$.
18. Show that $2(x - 3y) < 8$ is equivalent to $x < 4 + 3y$.
19. Show that $3(x + 2y) < 2x + 4$ is equivalent to $x < 4 - 6y$.
20. Show that $-2(x + 3) < 4x + 12$ is equivalent to $x > -3$.

Find the solutions of the following inequalities in one variable. Sketch their graphs.

21. $3x + 6 > 4$
22. $-2x \geq 5$
23. $5x + 3 \leq 13$
24. $3y + 2 < 11$
25. $3 - 4y > -5$
26. $2 - 3y < 17$
27. $2(x + 4) < x - 7$
28. $-3(2x + 2) < 6x + 5$
29. $2(x - 4) < 3(2x - 6)$
30. $4(3 - 2x) \geq 7 - 3x$
31. $3(4 - 2y) \leq y - 2$
32. $-2(3 - y) \geq y + 1$

Graphs of Two-Variable Inequalities

The **graph** of an inequality in two variables is the collection of all those points, and only those points, whose coordinates satisfy the inequality. In a number of situations the geometric view of an inequality provided by its graph is most useful. We shall find graphs of particular importance in our discussion of basic linear programming in chapter 3. In this connection especially, graphs of linear inequalities occupy the spotlight.

A two-variable linear equation is of the form

$$Ax + By = C$$

where A and B are not both zero. A **two-variable linear inequality** is an inequality that can be obtained from $Ax + By = C$ by replacing the equality condition ($=$) by any one of the inequality conditions $>$, $\geq$, $<$, or $\leq$. $2x - 3y \leq 6$, for example, is a two-variable linear inequality; it can be obtained from the linear equation $2x - 3y = 6$ by replacing $=$ by $\leq$.

The graph of $2x - 3y = 6$ is a line that divides the rest of the coordinate plane into two components (see Figure 1.44). All points on one side of this boundary line satisfy $2x - 3y < 6$, while all points on the other side satisfy $2x - 3y > 6$. To find which side corresponds to which inequality, it suffices to choose a test point on one side of $2x - 3y = 6$ and substitute it into $2x - 3y < 6$. If this test point satisfies $2x - 3y < 6$, then all points on the same side as this test point satisfy $2x - 3y < 6$, and all points on the other side of the boundary line satisfy $2x - 3y > 6$. If the test point does not satisfy $2x - 3y < 6$, then it satisfies $2x - 3y > 6$, as do all other points on the same side as the test point. The points on the other side of the boundary line satisfy $2x - 3y < 6$.

To illustrate, let us take $(0, 0)$ as our test point. (Any other point not on the boundary line $2x - 3y = 6$ would do just as well.) Substituting $(0, 0)$ into $2x -$

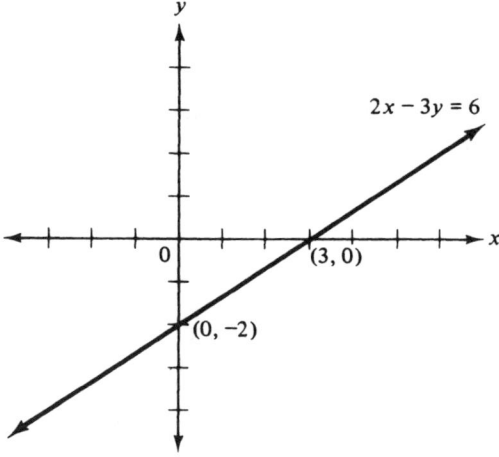

Figure 1.44

$3y < 6$ yields $0 < 6$, which is a correct statement. Thus $(0, 0)$ satisfies $2x - 3y < 6$, and all points on the same side as $(0, 0)$ (above $2x - 3y = 6$) satisfy $2x - 3y < 6$. The inequality $2x - 3y > 6$ is satisfied by all points below $2x - 3y = 6$ (see Figure 1.45).

More generally, the graph of $Ax + By = C$ is a line that divides the rest of the coordinate plane into two components. All points on one side of $Ax + By = C$ satisfy $Ax + By < C$, while all points on the other side satisfy $Ax + By > C$. Examples 5 and 6 further illustrate the graphing of linear inequalities.

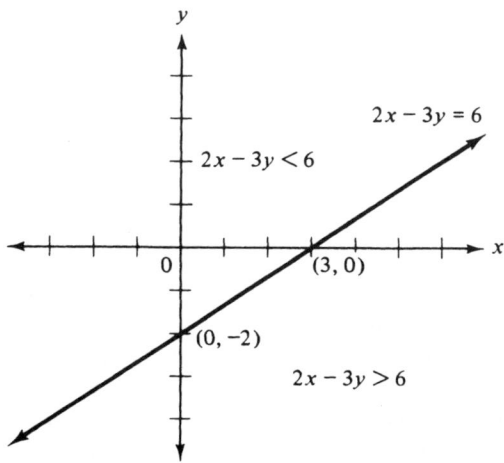

Figure 1.45

EXAMPLE 5

Sketch the graph of $5x + 2y \geq 10$.

We first graph the boundary line $5x + 2y = 10$. For $x = 0$, $y = 5$; for $y = 0$, $x = 2$. Thus $5x + 2y = 10$ passes through $(0, 5)$ and $(2, 0)$ (see Figure 1.46). We next

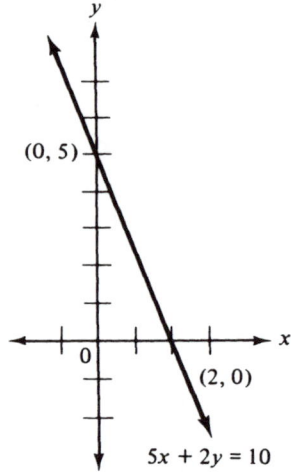

Figure 1.46

choose a test point not on $5x + 2y = 10$, $(0, 0)$, for example, and determine if it satisfies $5x + 2y > 10$. Substituting 0 for x and 0 for y in $5x + 2y > 10$ yields $0 > 10$, which is an incorrect statement. Since $(0, 0)$ does not satisfy $5x + 2y > 10$, the points that do satisfy $5x + 2y > 10$ are on the other side of (above) the boundary line $5x + 2y = 10$. The graphs of $5x + 2y \geq 10$ consists of all points on and above the boundary line $5x + 2y = 10$. This is indicated visually by drawing $5x + 2y = 10$ as a solid line and shading the region above it (see Figure 1.47).

EXAMPLE 6

Sketch the graph of $x - 2y < 4$.

We first graph the boundary line $x - 2y = 4$. For $x = 0$, $y = -2$; for $y = 0$, $x = 4$. Thus $x - 2y = 4$ passes through $(0, -2)$ and $(4, 0)$. Since this boundary line is not part of the graph of $x - 2y < 4$ (this inequality is a strict inequality), we draw a dashed line as shown in Figure 1.48. We next choose a test point not on $x - 2y = 4$, $(0, 0)$, for example, and determine if it satisfies $x - 2y < 4$. Substituting 0 for x and 0 for y in $x - 2y < 4$ yields $0 < 4$, which is a correct statement. Since $(0, 0)$ satisfies our inequality and is above the boundary line $x - 2y = 4$, the graph of $x - 2y < 4$ consists of all points above this boundary line (see Figure 1.49).

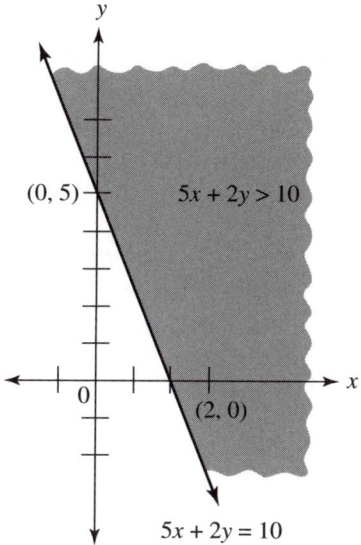

Figure 1.47

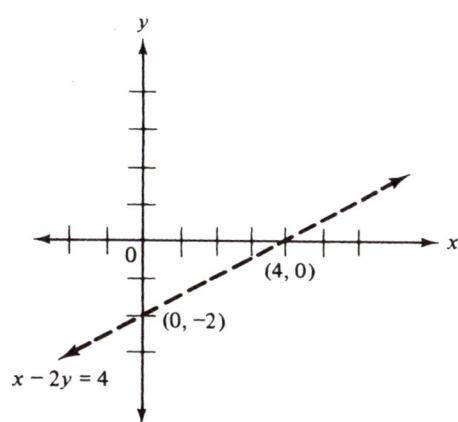

Figure 1.48

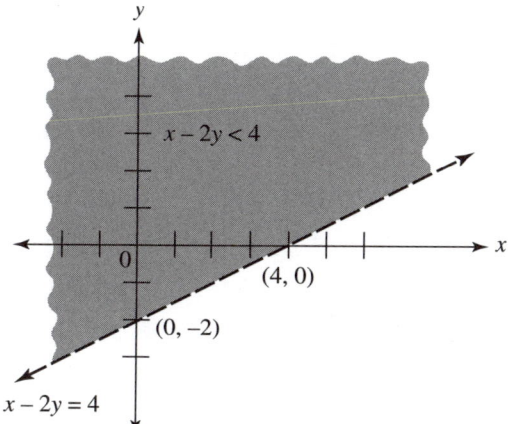

Figure 1.49

EXERCISES

Sketch the graphs of the following inequalities.

33. $x + y \geq 3$

34. $x + 2y \leq 6$

35. $x - y \geq 4$

36. $2x + y < 8$

37. $2x - y < 1$

38. $x + 3y < 6$

39. $x \geq 0$

40. $y \geq 0$

41. $3x - y \leq 3$

42. $4x - 3y > 12$
43. $x + y \geq 0$
44. $x - y < 0$
45. $-x + 4y \geq 0$
46. $4x + 5y \geq 75$
47. $2x + y \geq 23$

Systems of Linear Inequalities

Two or more linear inequalities involving the same variables, considered as a unit, are said to form a **system of linear inequalities**. For example,

$$2x - 3y \leq 6 \qquad (1.55)$$
$$5x + 2y \geq 10 \qquad (1.56)$$

and

$$x \geq 0 \qquad (1.57)$$
$$y \geq 0 \qquad (1.58)$$
$$x + 3y \leq 90 \qquad (1.59)$$
$$2x + y \leq 80 \qquad (1.60)$$

are systems of 2-variable linear inequalities while

$$x + 2y + z \leq 8 \qquad (1.61)$$
$$2x - y - \tfrac{1}{2}z \leq 6 \qquad (1.62)$$

and

$$x \geq 0 \qquad (1.63)$$
$$y \geq 0 \qquad (1.64)$$
$$z \geq 0 \qquad (1.65)$$
$$2x + 3y + z \leq 11 \qquad (1.66)$$
$$x + y + 3z \leq 10 \qquad (1.67)$$
$$2x + 2y + z \leq 10 \qquad (1.68)$$

are systems of three-variable linear inequalities.

The notion of solution of a system of inequalities is analogous to the notion of solution of a system of equations. For example, any ordered pair of numbers that satisfies all inequalities in the system

$$2x - 3y \leq 6 \qquad (1.55)$$
$$5x + 2y \geq 10 \qquad (1.56)$$

is said to be a solution of this two-variable system. Thus (2, 1) is a solution, since the substitution of 2 for x and 1 for y in (1.55) and (1.56) yields

$$1 \leq 6$$
$$12 \geq 10$$

both of which are correct statements. The ordered pair (1, 0) is not a solution since the substitution of 1 for x and 0 for y in (1.55) and (1.56) yields the statements

$$2 \leq 6$$
$$5 \geq 10$$

not both of which are correct.

Any ordered triple of numbers that satisfies all inequalities in the system

$$x \geq 0 \tag{1.63}$$
$$y \geq 0 \tag{1.64}$$
$$z \geq 0 \tag{1.65}$$
$$2x + 3y + z \leq 11 \tag{1.66}$$
$$x + y + 3z \leq 10 \tag{1.67}$$
$$2x + 2y + z \leq 10 \tag{1.68}$$

is said to be a solution of this three-variable system. Thus (1, 2, 1) is a solution, since it satisfies inequalities (1.63) through (1.68), whereas (2, 1, 3) is not a solution since it does not satisfy (1.67) (substitution yields $12 \leq 10$). More generally, a **solution** of an n-variable system of linear inequalities is any ordered collection of n numbers that satisfies all inequalities in the system.

EXERCISES

48. Which of (2, 3), (4, 1), (−1, 3), ($\frac{1}{2}$, 2), (6, −1), and (3, 1) are solutions of the following system?

$$2x + 4y \leq 15 \tag{1.69}$$
$$3x + y \leq 10 \tag{1.70}$$

49. Which of (8, 2), (8, 17), (20, 5), (9, 12), (5, 4), (12, 2), and (14, 7) are solutions of the following system?

$$x \geq 0 \tag{1.71}$$
$$y \geq 0 \tag{1.72}$$
$$x + y \leq 25 \tag{1.73}$$
$$-x + 4y \geq 0 \tag{1.74}$$
$$x \geq 8 \tag{1.75}$$

50. Which of (2, 1, 3), (1, 4, 1), (4, 3, 2), (3, 1, 1), (4, 2, 2,), (1, 5, 1), (1, 2, 1), and (0, 4, 3) are solutions of the following system?

$$x + 2y + 3z \leq 12 \tag{1.76}$$
$$5x - y + z \leq 9 \tag{1.77}$$

51. Which of (1, 1, 2), (0, 1, 3), (2, 1, 1), (3, −1, 2), (4, 1, 0), (2, 1, 2), (1, 3, 1), and (0, 2, 3) are solutions of the following system?

$$x \geq 0, y \geq 0, z \geq 0$$
$$2x + 3y + z \leq 11$$
$$x + y + 3z \leq 10$$
$$2x + 2y + z \leq 10$$

Graphs of Systems of Two-Variable Linear Inequalities

To graph a system of two-variable inequalities we graph the individual inequalities on the same coordinate system and then pick out the region which is common to them—that is, the overlap.

EXAMPLE 7

Sketch the graph of the system $x + 2y \leq 8$
$$3x + y \leq 9$$

We first graph $x + 2y \leq 8$. The boundary line $x + 2y = 8$ is shown in Figure 1.50. Since the test point (0.0) satisfies $x + 2y \leq 8$, the graph of $x + 2y \leq 8$ consists of all points on and below $x + 2y = 8$ as shown in Figure 1.51.

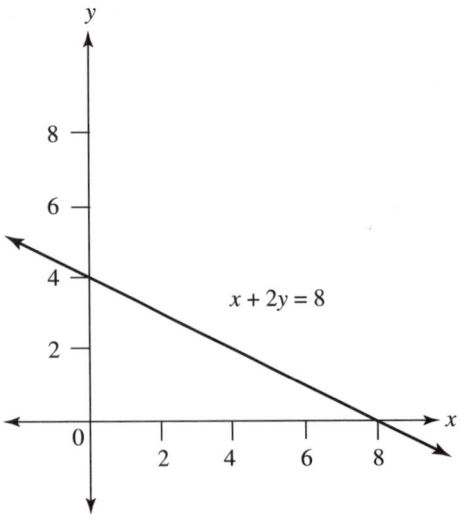

Figure 1.50

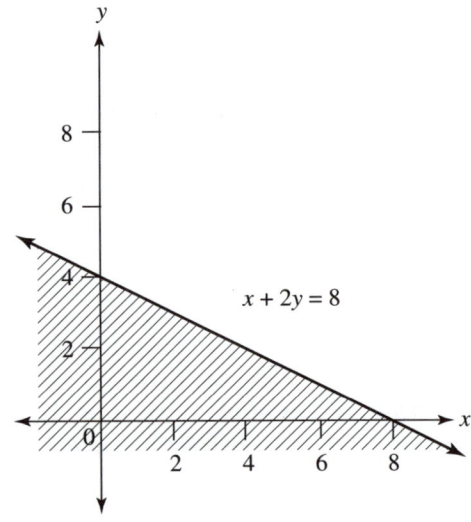

Figure 1.51

Sketching the graph of $3x + y \leq 9$ on the same coordinate system yields Figure 1.52. The overlap consists of all points on and below both boundary lines, shown in Figure 1.53.

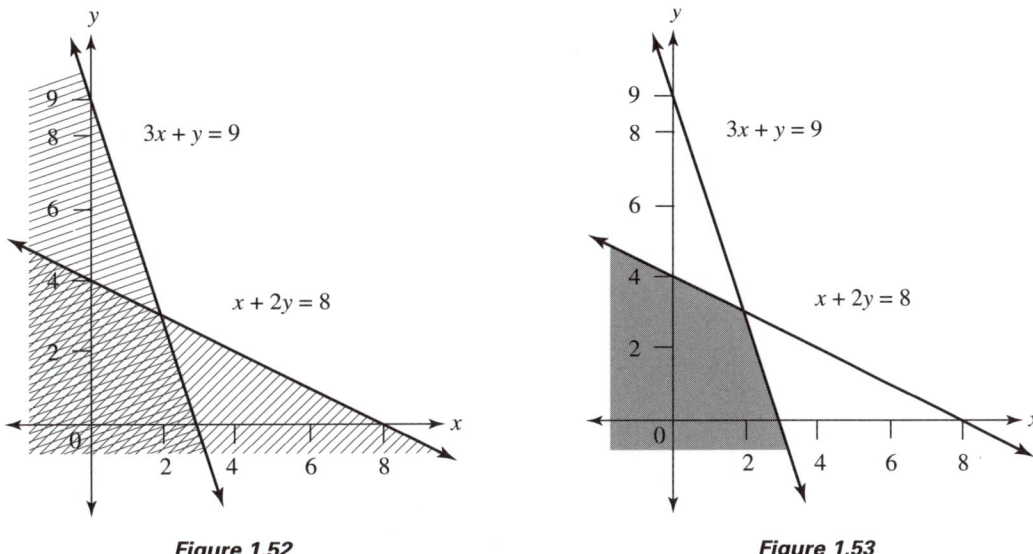

Figure 1.52

Figure 1.53

The nonnegativity conditions $x \geq 0$, $y \geq 0$ restrict us to the first quadrant. It is useful to make mental note of this because many problems of interest require such conditions. We encounter such problems in Chapter 3.

EXAMPLE 8

Sketch the graph of the system

$$x \geq 0$$
$$y \geq 0$$
$$x + 2y \leq 8$$
$$3x + y \leq 9$$

All we have to do is restrict the graph shown in Figure 1.53 to the first quadrant. The result is shown in Figure 1.54.

EXAMPLE 9

Sketch the graph of the system

$$x \geq 0$$
$$y \geq 0$$
$$2x + y \geq 23$$
$$4x + 5y \geq 75$$

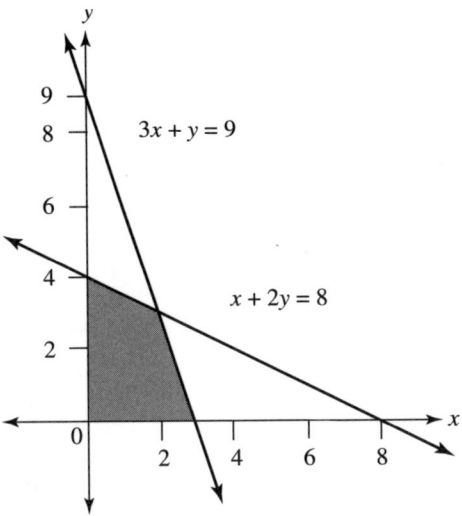

Figure 1.54

Since $x \geq 0$ and $y \geq 0$ restrict us to the first quadrant, it suffices to graph $2x + y \geq 23$ and $4x + 5y \geq 75$ restricted to the first quadrant. The graph of $2x + y \geq 23$ restricted to the first quadrant is shown in Figure 1.55(a), and the graphs of $2x + y \geq 23$ and $4x + 5y \geq 75$, restricted to the first quadrant, are shown in Figure 1.55(b). The graph of our system is the region in the first quadrant that is both shaded and

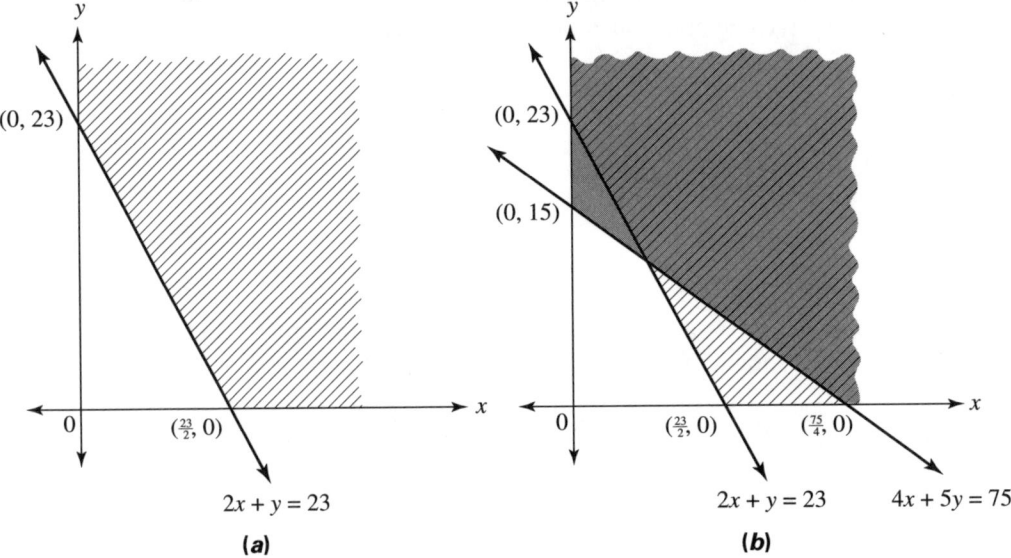

Figure 1.55

lined in (the region in the first quadrant above and including both boundary lines) and is shown in Figure 1.56.

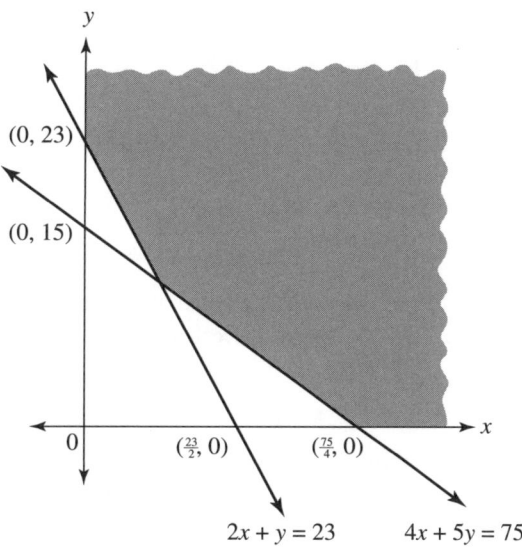

Figure 1.56

Sketch the graphs of the following systems of inequalities.

52. $x \geq 0, y \geq 0$
$x + y \geq 3$

53. $x \geq 0, y \geq 0$
$2x + y \geq 6$
$x \geq 3$

54. $x + y \leq 5$
$2x + 3y \leq 12$

55. $x \geq 0, y \geq 0$
$x + y \leq 5$
$2x + 3y \leq 12$

56. $2x + y \geq 10$
$x + 2y \leq 12$

57. $2x + y \leq 10$
$x + 2y \geq 12$

58. $x + y \geq 5$
$2x + 3y \geq 12$

59. $x + y \leq 5$
$2x + 3y \geq 12$

60. $x + 4y \leq 10$
$3x + y \leq 8$

61. $x \geq 0, y \geq 0$
$x + 2y \geq 10$
$3x + y \geq 15$

62. $x \geq 0, y \geq 0$
$x + 2y \geq 4$
$x - 2y \geq 2$

63. $x \geq 0, y \geq 0$
$2x + y \geq 4$
$x + y \geq 3$
$x + 2y \geq 4$

64. $x \geq 0, y \geq 0$
$x + y \leq 25$
$-x + 4y \geq 0$
$x \geq 8$

65. $x \geq 0, y \geq 0$
$5x + 4y \leq 120$
$x \geq 8, x \leq 16, y \geq 6$

66. $x \geq 0, y \geq 0$
$4x + 3y \leq 320$
$5x + 2y \leq 330$

67. $x \geq 0, y \geq 0$
    $2x + 3y \leq 1100$
    $5x + 3y \leq 1400$
    $4x + y \leq 756$
    $x \geq 25, y \geq 40$

68. $x \geq 0, y \geq 0$
    $8x + 5y \leq 2210$
    $3x + 2y \leq 860$
    $x \geq 50, y \geq 50$

1.8. PREFACE TO LINEAR PROGRAMMING

A **linear function in two variables**, x and y, say, is a function with the structure

$$F(x, y) = Ax + By + C$$

where A, B, and C are constants. Linear functions in n variables are defined in an analogous manner. Thus a **linear function in three variables,** x, y, and z, is one with the structure

$$F(x, y, z) = Ax + By + Cz + D$$

Of particular interest in linear programming (Chapter 3) are linear functions with domains of definition described by systems of linear inequalities.

EXAMPLE 1

Consider $F(x, y) = 5x + 10y$, where x and y must satisfy the following conditions:

$$x \geq 0 \qquad (1.78)$$
$$y \geq 0 \qquad (1.79)$$
$$2x + 3y \leq 24 \qquad (1.80)$$
$$x + 4y \leq 16 \qquad (1.81)$$

$F(x, y) = 5x + 10y$ is a linear function with $A = 5$, $B = 10$, and $C = 0$. Inequalities (1.78) through (1.81) describe the domain of definition of this function, and Figure 1.57 gives us a geometric view of this domain of definition. $F(x, y) = 5x + 10y$ is defined only for points that satisfy (1.78) through (1.81), that is, for points in the region shown in Figure 1.57. For example, (6, 2) satisfies (1.78) through (1.81); thus $F(6, 2)$ is defined and equals $5(6) + 10(2) = 50$. The point (2, 6), on the other hand, is not in the domain of definition, (1.81) is not satisfied, and thus $F(2, 6)$ is not defined.

Linear programming problems are concerned with finding the largest or smallest value of a linear function where the variables are required to satisfy linear conditions. In terms of Example 1 a linear programming problem emerges when we ask: What values of x and y satisfy (1.78) through (1.81), that is, are contained in the region shown in Figure 1.57, and yield the largest value for $F(x, y) = 5x + 10y$?

We apply the algebra developed in this chapter to develop a method for solving such problems in Chapter 3, and also examine a spectrum of realistic situations which lead to such problems. As a prelude to these developments we look at the nature of mathematical modeling in the next chapter.

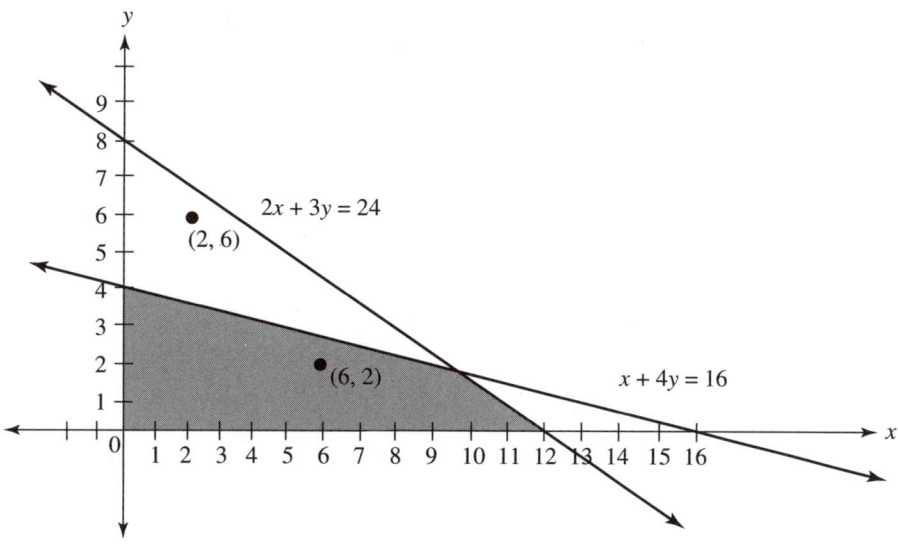

Figure 1.57

EXERCISES

Which of the following are linear functions? Explain.

1. $F(x, y) = 3x + \dfrac{1}{y}$

2. $F(x, y, z) = 3x + 4y + 2xz$

3. $F(x, y) = \frac{1}{2}x - 3y - 10$

4. $F(x, y, z) = 3x + 2y + z^2$

5. For $C(x, y) = 5x + 4y$, where x and y satisfy $x \geq 0$, $y \geq 0$, $2x + y \geq 23$, $4x + 5y \geq 75$, determine $C(0, 15)$, $C(0, 23)$, $C(\frac{23}{2}, 0)$, $C(\frac{20}{2}, \frac{29}{3})$, and $C(\frac{75}{4}, 0)$.

6. For $C(x, y) = 2x + 3y + 150{,}000$, where x and y satisfy $x \geq 0$, $y \geq 0$, $x + y \leq 15{,}000$, $x \leq 10{,}000$, $y \leq 12{,}000$, $x + y \geq 2000$, determine $C(0, 2000)$, $C(0, 12{,}000)$, $C(3{,}000, 12{,}000)$, $C(15{,}000, 0)$, and $C(10{,}000, 5000)$.

7. For $I(x, y) = 0.09x + 0.06y$, where x and y satisfy $x \geq 0$, $y \geq 0$, $x + y \leq 25$, $-x + 4y \geq 0$, $x \geq 8$, determine $I(8, 2)$, $I(25, 0)$, $I(8, 17)$, and $I(20, 5)$.

CHAPTER 2

Introduction to Mathematical Modeling

2.1. MATH MODELS FOR A VACATION TRIP

Ann's Model

Members of the Adams family were recently engaged in planning their trips from home in Brooklyn, New York to the vacation town of Kennebunkport, Maine. Ann planned to make the trip in late July with a major stop at Putnam, Connecticut for the annual picnic held there. Ann's problem was to set up a mathematical representation or model of the situation which would enable her to predict the total time required for the journey.

The setting of any such problem presents numerous features and characteristics, many of which are irrelevant or unessential to the focus of the problem. In developing her model Ann had to sort this out and decide on which features were negligible. This required discretion and judgment, the most controversial aspect of the math model development process; one person's essential might be another's irrelevancy.

Ann examined a map and laid out a route. She made assumptions about the traffic flow to be expected along various points, speeds that would be possible, and the number of rest stops to be made and their duration. These considerations led her to a model consisting of two line segments joining points representing Brooklyn, Putnam, and Kennebunkport, the sum of whose lengths is 350 miles, and the problem of determining how long it would take an object moving at an average speed of 50 miles an hour to travel this distance. She envisioned an approximate 2 hour stay at the picnic in Putnam before continuing to Kennebunkport.

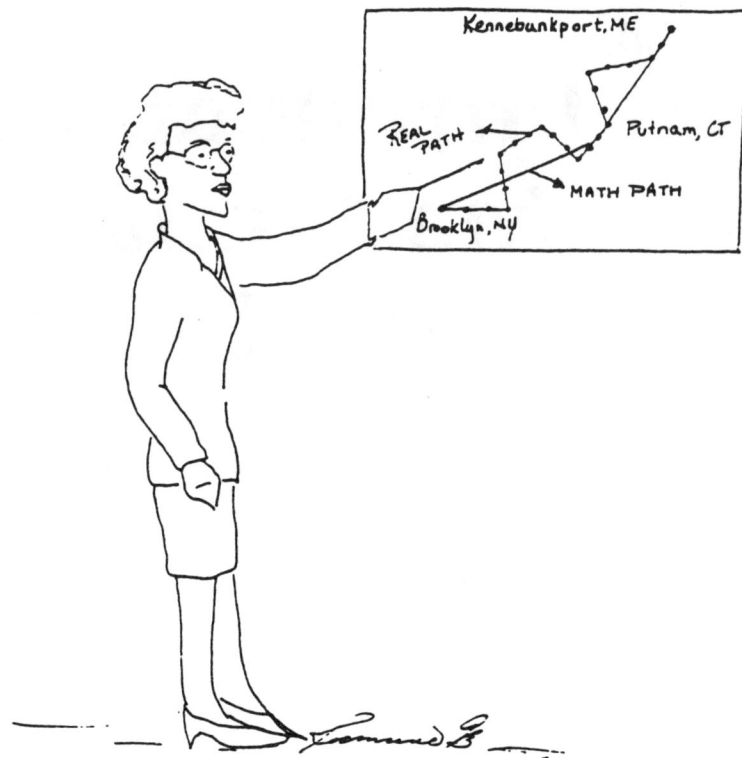

Ann's mathematical model is an idealized, abstract rendering of the real situation involving a trip from Brooklyn to Kennebunkport with a stop at Putnam. It is intended to capture the main features involved in taking such a trip and reflects these features as she sees them and the assumptions that she was led to make. It is possible that someone else planning such a trip would see things in another light and make very different assumptions.

By employing the mathematical operation division we obtain the valid conclusion that an object moving along the idealized path of Ann's model at an average speed of 50 miles per hour would take 350/50 = 7 hours to make the journey. If we add 2 hours for the picnic, we have a total of 9 hours for the travel time. This conclusion is a valid one with respect to Ann's model, valid in the sense that it is an inescapable consequence of the assumptions she was led to make. If we accept Ann's model as a starting point, then we must accept this conclusion as following from it in a deductive logical sense; it is inherent in the model. Mathematical methods, whether simple or the ultimate in technical sophistication, give valid conclusions from the mathematical models we set up. Valid conclusions obtained from a mathematical model are called **theorems** and the assumptions underlying the model are called **postulates** or **axioms**.

The interpretation of Ann's theorem is that if the trip were made under conditions realistically described by her postulates, it would take around 9 hours to complete the journey. What is the acid test of the accuracy of Ann's theorem—accuracy in the sense of how closely it describes the journey in the real world? Take the trip and see how long it takes. This is just what Ann did. It took her 9 hours and 10 minutes, which is close to the projected 9 hours for the journey. This established that her theorem is realistic in this case and, by reflection, was evidence in support of the realism of her model for the trip.

Andy's Model

Ann's son Andrius, called Andy by his friends, was planning his own trip to Kennebunkport to be taken in August. Andy did not intend to stop at Putnam and laid out a different model. Andy's assumptions about traffic flow, speeds that would be possible, and the duration of the one rest stop that he envisioned led him to a model consisting of a line 330 miles long joining points representing Brooklyn and Kennebunkport and the problem of determining how long it would take an object moving at an average speed of 55 miles per hour to cover this distance.

By again employing the mathematical operation division we obtain from Andy's model the valid conclusion that it would take 330/55 = 6 hours. This conclu-

Finite Mathematics, Models, and Structure

sion is valid with respect to Andy's model in the sense that it follows as an inescapable consequence of his model; it is inherent in Andy's model.

The interpretation of Andy's theorem is that if the trip were made under conditions realistically described by the postulates of his model, it would take around six hours to complete the journey. Andy took the trip in the middle of August and it took him 5 hours and 55 minutes, which is close to the projected time stated by his theorem. This established that his theorem is realistic in this case and, by reflection, was evidence in support of the realism of his model for the trip.

Rasa's Trip: Who's Right?

Andy's sister Rasa was planning to take a brief vacation trip to Kennebunkport during Labor Day weekend. Andy's model looked reasonable to her and she decided to follow the route it prescribed, expecting the journey to take around 6 hours.

Rasa took the trip on Labor Day weekend, but it took her 7 hours and 20 minutes. This actual trip time differs considerably from the projected 6 hour trip time of Andy's theorem so that something clearly went wrong; but what? "Your theorem

stinks," Rasa shouted at her brother in a somewhat agitated manner; "it's wrong; it's not valid," she continued.

Rasa's experience proved Andy's theorem wrong in terms of reality, but not validity. The distinction is a fundamental one. In the confrontation between what actually happened and what the theorem says should happen, what actually happened—reality—wins. Andy's theorem is a false statement as a description of the travel time to Kennebunkport on a Labor Day weekend, but it remains a theorem. It is an inescapable consequence of Andy's model—more specifically, his assumptions or postulates—and this is what makes a theorem a theorem. Mathematical methods, in this case the division of 330 by 55 yielding 6, did what it is capable of doing; it yielded a valid conclusion with respect to Andy's model. That valid conclusion may or may not be realistic. Validity is not the same as reality, but they are linked in this way: If the postulates of a model are realistic, so will be the valid conclusions obtained from it. If a valid conclusion is found to be false, this sends us a signal that some of the postulates of the model are unrealistic and require modification.

Andy's model, Rasa's experience showed, is not realistic for travel to Kennebunkport on a Labor Day weekend. In reexamining Andy's postulates we find that they do not realistically take into account unusually heavy traffic delays around the tollgates of the Whitestone Bridge characteristic of holiday weekends. Further examination of Rasa's actual trip shows that this is precisely where she had the difficulty.

We should keep in mind too that Andy did not design his model with a holiday weekend trip in mind. Major lesson: Look before you leap! That is, look at a mathematical model's assumptions before you use it.

Ann's and Andy's experiences showed that their theorems are realistic for their trips. By reflection, this is evidence in support of the realism of their models for the envisioned trips. This evidence does not, however, establish that their models are completely in accord with reality, as Rasa distressingly discovered when she employed her brother's model.

A mathematical model gives us a picture of a situation or phenomenon, but we cannot view the model as complete or the only possible picture. Other models are possible as well.

2.2. THE MATHEMATICAL MODEL BUILDING PROCESS

The development of a mathematical model for a real world problem or phenomenon basically consists of the following steps.

> **1.** Specify the real world problem or phenomenon to be investigated.

Let us reconsider, for example, Andy's desire to travel from home in Brooklyn, New York, to the vacation town of Kennebunkport, Maine. His problem is to determine how long the trip would take.

> **2.** Collect relevant data about the problem or phenomenon being studied and formulate an idealized representation for it, called a **mathematical model** (or more properly, the **hypothesis of a mathematical model**), from which valid conclusions, called **predictions,** can be obtained by mathematical methods.

The setting of a problem or phenomenon contains numerous features and characteristics, many of which are irrelevant or unessential to the focus of the study. In developing a model assumptions must be made as to which features of the setting are essential to the study and which are negligible. This calls for insight and discretion and is the most controversial aspect of the model development process; one person's essential might be another's irrelevancy.

Moreover, a delicate balance between realism and mathematical manageability must be struck. If we adopt a play-it-safe approach which favors incorporating into the model as many features as possible in the belief that we would thereby be less likely to leave out something important, then not only do we run the risk of clogging up the model with gibberish, but we might make it mathematically intractable; that is, we might not be able to apply existing mathematical methods to obtain useful theorems from the hypothesis of the model. Let us note that a **mathematical model** consists of its assumptions (or postulates) together with its theorems, although we will sometimes use the term model synonymously with its assumptions. If we dismiss too many features of the problem or phenomenon, or the wrong ones, as unessential to our study, then our model diverges too greatly from the phenomenon it is to represent and thus becomes unsuitable in this regard.

Concerning Andy's Brooklyn to Kennebunkport trip, for example, he began the data collection phase by examining a map and laying out a route. Estimates were obtained on the nature of the traffic flow along various parts of the route and the speeds that are possible. He collected data about the distances between various points on the route and introduced assumptions about the number of rest stops to be made and their duration. All such considerations led him to a model consisting of a line 330 miles long joining points representing Brooklyn and Kennebunkport and the problem of determining how long it takes an object going an average speed of 55 miles an hour to travel this distance. This mathematical model is an abstraction of the real problem of determining how long a trip from Brooklyn to Kennebunkport would take. How suitable this model is depends on how realistic are the assumptions on which the model is based.

> **3.** Apply mathematical methods to obtain valid conclusions (theorems) with respect to the hypothesis of the model.

For Andy's model for the Brooklyn to Kennebunkport trip the mathematical operation division yields the valid conclusion that a point going an average speed of 55 miles an hour along a straight line path 330 miles long will travel this distance in $330/55 = 6$ hours.

> **4.** Interpret the theorems obtained from the hypothesis of the model in terms of the real world phenomenon the model represents.

For Andy's Brooklyn to Kennebunkport trip we would expect the journey to take somewhere in the neighborhood of 6 hours, provided that the conditions under which the actual trip takes place are realistically represented in his model.

> **5.** Test the accuracy of the model; that is, compare the interpreted conclusions of the model with the results obtained from reality within some desired degree of accuracy.

If the interpreted conclusions can be shown to closely approximate reality within some degree of accuracy, that is, to be true within some desired degree of accuracy (by means of observation which may be coupled with experimentation), then this increases our confidence in the mathematical model as a realistic description of the phenomenon under study.

If in making the trip from Brooklyn to Kennebunkport Andy found that the time taken is close to 6 hours, then this would serve as an acid test which confirms the model as a realistic representation of his undertaking.

However, a model developed for a phenomenon may not be the only representation for the phenomenon. Other models are possible as well; we cannot entertain uniqueness in model representation. Andy, for example, may well have put together another model for his envisioned trip based on another route and assumptions.

Suppose alternative models are available for a phenomenon; how do we choose between them? The decisive verdict is rendered by reality. The crucial question is, are the model's theorems in accord with reality? When some of the model's theorems, suitably interpreted, are found to be in disagreement with the findings of experimentation and observation, this tells us that some part of the model's hypothesis is incomplete or unrealistic and that the model must be refined or abandoned as a description of the phenomenon in question. A mathematical model lives as a description of a phenomenon as long as it avoids head on collision with some contrary fact of observation or experiment.

72   Finite Mathematics, Models, and Structure

A theorem found to be false (or true for that matter) does not lose its status as a theorem. It remains a valid consequence of the hypothesis of the model, correct in the sense of validity, incorrect in the sense of reality. It is unfortunate that many writers on science and mathematics use the terms valid and true synonymously. (Sometimes this is done out of sloppiness, sometimes out of ignorance.)

Suppose, as Rasa found, much to her displeasure, that the Brooklyn to Kennebunkport trip, made on Labor Day weekend, took 7 hours and 20 minutes instead of the 6 hours stated in the model's theorem. The theorem remains a theorem of Andy's model, but it now acquires the status of false theorem for travel on a holiday weekend. His model is unsuitable for holiday weekend travel and its hypothesis would have to be modified for such a purpose.

The following characteristics are also important for mathematical models. A model should bring in only observable entities connected with the phenomenon it is to describe. Although a model has its origins in data and observations, it must go beyond them in leading to significant new predictions and results; the theorems of the model should provide us with significant new insights into the phenomenon in question.

> **6.** Refine the model as evidence obtained from experimentation and observation make necessary.

The model Rasa employed for her Labor Day weekend trip from Brooklyn to Kennebunkport proved unsatisfactory because its predicted travel time of 6 hours differed considerably from the actual 7 hours and 20 minutes needed for the journey. In reexamining the hypothesis of Andy's model she found that it did not take into account unusually heavy traffic delays around the tollgates of the Whitestone bridge characteristic of holiday weekends. This would lead her to suitably refine the hypothesis of Andy's model to make it applicable to holiday weekend trips.

Figure 2.1 summarizes the basic steps in the development of a mathematical model in diagramatic form.

EXERCISES

1. The Asta Company makes shoes. The Company plans to introduce two new styles, designated by A-18 and A-21, for the fall season and management wants to know how to set its monthly production schedule so as to maximize profit. Two consulting firms were hired to study this problem and to make recommendations. Each consulting firm formulated a mathematical model, designated by M1 and M2 for convenience. A conclusion obtained from M1 states that to maximize profit 50,000 A-18 pairs and 35,000 A-21 pairs should be produced monthly with an anticipated

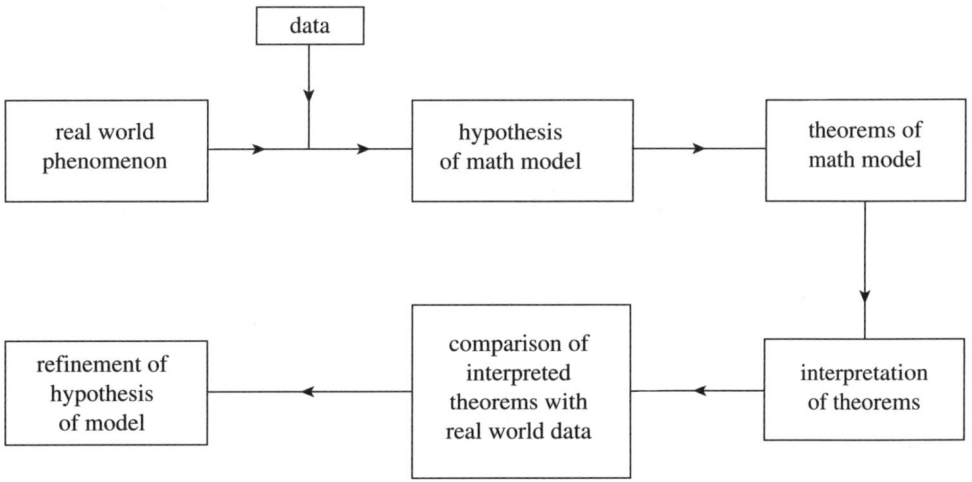

Figure 2.1

monthly profit of $230,000. A conclusion obtained from M2 states that to maximize profit 40,000 A-18 and 50,000 A-21 pairs should be produced monthly with an anticipated profit of $200,000.

(a) If mathematics is the precise subject it is reputed to be, should there not be one solution to the problem of developing a production schedule to maximize profit instead of two? Explain.

(b) Since two solutions emerge, does it follow that not both are valid? Explain.

(c) Would you implement model M1 because it projects a larger monthly profit than M2? Explain.

(d) It came to pass that M1 was implemented. Subsequently, the Asta Company sustained a consistent monthly loss on the A-18 and A-21 styles and discontinued them. How can one explain these developments?

2. While on your way to your Aunt Alice's birthday celebration, her hundredth, it has been suggested, the world seemed to be in conspiracy against you. It was snowing, traffic was bumper to bumper, and then someone in an old oldsmobile sideswiped your new buick and took off. You finally arrive at your aunt's celebration, but, needless to say, not in very good humor. Before you know it you find yourself in a social group being "educated" by Uncle George. George thinks he knows everything about everything and on this occasion his subject is mathematics, specifically geometry. "Since their truth was established by the precise mathematical reasoning for which the ancient Greek mathematicians are justly famous, the truth of the theorems of Euclidean geometry is beyond question," bellows George in his most authoritative sounding tone. Usually your attitude toward George is one of toleration, but this time you're ready for bear. What reply would give to his "profound observation"?

CHAPTER 3

Linear Program Models

3.1. THE BIRTH AND REBIRTH OF LINEAR PROGRAMMING

The seed from which the discipline called linear programming first germinated was planted in the late 1930's when the Leningrad Plywood Trust approached the Mathematics and Mechanics Department of Leningard University for help in solving a production scheduling problem of the following nature. The Plywood Trust had different machines for peeling logs for the manufacture of plywood. Various kinds of logs were handled and the productivity of each kind of machine (that is, the number of logs peeled per unit of time) depended on the wood being worked on. The problem was to determine how much work time each kind of machine should be assigned to each kind of log so that the number of peeled logs produced is maximized. A basic condition which had to be satisfied is that if logs of a given type of wood make up a specified percent of the input, then peeled logs of that type would also constitute that percent of the output.

The germination of this seed is due to Leonid Kantorovich who saw that it together with a wide variety of economic planning problems can be formulated in terms of what are today called linear program models. These problems involved the optimum distribution of worktime of machines, minimization of scrap in manufacturing processes, best utilization of raw materials, optimum distribution of arable land, optimal fulfillment of a construction plan with given construction materials, and the minimal cost plan for shipping freight from given sources to given destinations.

In 1939 Kantorovich published a report* on his discoveries that included a method sufficient for solving all the linear program models he had formulated for the aforenoted problems. The chaos of the Second World War and the postwar intellectual climate in the Soviet Union were highly unfavorable for the development and implementation of Kantorovich's linear programming methods in the Soviet economic scene. Independently of the Soviet scene, linear programming methods were developed in the United States and Western Europe in the late 1940's, and the 1950's and 60's saw the development of a wide variety of linear program models for problems arising in such areas as economic planning, accounting, banking, finance, industrial engineering, and marketing. The thaw in the Soviet Union's intellectual climate which followed Joseph Stalin's death in 1953 saw the development and implementation of Kantorovich's ideas in the economic life of the U.S.S.R.

In 1975 Kantorovich was a co-recipient of the Nobel Prize in economics for his development of linear programming methods and their application in economics.

3.2. A TALE OF TWO LINEAR PROGRAMS

The Austin Company, a producer of high quality electronic home entertainment equipment, has decided to enter the digital tape player market by introducing two models, DT-1 and DT-2. Their problem is to determine the number of units of each model to be produced to maximize profit.

The Company's operations research department was asked to study the situation and make recommendations. The OR department began their analysis by collecting data. They divided the manufacturing process into three phases; construction, assembly, and finishing. The data collected and their analysis led them to make the following assumptions.

(a) In the construction phase each DT-1 unit requires 2 hours of labor and each DT-2 unit requires 3 hours of labor. At most 1,100 hours of construction time are available per week.

(b) In the assembly phase each DT-1 unit requires 5 hours of labor and each DT-2 unit requires 3 hours of labor. At most 1,400 hours of assembly time are available per week.

(c) In the finishing phase each DT-1 unit requires 4 hours of labor and each DT-2 unit requires 1 hour of labor. At most 756 hours of finishing time are available per week.

(d) After taking cost and revenue factors into consideration the anticipated profit for each DT-1 unit is $150 and the anticipated profit for each DT-2 unit is $120. In order for these unit profit values to hold the Company must produce at least 25 DT-1 and 40 DT-2 units per week.

(e) There is an unlimited market for the DT-1 and DT-2 models.

---

*L. V. Kantorovich, *Mathematical Methods of Organizing and Planning Production,* Leningrad University, 1939. For an English translation, see *Management Science,* vol. 6, no. 4 (July 1960), pp. 363–422; or V. S. Nemchinov, ed., *The Use of Mathematics in Economics* (Cambridge, Mass.: MIT Press, 1964).

(f) All factors other than the ones considered in the analysis of the production of the DT-1 and DT-2 models are negligible.

Its next task was to translate these assumptions into mathematical form, being careful to include everything stated in the assumptions and not go beyond them. The OR department began by introducing variables for the quantities it sought to determine; let x denote the number of DT-1 and y the number of DT-2 units to be made weekly. There is a fair amount of data contained in the assumptions and it is useful to make it available at a glance in tabular form. This is done in Table 3.1.

Table 3.1

	No. of Units to be made	Profit per unit	Construction time per unit (hrs)	Assembly time per unit (hrs)	Finishing time per unit (hrs)
DT-1	x	$150	2	5	4
DT-2	y	$120	3	3	1

The key to expressing profit and the conditions that have emerged in terms of x and y is the information stated on unit profits and unit construction, assembly and finishing times for DT-1 and DT-2. We have:

$$profit = \begin{bmatrix} profit\ on \\ DT-1 \end{bmatrix} + \begin{bmatrix} profit\ on \\ DT-2 \end{bmatrix}$$

$$= \begin{bmatrix} profit\ on \\ one\ DT-1 \\ unit \end{bmatrix} \cdot \begin{bmatrix} no.\ of \\ units \\ made \end{bmatrix} + \begin{bmatrix} profit\ on \\ one\ DT-2 \\ unit \end{bmatrix} \cdot \begin{bmatrix} no.\ of \\ units \\ made \end{bmatrix}$$

$$= 150x + 120y$$

The profit obtained by making x DT-1 and y DT-2 units per week is expressed by the linear function:

$$P(x, y) = 150x + 120y$$

As to the conditions that x and y must satisfy, since the number of units made must be non-negative, we have:

$$x \geq 0$$
$$y \geq 0$$

The construction time condition is that

$$(\text{total construction time used}) \leq 1100.$$

In terms of unit construction times, 2 hours are needed for one unit of DT-1 and 3 hours are needed for one unit of DT-2, $2x$ hours are needed for x DT-1 units and $3y$ hours are needed for y DT 2 units. The total construction time utilized is expressed by $2x + 3y$. Thus, the construction time utilized is expressed by $2x + 3y$. Thus, the construction time condition is:

$$2x + 3y \leq 1100$$

Similarly, the assembly and finishing time conditions are stated by the inequalities:

$$5x + 3y \leq 1400$$
$$4x + y \leq 756$$

The conditions that at least 25 DT-1 and 40 DT-2 units must be produced weekly are expressed by the inequalities:

$$x \geq 25$$
$$y \geq 40$$

We thus emerge with the following mathematical structure, called linear program model LP-1, as a translation of the assumptions made by the OR department of the Austin Company.

$$\text{Maximize } P(x, y) = 150x + 120y$$

subject to

$$x \geq 0, \quad y \geq 0$$
$$2x + 3y \leq 1100$$
$$5x + 3y \leq 1400$$
$$4x + y \leq 756$$
$$x \geq 25, y \geq 40$$

Here x represents the number of DT-1 and y the number of DT-2 units to be made weekly.

More generally, a **linear program** is a mathematical problem with the following structure: there is specified a linear function of a number of variables that are required to satisfy linear conditions described by some mixture of linear inequalities and linear equations, called **constraints**. The problem is to find values for these variables which satisfy the constraints and yield the maximum, or minimum, value of the function, which is called an **objective function.** LP-1 is a 2-variable linear program, but the same kind of problems may involve 200, or 2000, or even 200,000 or more variables.

The Austin Company also hired the Marks Company, a consulting operations research firm, to independently study the digital tape player situation and make recommendations. The Marks OR group divided the manufacturing process into two phases: construction (which included assembly) and finishing. The data collected and their analysis led them to make the following assumptions:

(a) In the construction phase each DT-1 unit requires 8 hours of labor and each DT-2 unit requires 5 hours of labor. At most 2,210 hours of construction time are available per week.
(b) In the finishing phase each DT-1 unit requires 3 hours of labor and each DT-2 unit requires 2 hours of labor. At most 860 hours of finishing time are available per week.
(c) The anticipated profit for each DT-1 unit is $140 and the anticipated profit for each DT-2 unit is $150. In order for these unit profit values to hold the company must produce at least 50 DT-1 and 50 DT-2 units per week.
(d) There is an unlimited market for the DT-1 and DT-2 models.
(e) All factors other than the ones considered in the analysis of the production of the DT-1 and DT-2 models are negligible.

The same sort of analysis that leads to LP-1 from the assumptions made by the Austin Company's operation research department leads to the Marks OR group's linear program model LP-2:

$$\text{Maximize } P(x, y) = 140x + 150y$$

subject to

$$x \geq 0, \quad y \geq 0$$
$$8x + 5y \leq 2210$$
$$3x + 2y \leq 860$$
$$x \geq 50, y \geq 50,$$

where x represents the number of DT-1 and y the number of DT-2 units to be made weekly.

We turn our attention to developing a systematic approach to solving such problems in the next section.

EXERCISES

1. Set up LP-2 from the assumptions formulated by the Marks Company.

3.3. THE CORNER POINT SOLUTION METHOD

We next address the problem of solving LP-1 and LP-2. To do this we develop a simple method for solving linear programs called the **corner point method.** It has an appealing geometric flavor and is effective for 2-variable linear programs.

As a working vehicle consider the linear program:

$$\text{Maximize } F(x, y) = 5x + 8y$$

subject to

$$x \geq 0, \quad y \geq 0$$
$$x + 2y \leq 8$$
$$3x + y \leq 9.$$

The points which satisfy the constraints of a linear program, called **feasible points,** are the points that the objective function to be optimized may be applied to.

Our problem here is to determine that feasible point (or those feasible points, if there is more than one) which yields the maximum value for the objective function $F(x, y) = 5x + 8y$.

Our first step in solving this problem is to obtain a geometric representation of the feasible points by graphing the constraints. This is done in Example 8 of Section 1.7 (p. 60).

The intersection points of at least two boundary lines which come out of the equality conditions of the constraints of a 2-variable linear program are called **corner points.** The corner points of our linear program are (0, 0), (0, 4), (3, 0) and (2, 3) – obtained by solving the system of equations $x + 2y = 8$ and $3x + y = 9$; they are shown in Figure 3.1.

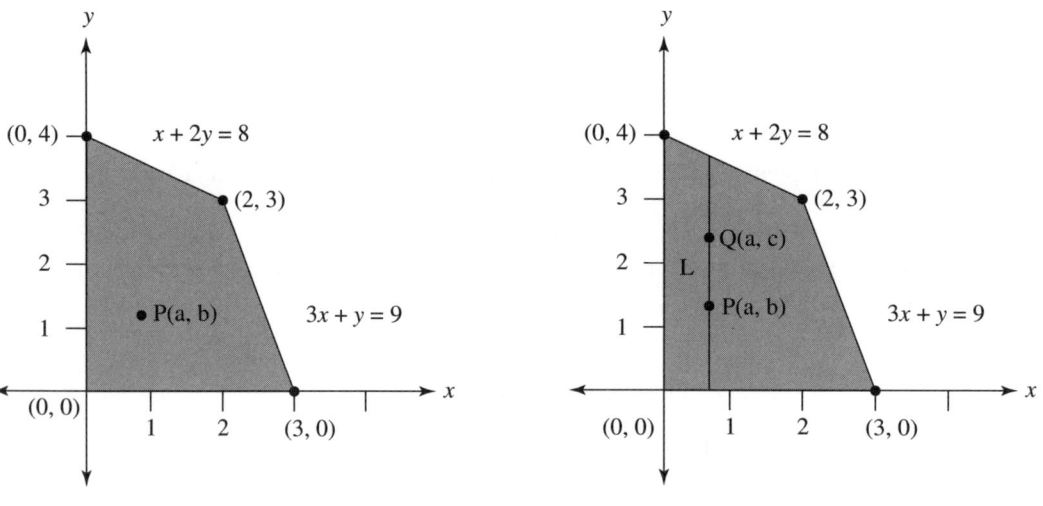

Figure 3.1 Figure 3.2

The significance of corner points is made clear by the following argument. As a starting point consider any feasible point $P(a, b)$ in the region of feasible points of our linear program; see Figure 3.1. If we move up the vertical line L passing through $P(a, b)$ to $Q(a, c)$, (see Figure 3.2), our objective function $F(x, y) = 5x + 8y$

increases in value from $5a + 8b$ to $5a + 8c$ since c is larger than b. By taking feasible points higher and higher on L we increase $F(x, y) = 5x + 8y$. Since we must remain within the set of feasible points in taking points higher and higher on L, we can go as far as $R(a, d)$ on the boundary line $x + 2y = 8$ (Figure 3.3). From there we can move in one of two directions on the boundary line until we come to a corner point (see Figure 3.4).

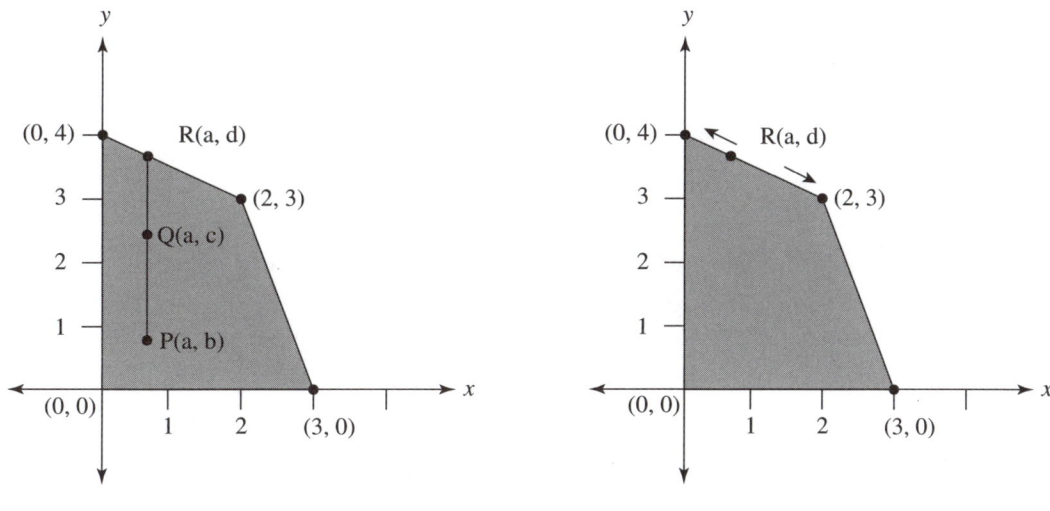

Figure 3.3   Figure 3.4

The question that this raises is, how does a linear function behave as we take points in one direction or the other along a boundary line? We shall not do so here, but one can prove that one of two things happens: (a) The linear function increases in value as we take points in one direction along the boundary line and decreases as we take points in the other direction, or (b) the linear function has the same value at all the points on the boundary line. In either case we are led to a corner point. This argument suggests the following theorem.

Corner Point Theorem: If a 2-variable linear program has an optimal value (maximum or minimum value, depending on the nature of the linear program), then a solution yielding this optimal value can be found from among the corner points of the linear program.

The corner point theorem's hypothesis presupposes that the linear program under consideration has a solution. It does not say that a solution cannot occur at a feasible point which is not a corner point; this does happen with some linear programs. We are, however, assured of a solution at a corner point, assuming that the linear program has a solution to begin with.

Implementation of the corner point method to solve a 2-variable linear program involves the following sequence of steps:

> **Corner Point Method Steps**
>
> 1. Graph the feasible points of the linear program.
> 2. Locate its corner points on the graph.
> 3. Determine the coordinates of all corner points. For a corner point which is not on either of the coordinate axes this is done by solving the system of equations which describe a pair of boundary lines which intersect at the corner point.
> 4. Compute the value of the objective function at each corner point.
> 5. From these values pick out the largest or smallest value, depending on the nature of the linear program, and the solution(s) which yields it.

To illustrate these procedures we return to our vehicle:

$$\text{Maximize } F(x, y) = 5x + 8y$$

subject to

$$x \geq 0, \quad y \geq 0$$
$$x + 2y \leq 8$$
$$3x + y \leq 9$$

The graph of its feasible points is reproduced as Figure 3.5. The corner points, displayed in Figure 3.6, are $(0, 0)$, $(3, 0)$, $(2, 3)$ and $(0, 4)$; as was previously noted, $(2, 3)$ is obtained by solving the system of boundary line equations $3x + y = 9$ and $x + 2y = 8$.

The computation of the value of the objective function $F(x, y) = 5x + 8y$ at the corner points yields the results summarized in Table 3.2, from which we see that 34 is the maximum value and $(2, 3)$ is the solution.

EXAMPLE 1

Solve LP-1, the linear program model obtained by the operations research department of the Austin Company.

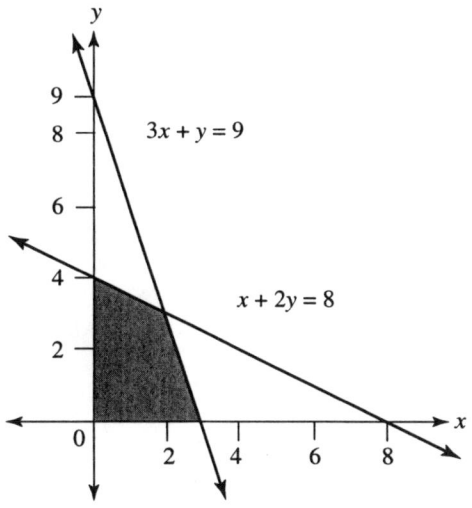
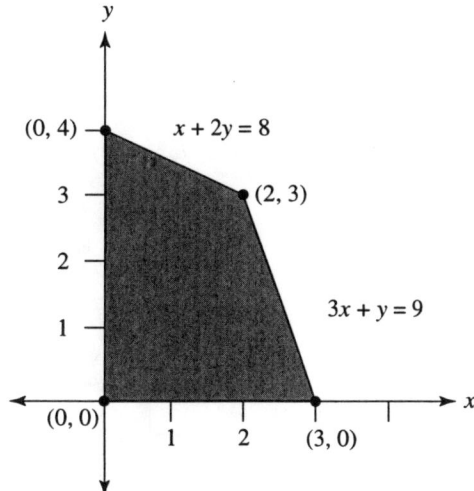

Figure 3.5   Figure 3.6

Table 3.2

Corner Point	$F(x, y) = 5x + 8y$
(0, 0)	0
(3, 0)	15
(2, 3)	34
(0, 4)	32

Maximize $P(x, y) = 150x + 120y$

subject to

$$x \geq 0, \quad y \geq 0$$
$$2x + 3y \leq 1100$$
$$5x + 3y \leq 1400$$
$$4x + y \leq 756$$
$$x \geq 25, y \geq 40$$

Our first step is to sketch the graph of the feasible points. For this see Exercise 67 of Section 1.7 (p. 62).

Locate the corner points on the graph and solve the appropriate systems of equations to determine their coordinates. There are five corner points, shown in Figure 3.7: (25, 40), (25, 350), (100, 300)—obtained by solving $2x + 3y = 1100$ and $5x + 3y = 1400$, (124, 260)—obtained by solving $4x + y = 756$ and $5x + 3y = 1400$, and (179, 40).

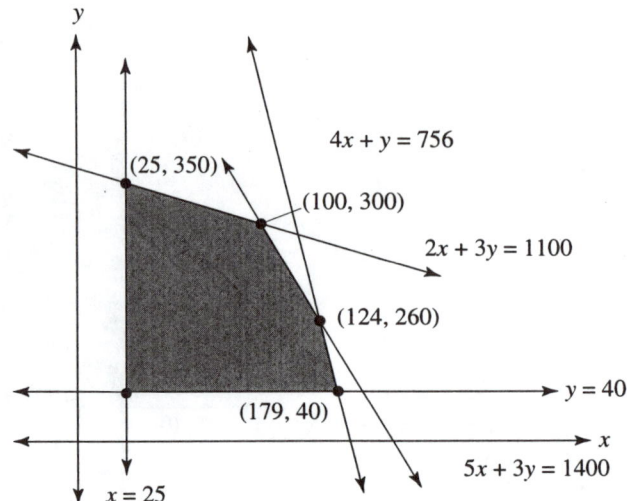

Figure 3.7

The computation of $P(x, y) = 150x + 120y$ at the corner points, summarized in Table 3.3, yields the solution (100, 300) with maximum value 51,000.

Table 3.3

Corner Point	$P(x, y) = 150x + 120y$
(25, 40)	8,550
(25, 350)	45,750
(100, 300)	51,000
(124, 260)	49,800
(179, 40,)	31, 650

EXAMPLE 2

Solve LP-2, the linear program model obtained by the Marks Company for the Austin Company:

$$\text{Maximize } P(x, y) = 140x + 150y$$

subject to

$$x \geq 0, y \geq 0$$
$$8x + 5y \leq 2210$$
$$3x + 2y \leq 860$$
$$x \geq 50, y \geq 50$$

Our first step is to sketch the graph of the feasible points. For this see Exercise 68 of Section 1.7 (p. 62).

Locate the corner points on the graph and solve the appropriate systems of equations to determine coordinates. There are four corner points, shown in Figure 3.8: (50, 50), (245, 50), (120, 250), and (50, 355).

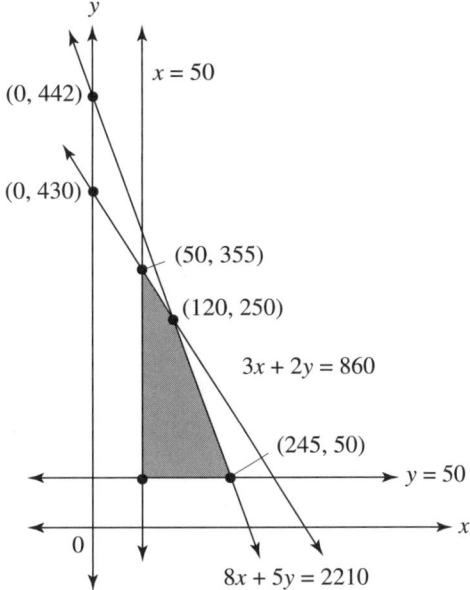

Figure 3.8

The computation of $P(x, y) = 140x + 150y$ at the corner points, summarized in Table 3.4, yields the solution (50, 355) with maximum value 60, 250.

Table 3.4

Corner Point	$P(x, y) = 140x + 150y$
(50, 50)	14,500
(245, 50)	41,800
(120, 250)	54,300
(50, 355)	60,250

Model LP-1 has solution (100, 300) with maximum value 51,000 whereas model LP-2 has solution (50, 355) with maximum value 60,250. In terms of the Austin Company's situation, to implement LP-1's conclusion the production schedule would have to be set to manufacture 100 DT-1 and 300 DT-2 units per week with an anticipated weekly profit of $51,000; to implement LP-2's conclusion the production schedule would have to be set to manufacture 50 DT-1 and 355 DT-2 units per week with an anticipated weekly profit of $60,250.

Which solution should be implemented? It may seem foolish even to ask, but is it? We explore this issue in the next section.

EXERCISES

Solve, if possible, the following linear programs.

1. Max. $F(x, y) = 5x + 4y$
   subject to
   $x \geq 0, y \geq 0$
   $2x + 3y \leq 15$
   $x + y \leq 6$

2. Min. $G(x, y) = 10x + 12y$
   subject to
   $x \geq 0, y \geq 0$
   $3x + 5y \leq 12$
   $x + y \geq 3$

3. Min. $C(x, y) = 1.50x + 1.10y$
   subject to
   $x \geq 0, y \geq 0$
   $2x + y \geq 4$
   $x + y \geq 3$
   $x + 2y \geq 4$

4. Min. $C(x, y) = 8x + 12y$
   subject to
   $x \geq 0, y \geq 0$
   $x + y \geq 250,000$
   $2x + y \leq 400,000$

5. Max. $I(x, y) = 0.10x + 0.08y$
   subject to
   $x \geq 0, y \geq 0$
   $x + y \leq 60$
   $-x + 3y \geq 0$
   $x \geq 15$

6. Max. $F(x, y) = 3x + 2y$
   subject to
   $x \geq 0, y \geq 0$
   $x + 2y \geq 4$
   $x - 2y \geq 2$

7. Max. $I(x, y) = 110x + 70y$
   subject to
   $x \geq 0, y \geq 0$
   $x \leq 25$
   $2x + 4y \leq 120$
   $4x + 2y \leq 110$

8. Max. $C(x, y) = 39,600x + 54,000y$
   subject to
   $x \geq 0, y \geq 0$
   $2x + 3y \leq 50$
   $x + y \geq 30$

9. Max. $P(x, y) = 180x + 120y$
   subject to
   $x \geq 0, y \geq 0$
   $4x + 3y \leq 320$
   $5x + 2y \leq 330$

10. Max. $P(x, y) = 190x + 110y$
    subject to
    $x \geq 0, y \geq 0$
    $5x + 2y \leq 330$
    $3.25x + 2y \leq 225$
    $4x + 3y \leq 320$

11. Min. $C = -5x + 12,795$
    subject to
    $x \geq 0, y \geq 0$
    $x + y \leq 100$
    $x + y \geq 15$
    $x \leq 75, y \leq 90$

12. Min. $G(s, t) = 90s + 80t$
    subject to
    $s \geq 0, t \geq 0$
    $s + 2t \geq 10$
    $3s + t \geq 15$

3.4. WHICH SOLUTION SHOULD BE IMPLEMENTED?

It's always easier when you have a choice of one; take it, or leave it. But a choice of two is another matter. Bottom-line Bob, chairman of the ten member board charged with making a decision on how to implement the Company's entry into the digital tape player market, argued that it's obvious what we should do. "Implementation of LP-2 brings us a weekly profit of $60,250, whereas implementation of LP-1 brings us a weekly profit of $51,000. Since we want the largest possible return, we should go with LP-2. You can't make it any simpler." The board voted nine to one to implement LP-2.

Alas, the $60,250 weekly profit was far from being realized after LP-2 was implemented, and two years later the Austin Company's venture into the digital tape player market had to be written off as a disaster.

Bottom-line Bob, who presided over this disaster was confused, upset, and out of a job. He went to Reflective Ramuné, the chair of the new board and the one person who had voted against implementation of LP-2, with some questions: "We were ultra-cautious and obtained the additional services of the Marks OR group to make recommendations which, subsequently, had disastrous consequences for us; what

went wrong? How is it that mathematics failed us? Why did you vote against implementation of LP-2?" "Well Bob, as I pointed out at the board meeting, I voted against implementation of LP-2 because I was not convinced that its promise of a $60,250 weekly profit was realistic. As you yourself pointed out, 'the promise of LP-2 is $9,250 better than that promised by LP-1,' but promises may not be realizable if they are founded on unrealistic assumptions. The conclusion reached from LP-2 was indeed tempting, and in fact proved too tempting for my colleagues on the board, but since it came from a linear program founded on assumptions which I viewed as unrealistic, I resisted temptation. We have no quarrel with mathematics; mathematics gave us a valid conclusion from LP-2, which is all that we can legitimately expect. Unfortunately that conclusion proved to be unrealistic."

Is Mathematics Precise?

"Ramuné, I don't understand this. I always liked math in high school and college. Solving those equations, factoring those expressions, differentiating those functions, throwing the data into the computer and letting it do its thing, that was real fun. What I like most about math is its precision. You don't get ten sides to a story. You get one answer and that's that; no baloney"."Bob, I think your math

courses may have focused too much on technique and not enough on perspective. Technique can be fun to a point, but without a perspective on its place in the over-all role of mathematics in applications, we see only a small tip of the mathematical iceberg. Mathematics is precise in the sense that it gives us valid conclusions based on the assumptions made, which is where technique—factoring, differentiating functions, and the like—plays its major role. Whether the assumptions made are realistic or not is another matter which technique can't help us with. The question of how to formulate these assumptions and reach a judgment on their realism may indeed yield ten sides to the story. I'm afraid that those who find mathematics attractive because of what they perceive to be its absolutist nature have misunderstood the meaning of mathematical precision."

EXERCISES

1. ZKB Electronics puts out two kinds of personal computers, model ZKB-47 and model ZKB-82. The management of ZKB called in the Rex Consulting Firm to determine how many units of each model should be made daily so as to maximize profit. The consulting firm set up a linear program model for the electronics company's production problem and, by applying the corner-point method, reached the conclusion that 300 ZKB-47 units and 250 ZKB-82 units should be made daily to maximize profit. Before implementing this conclusion, the management of ZKB put the following questions to the director of the consulting firm.

   (a) Does use of the corner-point method guarantee that profit will be maximized when 300 ZKB-47 units and 250 ZKB-82 units are made daily and sold?

   (b) What is your basis for recommending that we implement your conclusion?

   How would you reply to these questions?

2. The Onute Corporation plans to introduce two high resolution TV models, T20 and T24, to the market. Its own operations research group was led to introduce the following M1 model to determine the optimal production schedule for maximizing profit:

   $$\text{Maximize } P(x, y) = 180x + 120y$$
   subject to
   $$x \geq 0, y \geq 0$$
   $$4x + 3y \leq 320$$
   $$5x + 2y \leq 330,$$

   where, x and y denote the number of T20 and T24 units, respectively, to be made daily. Its solution is (50, 40) with maximum value 13,800 (see Exercise 9, Sec. 3.3).

   The Alexis company was also hired to study the Onute Corporation's production scheduling problem. It was led to introduce the following M2 model to determine the optimal production schedule for maximizing profit:

$$\text{Maximize } P(x, y) = 190x + 110y$$

subject to

$$x \geq 0, y \geq 0$$
$$5x + 2y \leq 330$$
$$3.25x + 2y \leq 225$$
$$4x + 3y \leq 320,$$

where x and y denote the number of T20 and T24 units, respectively, to be made daily. Its solution is (60, 15) with maximum value 13,050 (see Exercise 10, Sec. 3.3).

The following questions have arisen. How would you answer them?

(a) If mathematics is the precise subject that it is reputed to be, should there not be one solution to this problem rather than two?

(b) Since two solutions emerge, does it follow that not both are valid? Explain.

(c) Before making a decision about whether to implement M1 or M2, what questions would you put to the two operations research groups?

(d) Which model, if either, would you adopt and implement? Why? Is it possible that you would not adopt either model?

3.5. HOW COULD IT BE WRONG? I USED A COMPUTER

"Ramuné, I still don't fully understand what went wrong. The company just spent millions to update its computer system. I had access to the latest and the best. Why didn't this save us from disaster." "Bob, Henry Clay's observation that 'statistics are no substitute for judgment' applies equally well to the computer.

We cannot expect the computer to employ technological alchemy and convert unrealistic assumptions into golden truths. Keep in mind the GIGO principle; if garbage in, then garbage out. Indiscriminate use of computer technology has made possible the generation of more and more nonsense more quickly than ever before by more people having less and less understanding of what they are doing."

The Computer's Right of Way

"If what you say is true Ramuné, then what good is this super computer technology to us?" "For number crunching and delivering results quickly and efficiently, the computer is without equal, Bob.

In this dimension it is the undisputed master of the field. The mathematical model building process and computers have developed a symbiotic relationship in that computers have made it possible for us to solve previously unapproachable large scale problems that come out of mathematical models, while the accessibility of such

problems to computer solution has made possible the use of such complex models. Alas, none of this overrides the GIGO principle."

3.6. THE SCOPE OF LINEAR PROGRAMMING APPLICATIONS

Linear programming has turned out to have a wide spectrum of applications. To obtain some sense of this spectrum we look at six case studies. The case studies are all realistic, but are presented in miniature for the sake of manageability. Actual real-life situations that emerge have the same structure and tone, but are more complex in that more factors are generally considered and more variables are required.

We view all of these situations through the eyes of others in much the same way that we see events through the eyes of a reporter or observer by reading his account of them in a newspaper, journal or book. Just as the reporter has selected what he believes are important features surrounding the events and has omitted those he considers unessential, we too are looking at features considered crucial to the situations we examine as seen by someone who has made such a selection. This selection reflects assumptions that have been made. To maintain a proper perspective on this it is important to keep in mind that other analysts, as other reporters, might see things in a different light and, accordingly, make other assumptions.

Case 1 Production Planning

The Austin Company's problem of determining the number of DT-1 and DT-2 digital tape players to be made per week so as to maximize profit, considered in Section 3.2, is a production scheduling problem which we expressed in linear program terms under the assumptions introduced. The background that led to the Austin Company's linear programs is illustrative of situations with the following general features: A firm makes a number of products or models of a product and utilizes a number of resources in their manufacture, such as raw materials, labor, capital, different machines, storage facilities. It is assumed that for each product made a fixed amount of each resource is required to make a unit of that product. Within the production time frame a fixed amount of each resource is available and cannot be exceeded. It is also assumed that for a range of possible output levels there is a fixed profit per unit of each product which does not depend on the number of units produced. Under these conditions the problem of determining output levels of the products produced so as to maximize total profit can be formulated in terms of a linear program.

As a further illustration of a production planning situation we turn to the Ramuné Company's problem.

The Ramuné Company makes stereo systems. Two new models, RA5 and RA9, are to be mass produced. Both models pass through assembly and finishing plants of the company. In the assembly plant an RA5 unit is worked on for 1 hour; an RA9 unit is worked on for 3 hours. In the finishing plant an RA5 unit is worked on for 2 hours; an RA9 unit is worked on for 1 hour. At most, 90 hours of assembly time and 80 hours of finishing time are available per week. The anticipated profit on an RA5 unit is $10 and the anticipated profit on an RA9 unit is $15.

The problem is to determine, with respect to the given assumptions, how many RA5 and RA9 units should be made per week so as to maximize profit.

We begin by introducing variables to stand for the quantities we wish to determine. Let x denote the number of RA5 units to be made and let y denote the number of RA9 units to be made. To make needed information available at a glance, we express the basic data in tabular form, as shown in Table 3.5.

Table 3.5

	No. of Units to Be Made	Profit per Unit	Assembly Time per Unit (hours)	Finishing Time per Unit (hours)
Model RA5	x	$10	1	2
Model RA9	y	$15	3	1

Since profit is to be maximized, we must express profit in terms of x and y. It is useful to note that

$$\text{profit} = \begin{bmatrix} \text{profit on} \\ \text{model RA5} \end{bmatrix} + \begin{bmatrix} \text{profit on} \\ \text{model RA9} \end{bmatrix}$$

$$\text{profit} = \begin{bmatrix} \text{profit on} \\ \text{one RA5} \\ \text{unit} \end{bmatrix} \cdot \begin{bmatrix} \text{no. of} \\ \text{units} \\ \text{made} \end{bmatrix} + \begin{bmatrix} \text{profit on} \\ \text{one RA9} \\ \text{unit} \end{bmatrix} \cdot \begin{bmatrix} \text{no. of} \\ \text{units} \\ \text{made} \end{bmatrix}$$

$$\text{profit} = 10x + 15y$$

The profit obtained by making x RA5 units and y RA9 units is expressed by the linear function

$$P(x, y) = 10x + 15y$$

Our next task is to describe the conditions that x and y must satisfy. Since the number of units made is nonnegative, we have

$$x \geq 0$$
$$y \geq 0$$

To express assembly plant operation time in terms of x and y, we note that

$$\begin{bmatrix} \text{assembly} \\ \text{plant} \\ \text{time} \end{bmatrix} = \begin{bmatrix} \text{assembly} \\ \text{time on} \\ \text{RA5} \end{bmatrix} + \begin{bmatrix} \text{assembly} \\ \text{time on} \\ \text{RA9} \end{bmatrix}$$

$$\begin{bmatrix} \text{assembly} \\ \text{plant} \\ \text{time} \end{bmatrix} = \begin{bmatrix} \text{assembly} \\ \text{time for} \\ \text{1 RA5 unit} \end{bmatrix} \cdot \begin{bmatrix} \text{no. of} \\ \text{RA5 units} \\ \text{made} \end{bmatrix} + \begin{bmatrix} \text{assembly} \\ \text{time for} \\ \text{1 RA9 unit} \end{bmatrix} \cdot \begin{bmatrix} \text{no. of} \\ \text{RA9 units} \\ \text{made} \end{bmatrix}$$

$$\begin{bmatrix} \text{assembly} \\ \text{plant} \\ \text{time} \end{bmatrix} = 1 \cdot x + 3 \cdot y$$

Since at most 90 hours of assembly time per week is available, we have

$$x + 3y \leq 90$$

Similarly, the condition that at most 80 hours of finishing time are available per week is expressed in terms of x and y by the inequality

$$2x + y \leq 80$$

In summary, the assumptions made lead to the linear program

Maximize $P(x, y) = 10x + 15y$

subject to

$$x \geq 0$$
$$y \geq 0$$

$$x + 3y \leq 90$$
$$2x + y \leq 80,$$

where x and y denote the number of RA-5 and RA-9 units to be made per week, respectively.

To solve the Ramuné Company's linear program we first sketch the graph of its feasible points, shown in Figure 3.9.

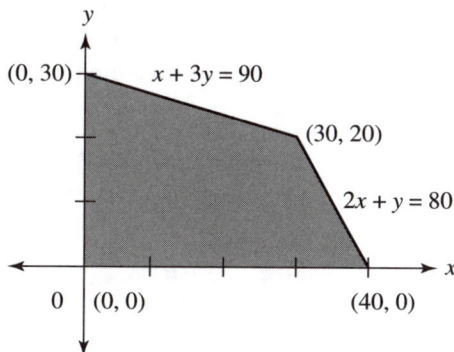

Figure 3.9

From Table 3.6 we see that (30, 20) is the solution of the Ramuné Company's linear program and 600 is its maximum value.

Table 3.6

Corner Point	$P(x, y) = 10x + 15y$
(0, 0)	0
(0, 30)	450
(40, 0)	400
(30, 20)	600

Implementation of this solution by the Ramuné Company calls for making 30 RA5 and 20 RA9 units per week. Whether this solution should be implemented or not is, of course, another issue. If a careful review of this linear program's assumptions confirms the assessment that they are realistic, then implementation of its solution makes good sense.

Case 2   Diet Problems

A sack of animal feed is to be put together from linseed oil meal and hay. It is required that each sack of feed contain at least 2 pounds of protein, 3 pounds of fat, and 8 pounds of carbohydrate. It is estimated that each unit (a unit is 30 pounds) of linseed oil meal contains 1 pound of protein, 1 pound of fat, 2 pounds of carbohydrate, and that each unit of hay contains 1/2 pound of protein, 1 pound of fat, and 4 pounds of carbohydrate. Linseed oil meal costs $1.50 per unit and hay costs $1.10 per unit.

The problem is to determine how many units of linseed oil meal and hay should be used to make up a sack of animal feed that satisfies the nutritional requirements at minimal cost.

To translate this problem and the assumptions which underlie it into a linear program, let x and y denote the number of units of linseed oil meal and hay, respectively, to be used in making up a sack of animal feed. The basic data are summarized in Table 3.7.

Table 3.7

	No. units used	Cost per unit	Protein per unit (lbs)	Fat per unit (lbs)	Carbohydrate per unit (lbs)
Linseed oil meal	x	$1.50	1	1	2
Hay	y	$1.20	1/2	1	4

Since the total cost equals the cost of linseed oil meal, $1.50x$, plus the cost of hay, $1.10y$, the cost function to be minimized is

$$C(x, y) = 1.50x + 1.10y.$$

The number of pounds of protein in the mixture equals the amount contributed by the linseed oil meal, x pounds, plus the amount contributed by the hay, $(1/2)y$ pounds. The sack of cattle feed must contain at least 2 pounds of protein, which translates to

$$x + (1/2)y \geq 2,$$

or equivalently

$$2x + y \geq 4.$$

Similarly, the fat and carbohydrate requirements are expressed by the constraints

$$x + y \geq 3$$

and

$$2x + 4y \geq 8,$$

or equivalently

$$x + 2y \geq 4.$$

We thus obtain the linear program model

$$\text{Minimize } C(x, y) = 1.50x + 1.10y$$

subject to

$$x \geq 0, y \geq 0$$
$$2x + y \geq 4$$
$$x + y \geq 3$$
$$x + 2y \geq 4,$$

which has solution (1, 2) and minimum value 2.7 (see Section 3.3, Exercise 3, p. 88).

To implement this valid conclusion of the model we would use 1 unit of linseed oil meal (30 pounds) and 2 units of hay (60 pounds). Based on the assumed costs of the ingredients, the anticipated cost of a sack of animal feed is $2.70.

The problem considered illustrates "diet problems" with the following general features: A diet, or food substance, is to be put together from a number of available foods. It is required that the diet be balanced in the sense that it must contain minimal amounts of stated nutrients—proteins, fats, carbohydrates, minerals, vitamins, etcetera. It is assumed that each food unit contains a known fixed amount of each nutritional unit and that the unit prices of the food items are known and fixed within the time period considered. The problem is to determine the minimal cost diet which satisfies the prescribed nutritional requirements.

Case 3 Environmental Protection

The Saxon Company must produce at least 250 thousand tons of paper annually. From the current operating system 10 pounds of chemical residue is deposited into a neighboring water system for each ton of paper produced. The resulting pollution has become a problem of serious concern, and to remain eligible for state tax benefits the Saxon Company must restrict the chemical residue emitted into the state's water system to not exceed 200 tons per year. Two filtration systems, Delta and Beta, have emerged for consideration. It is estimated that the installation of the Delta system would reduce emissions to 2 pounds for each ton of paper produced, and installation of the Beta system would reduce emissions to 1 pound for each ton of paper produced. Capital and operating costs for the Delta and Beta systems have been estimated at $8 and $12, respectively, per ton of paper produced.

The problem is to determine how many tons of paper should be produced subject to the Delta system and how many should be produced subject to the Beta system so that the emissions standard is met at minimal cost.

Let x and y denote the number of tons of paper to be produced annually subject to the Delta and Beta systems, respectively. The cost function to be minimized is

$$C(x, y) = 8x + 12y.$$

The condition that the Saxon Company must produce at least 250 thousand tons of paper annually is expressed by

$$x + y \geq 250{,}000.$$

The total amount of chemical residue produced annually is the number of pounds produced through use of the Delta system, $2x$ pounds, plus the amount produced through use of the Beta system, y pounds. Since this cannot exceed 200 tons, we have

$$2x + y \leq 400{,}000.$$

We thus emerge with the linear program model

$$\text{Minimize } C(x, y) = 8x + 12y$$

subject to

$$x \geq 0,\ y \geq 0$$
$$x + y \geq 250{,}000$$
$$2x + y \leq 400{,}000,$$

which has solution (150000, 100000) and minimum value 2,400,000 (see Section 3.3, Exercise 4, p. 88).

To implement this result the Saxon Company would have to produce 150,000 tons of paper annually subject to the Delta system and 100,000 tons of paper annually subject to the Beta system. The anticipated cost would be $2.4 million.

Case 4   Bank Portfolio Management

The Charles National Bank has assets in the form of loans and negotiable securities which, it is assumed, bring returns of 10 and 8 percent, respectively, in a certain time period. The bank has a total of $60 million to allocate between loans and securities. To meet unanticipated deposit withdrawals the bank maintains a securities balance greater than or equal to 25 percent of total assets. Lending is the bank's most important activity and to satisfy its clients it requires that at least $15 million be available for loans.

The bank wishes to determine, under these conditions, how funds should be allocated to maximize total investment income.

Let x and y denote the amount, in millions of dollars, to be allocated for loans and securities, respectively. The income function to be maximized is

$$I(x, y) = 0.10x + 0.08y.$$

The following constraints emerge:

$x + y \leq 60$: $60 million is available for investment in loans and securities.

$y \geq (1/4)(x + y)$,   or equivalently, $-x + 3y \geq 0$: A securities balance greater than or equal to 25% of total assets must be maintained. Note, total assets is defined as the sum of the amounts invested in loans and securities, which is $x + y$.

$x \geq 15$:   at least $15 million must be available for loans.

We thus obtain the linear program model

$$\text{Maximize } I(x, y) = 0.10x + 0.08y$$

subject to

$$x \geq 0, y \geq 0$$
$$x + y \leq 60$$
$$-x + 3y \geq 0$$
$$x \geq 15,$$

which has solution (45, 15) and maximum value 5.7 (see Section 3.3, Exercise 5, p. 88).

To implement this result the Charles Bank would have to allocate $45 million to loans and $15 million to securities. The anticipated interest on investment is $5.7 million.

Of Interest

A. Broaddus, "Linear programming: A New Approach to Bank Portfolio Management," *Federal Reserve Bank of Richmond: Monthly Review,* vol. 58, No. 11 (Nov. 1972), pp. 3–11. This article provides an introductory nontechnical discussion of linear programming for bank portfolio management.

K. J. Cohen and F. S. Hammer, "Linear Programming and Optimal Bank Asset Management Decisions," *Journal of Finance,* vol. 22 (May 1967), pp. 147–165. This paper describes a linear program model that had been used for several years by Bankers Trust Company in New York to assist in reaching portfolio decisions.

Case 5 Transportation

Heavy duty transformers made by the Thomas Company are to be sent from their plants in Dobsville and Watertown to distribution centers in New York, Chicago and Detroit. There are 100 transformers in Dobsville and 150 transformers in Watertown. The distribution centers in New York, Chicago and Detroit are to receive 75, 90 and 85 transformers, respectively. It costs $50, $52 and $54 to ship a transformer from Dobsville to New York, Chicago and Detroit, respectively. It costs $51, $48 and $50 to ship a transformer from Watertown to New York, Chicago and Detroit, respectively.

The problem is to determine how many transformers should be sent from each plant to each distribution center so that total cost is minimized.

This situation has the special feature that the total number of transformers to be sent from the plants (250) is equal to total number to be received by the destinations. This allows us to analyze the problem in terms of two variables as opposed to six variables (one linking each source to each destination) which would be needed if this equilibrium condition were not satisfied.

Let x and y denote the number of transformers to be sent from Dobsville to New York and Chicago, respectively. Send what remains at Dobsville, $100 - x - y$ transformers, to Detroit. From Watertown we send to New York, Chicago and Detroit the difference between what they should receive and what they have been sent from Dobsville. Thus, from Watertown we send $75 - x$ transformers to New York, $90 - y$ transformers to Chicago, and $85 - (100 - x - y) = x + y - 15$ transformers to Detroit. In summary, we have the shipping schedule shown in Figure 3.10.

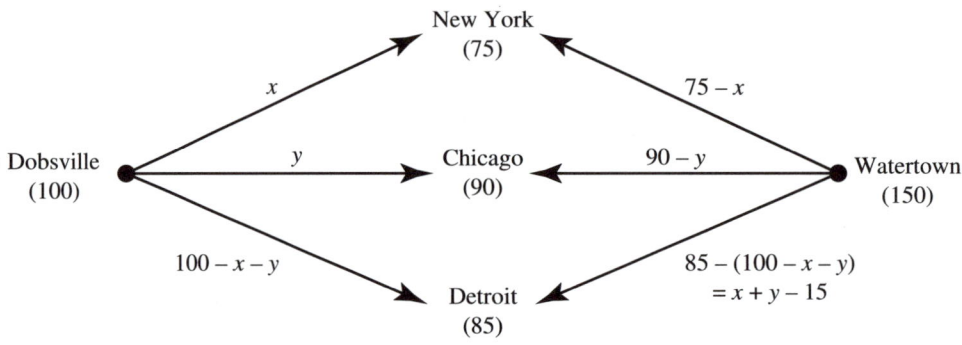

Figure 3.10

The cost of shipping the transformers from Dobsville to New York is $50x$ dollars, the cost of shipping one transformer, $50, times the number being sent, x. Similarly, the cost of shipping the transformers from Dobsville to Chicago and Detroit is $52y$ and $54(100 - x - y)$ dollars, respectively. The cost of shipping the transformers from Watertown to New York, Chicago and Detroit is $51(75 - x)$, $48(90 - y)$ and $50(x + y - 15)$ dollars, respectively.

The total cost function $C(x, y)$, obtained by adding up the costs of shipping the transformers from the sources to the destinations is

$$C(x, y) = -5x + 12{,}795.$$

Since we have taken into account the number of transformers at each source and the number to be received by each destination, the only condition that remains to be stated is that the direction of the flow is from the sources to the destinations; no backflow. Doing so yields the linear program model

Minimize $C(x, y) = -5x + 12{,}795$

subject to

$$x \geq 0, y \geq 0$$
$$100 - x - y \geq 0$$
$$75 - x \geq 0$$
$$90 - y \geq 0$$
$$x + y - 15 \geq 0,$$

or equivalently,

Minimize $C(x, y) = -5x + 12{,}795$

subject to

$$x \geq 0, y \geq 0$$
$$x + y \leq 100$$
$$x + y \geq 15$$
$$x \leq 75, y \leq 90,$$

which has solutions (75, 0) and (75, 25) and minimum value 12,420. Moreover, all points of the form (75, y), where y is an integer value between 0 and 25, are also solutions of this linear program which could be implemented as a shipping schedule (see Section 3.3, Exercise 11, p. 88).

Implementation of the solution (75, 0), for example, requires that 75, 0 and 25 transformers be sent from Dobsville to New York, Chicago, and Detroit, respectively, and that 0, 90 and 60 transformers be sent from Watertown to New York, Chicago, and Detroit, respectively. The anticipated total cost is $12,420.

More generally, a transportation problem has the following features. Given amounts of a commodity are available at a number of sources of supply, such as warehouses. Specified amounts are required by various destinations, such as retail outlets. The total amount required by the destinations may or may not be equal to the total amount available at the sources. Estimates (assumptions) are available on the cost of sending one unit of the commodity from each source to each destination. The problem is to determine the least cost shipping schedule. When the problem involves very few sources and destinations, as in Case 5, it can be solved by inspection. When the number of sources and destinations is large, inspection will not do, and linear programming provides us with a systematic approach to such problems. We discuss this further in the context of the Vytis Publishing Company's shipping problem in Section 3.9 (p. 122).

Case 6   An Assignment Problem

The Rasa Publishing Company has two positions to fill, editor of the mathematics list (job 1) and editor of the social science list (job 2), and is considering three candidates, Albert Roberts (candidate 1), Rita O'Brien (candidate 2), and Martin Thorp (candidate 3). After considering resumes, letters of recommendation, and conducting interviews the editorial board of the company has assigned a numerical rating to each person's qualifications for each position as stated in Table 3.8. These ratings

Table 3.8

Candidate	Position	
	Math, Editor (job 1)	Soc. Sci. Editor (job 2)
Roberts (candidate 1)	8	8
O'Brien (candidate 2)	7	9
Thorp (candidate 3)	9	8

serve as a quantitative measure of each candidate's potential for each position as seen by the editorial board. The editorial board wishes to assign candidates to positions in such a way that total potential is maximized.

To relate the candidates to the jobs we introduce X_{ij} to relate candidate i to job j. X_{ij} can assume one of two values, 0 if candidate i is not assigned job j, and 1 if candidate i is assigned job j. In summary, we emerge with Table 3.9.

Table 3.9

Candidate	Position	
	Math, Editor (job 1)	Soc. Sci. Editor (job 2)
Roberts (candidate 1)	X_{11}	X_{12}
O'Brien (candidate 2)	X_{21}	X_{22}
Thorp (candidate 3)	X_{31}	X_{32}

The function

$$P = 8X_{11} + 8X_{12} + 7X_{21} + 9X_{22} + 9X_{31} + 8X_{32},$$

obtained by multiplying the variable that relates a candidate to a job by the candidate's potential for the job and adding, is the potential function to be maximized subject to two conditions:

> **1.** Each candidate is assigned to at most one job.

The variables X_{11} and X_{12} (row 1 of Table 3.9) relate candidate 1 to the available jobs 1 and 2. The constraint

$$X_{11} + X_{12} \leq 1$$

expresses the requirement that candidate 1 be assigned to at most one job since it makes it impossible for candidate 1 to be assigned job 1 ($X_{11} = 1$) and job 2 ($X_{12} = 1$). This constraint comes from row 1 of Table 3.9, and in general the condition that each candidate be assigned at most one job is expressed by requiring that the sum of the variables in each row of Table 3.9 be less than or equal to one. Rows 2 and 3 yield the same condition for candidates 2 and 3:

$$X_{21} + X_{22} \leq 1$$
$$X_{31} + X_{32} \leq 1$$

> **2.** Each job is filled by at most one person.

The variables X_{11}, X_{21}, and X_{31} in column 1 of Table 3.9 relate candidates 1, 2, and 3 to job 1. The constraint

$$X_{11} + X_{21} + X_{31} \leq 1,$$

obtained by requiring that the sum of the variables in the first column 1 of Table 3.9 be less than or equal to one, expresses the requirement that job 1 be filled by at most one candidate since it makes it impossible for any two or all three candidates to be assigned to job 1 ($X_{11} = 1$, $X_{21} = 1$, $X_{31} = 1$). Column 2 yields the same condition for job 2:

$$X_{12} + X_{22} + X_{32} \leq 1$$

We thus obtain the following linear program: Find nonnegative integers (zeros and ones) that

$$\text{Maximize } P = 8X_{11} + 8X_{12} + 7X_{21} + 9X_{22} + 9X_{31} + 8X_{32}$$

subject to

$$X_{11} + X_{12} \leq 1$$
$$X_{21} + X_{22} \leq 1$$
$$X_{31} + X_{32} \leq 1$$
$$X_{11} + X_{21} + X_{31} \leq 1$$
$$X_{12} + X_{22} + X_{32} \leq 1.$$

By inspection we can see from Table 3.8 that the potential function P is maximized when candidate 3 (Thorp) is assigned to job 1 (Math Editor) and candidate 2 (O'Brien) is assigned to job 2 (Social Science Editor). That is, $X_{11} = 0$, $X_{12} = 0$, $X_{21} = 0$, $X_{22} = 1$, $X_{31} = 1$, $X_{32} = 0$ maximizes potential. When the number of candidates and jobs is large such problems cannot be handled by inspection, but can be handled by linear programming methods.

Assignment problems may involve the most efficient assignment of people to jobs, machines to tasks, project leaders to projects, police cars to city sectors, departments to store locations, sales people to territories, and so on. The objective might involve maximizing effectiveness in some sense or minimizing cost or travel time.

EXERCISES

The situations presented in the following exercises reflect assumptions made by an individual or group. Set up linear programs for the problems that arise, if possible solve them, and state how the solutions obtained would be implemented. What concerns would you want to have satisfactorily addressed before implementing a solution obtained?

1. The Algis Company makes radios. Two new portable stereo models, K15, and K31 are to be introduced. Both models pass through assembly and finishing plants of the company. In the assembly plant a K15 unit is worked on for 2 hours; a K31 unit is worked on for 3/2 hours. In the finishing plant a K15 unit is worked on for 5/2 hours; a K31 unit is worked on for 1 hour. At most 160 hours of assembly time and 165 hours of finishing time are available per week. The anticipated profit is $180 per K15 unit and $120 per K31 unit.

    The problem is to determine, under such conditions, how many K15 and K31 units should be made weekly so as to maximize profit.

2. A fruit juice is to be made from orange juice concentrate and apricot juice concentrate. Particular attention is being paid to the vitamin A, C, and D content of the fruit juice. Each container of fruit juice is to contain at least 120 units of vitamin A, 150 units of vitamin C, and 55 units of vitamin D. One ounce of orange juice concentrate contains 2 units of vitamin A, 3 units of vitamin C, and 1 unit of vitamin D. One ounce of apricot juice concentrate contains 3 units of vitamin A, 2 units of vitamin C, and 1 unit of vitamin D. Orange juice concentrate costs 3¢ per ounce and apricot juice concentrate costs 2¢ per ounce.

    The problem is to determine how many ounces of each concentrate should be used to make a least-cost container of juice that satisfies the vitamin requirements.

3. Cans of meat are to be mass produced and made available for distribution in emergency situations (such as earthquakes and floods). Each can of meat is to be a mixture of pork and beef and must contain at least 12 ounces of protein and 9 ounces of fat. It is estimated that a pound of the beef to be used contains 5 ounces of protein and 3 ounces of fat while a pound of pork contains 3 ounces of protein and 3 ounces of fat.

    If beef costs 50¢ a pound and pork costs 45¢ a pound, how many pounds of each should be used to make up a can of meat so that the nutritional requirements are met and the cost is minimal? What is the minimal cost?

4. The Saturn Company makes refrigerators and air-conditioners. Two plants, I and II, are used. The assembly work is done in plant I, and it is estimated that 5 labor-hours of work are required to produce a refrigerator and 2 labor-hours of work are required to produce an air-conditioner. The finishing work is done in plant II, and it is estimated that 3 labor-hours of work are needed to finish a refrigerator and 2 labor-hours are needed to finish an air-conditioner. Plant I has 220 labor-hours per week available and plant II has 180 labor-hours per week available. A market survey indicates that there is an unlimited market for these products.

    If the Saturn Company makes a profit of $50 on each refrigerator and $30 on each air-conditioner, how many of each should be produced weekly so as to maximize profit?

5. The Andrius Bank has assets in the form of loans and negotiable securities. For a certain time period it is assumed that loans and securities bring returns of 9 and 6 percent, respectively. The bank has a total of $25 million, provided by demand deposit accounts and time deposit accounts, to allocate between loans and securities. To meet unanticipated deposit withdrawals the bank always maintains a securities balance equal to or greater than 20 percent of total assets. Since lending is the bank's most important activity, it imposes certain restrictions on its loan balance to satisfy its principal clients. Specifically, it requires that at least $8 million be available for loans.

   Under the given conditions, how should the bank allocate funds between loans and securities so that total investment income is maximized?

6. The Jay Toy Store plans to invest up to $2200 in buying and stocking two popular children's toys. The first toy costs $4 per unit and occupies 5 cubic feet of storage space; the second toy costs $6 per unit and occupies 3 cubic feet of storage space. The store has 1400 cubic feet of storage space available. The owner expects to make a profit of $1.50 on each unit of the first toy he buys and stocks and a profit of $2.00 on each unit of the second toy.

   How many units of each should be bought and stocked so that profit is maximized?

7. At Ecap University discussion has centered on determining the number of openings, called slots, to be made available in the forthcoming year at the associate and full-professor ranks. Each person promoted to associate professor is to receive a merit increment of $500, and each person promoted to full professor is to receive a merit increment of $1000. At most $15,000 is available for merit increments. A long-standing guideline is that the number of full-professor slots is not to exceed one fourth the number of associate professor slots. The university senate has recommended that at least 3 slots at the full-professor rank be established and that not more than 22 slots at the associate professor rank be established.

   Of interest to the faculty council is the question of how many slots should be established at each rank so as to maximize the total number of promotions. Administration has raised the question of how many slots at each rank should be established so as to minimize the total cost of increments.

8. The Brooks and Darius mines of Lexington Mines, Inc., produce high-grade and medium-grade silver ore. The Brooks mine yields 1 ton of high-grade ore and 4 tons of medium-grade ore per hour. The Darius mine yields 2 tons of high-grade ore and 3 tons of medium-grade ore per hour. To meet its commitments the company needs at least 40 tons of high-grade and 100 tons of medium-grade ore per hour. It costs $500 per hour to operate the Brooks mine and $700 per hour to operate the Darius mine.

   Lexington Mines, Inc., would like to determine how many hours per day each mine should be operated if their ore requirements are to be met at minimal cost.

9. The Petrovski Steel Company produces 2 million tons of steel annually. In the current operation of the blast and open-heart furnaces 50 pounds of particulate matter is emitted into the atmosphere for every ton of steel produced. The resulting air pollution has become a problem of serious concern and efforts are being directed at curbing the emissions. On the basis of studies that have been conducted, it is estimated that installation of the F14 filter system would reduce emissions to 20 pounds of particulate matter per ton of steel produced, and installation of the F24 filter system would reduce emissions to 18 pounds of particulate matter per ton of steel produced. Capital and operating costs for the F14 and F24 filter systems are estimated at $1.2 and $1.8, respectively, per ton of steel produced. It is desired that particulate emissions be reduced by 62,400,000 pounds or better per annum. At the same time cost is an important factor if the company is to remain competitive.

The problem is to determine how many tons of steel should be produced annually subject to the F14 system and how many tons should be produced subject to the F24 system so that the desired reduction in particulate emissions is achieved at minimal total cost.

10. The Rasa Publishing Company has warehouses in Dallas and Kansas City. Copies of its newly published best seller are to be sent to dealers in New York, San Francisco, and Atlanta. The Dallas warehouse has 15,000 copies and the

Kansas City warehouse has 20,000 copies. Orders for 10,000 copies, 12,000 copies, and 13,000 copies have been placed by New York, San Francisco, and Atlanta book dealers, respectively. It costs 5¢, 7¢, and 4¢ per copy to ship the books from Dallas to New York, San Francisco, and Atlanta, respectively; it costs 4¢, 5¢, and 5¢ per copy to ship the books from Kansas City to these respective destinations.

How should the books be shipped so as to minimize the total shipping cost?

11. The Linnus Company is planning to put into production three minicomputer models, C15, C24, and C51. Each unit must pass through assembly, finishing, and inspection plants of the company. In the assembly plant, a C15 unit is worked on for 2 hours, a C24 unit is worked on for 3 hours, and a C51 unit is worked on for 1 hour. In the finishing plant, it takes 1 hour to finish a C15 unit, 1 hour to finish a C24 unit, and 3 hours to finish a C51 unit. In the inspection plant, it takes 2 hours to inspect a C15 unit, 1 hour to inspect a C24 unit, and 1 hour to inspect a C51 unit. At most 16, 17, and 12 hours of assembly time, finishing time, and inspection time, respectively, are available per day. The anticipated profit is $24, $30, and $26 per unit of C15, C24, and C51, respectively.

The problem is to determine, under these conditions, how many units of each model should be made daily so as to maximize profit.

12. Soybeans are to be shipped from New York and New Orleans to Dakar, Marseille, and Odessa. 8000 tons are in New York and 11,000 tons are in New Orleans. 6000 tons are to be sent to Dakar, 7000 tons to Marseille, and 6000 tons to Odessa. The cost (in dollars) of shipping one ton of soybeans from each distribution point to each destination is given in Table 3.10.

Table 3.10

	Dakar	Marseille	Odessa
New York	10	11	12
New Orleans	10.5	11.2	12.3

The problem is to determine how many tons should be shipped from each distribution center to each destination so that the total shipping cost is minimized.

13. Up to $50,000 of a client's money is to be invested by the Onute Investment Company in low-risk bonds with a 10 percent yield, and high-risk speculative stock with an anticipated 18 percent yield. The client has insisted that the amount invested in stock be no greater than one fourth the amount invested in bonds. The managers of the investment company feel that at most $41,000 should be invested in bonds and that at least $5000 should be invested in stock.

The problem is to determine the investment plan that satisfies these conditions and maximizes return on investment.

14. The Selby and Turchin plants of the Asta Paper Company produce three grades of paper. The Selby plant produces 5 tons of grade A paper, 3 tons of grade B paper, and 7 tons of grade C paper per hour. The Turchin plant produces 3 tons of grade A paper, 5 tons of grade B paper, and 6 tons of grade C paper per hour. To satisfy existing contracts, the company must produce at least 95 tons of grade A paper, 88 tons of grade B paper, and 160 tons of grade C paper per day. It costs $200 per hour to operate the Selby plant and $150 per hour to operate the Turchin plant.

    The Asta Paper Company would like to determine the number of hours per day that each plant should be kept in operation if their paper needs are to be met at minimum cost.

15. A railroad company has freight cars of two types. Model AA1 has 8000 cubic feet of refrigerated space and 10,000 cubic feet of nonrefrigerated space; model AA2 has 6000 cubic feet of refrigerated space and 15,000 cubic feet of nonrefrigerated space. Audre Food Distributors deal in meat products, which require refrigeration, and a variety of nonrefrigerated foods. 60,000 cubic feet of meat products and 120,000 cubic feet of nonrefrigerated foods are to be shipped from Chicago to Louistown. A model AA1 car rents for $0.8 per mile, and a model AA2 car rents for $1.25 per mile.

    The problem is to determine the number of freight cars of each type that should be rented so that the total cost of shipping the foods is minimized.

16. The Stillwell Company produces lubricating oil for machine tools (BV3 oil) and steel mills (BV7 oil). The production of a unit of BV3 oil requires 1 hour of refining and 0.5 hours of blending; the production of a unit of BV7 oil requires 1.2 hours of refining and 1 hour of blending. The refinery can be kept in operation at most 12 hours per day, and the blending plant can be kept in operation at most 8 hours per day. BV3 oil brings a profit of $15 per unit, and BV7 oil brings a profit of $20 per unit.

    The problem is to determine the number of units of each type of oil that should be produced daily to maximize profit.

17. The Clinton Company produces a portable X-ray unit that it sells to two types of outlets, hospitals and medical supply houses. The profit margin varies between the two types of outlets owing to differences in order sizes, selling costs, and credit policies. It is estimated that the profit per unit is $100 and $120, respectively, for the hospitals and medical supply houses. Sales promotion is carried out by personal sales force calls and media advertising. The company has eight salespersons on its marketing staff, representing 12,000 hours of available customer contact time during the next year. $30,000 has been allocated for media advertising during the next year. An examination of past data indicates that a unit sale to a hospital requires about a 1/2-hour sales call and $2 worth of advertising; a unit sale to a medical supply house requires a 1-hour sales call and $1 worth of advertising. The company would like to achieve sales of at least 5000 units in each customer segment.

    The problem is to determine what sales volume (in product units) it should seek to develop in each segment in order to maximize total profit.

3.7. A LINEAR PROGRAMMING SHORTFALL

EXAMPLE 1 ADVERTISING MEDIA SELECTION

To advertise its new best seller, the Rasa Publishing Company is planning to buy morning and afternoon time on radio station WQRX. Morning time costs $1000 per minute and afternoon time costs $800 per minute. It is estimated that morning commercials reach 0.9 million listeners and afternoon commercials reach 0.6 million listeners. At most, 16 minutes of morning time is available in the month in which the advertising campaign is to run. The advertising department of the Rasa Company feels that at least 8 minutes of morning time and at least 6 minutes of afternoon time should be purchased. The advertising budget for this campaign is $24,000.

How much morning and afternoon time should be purchased so as to maximize total listener exposure time to the ads in the month in which the advertising campaign is to run?

Let x denote the number of minutes of morning time and y the number of minutes of afternoon time to be purchased. Total listener exposure time to the ads is expressed by

$$F(x, y) = 0.9x + 0.6y$$

The inequalities

$$x \geq 8, \quad y \geq 6$$

express the requirement that at least 8 minutes of morning time and 6 minutes of afternoon time are to be purchased. That at most 16 minutes of morning time is available is expressed by

$$x \leq 16$$

$1000x + 800y$ expresses the cost of x minutes of morning time and y minutes of afternoon time, and since this cost cannot exceed $24,000, we have

$$1000x + 800y \leq 24{,}000$$

Therefore, we obtain the following linear program:

Maximize $F(x, y) = 0.9x + 0.6y$

subject to

$$x \geq 0$$
$$y \geq 0$$
$$x \geq 8$$
$$y \geq 6$$
$$x \leq 16$$
$$1000x + 800y \leq 24{,}000$$

This is an example of a general class of problems called advertising media-selection problems. The general advertising media-selection problem is to choose from various media capable of carrying an advertisement a selection that is, in some sense, best. Specific choices within a given medium as well as given media are included in the alternatives. The constraints in media selection include the size of the advertising budget, the minimum and maximum usages of specific media categories, and the desired minimum exposure rate to envisioned buyers. A number of approaches have been developed for the media-selection problem,[1] and in the early 1960's hopes ran high in the world of advertising for the use of linear programming. An early linear-programming model for media selection was the one developed by James Engel and Martin Warshaw[2] for the McGraw-Edison Company. The Pennsylvania Transformer division of McGraw-Edison manufactures transformers for use by industrial plants, schools, hospitals, commercial construction projects, and so on. Ten trade publications were considered for advertising purposes, and $25,000 was allocated for industrial advertising for a period of 1 year. Since the purchase decision is usually made by the plant engineer, the objective posed was to maximize the number of plant engineers reached. The following linear programming model was developed:

$$\text{Maximize } f = 0X_1 + 15.15X_2 + 32.87X_3 + 49.44X_4 + 56.65X_5 + 17.54X_6 + 58.20X_7 + 0X_8 + 23.53X_9 + 40.00X_{10}$$

subject to nonnegativity of the variables ($X_1 \geq 0$, $X_2 \geq 0$, etc.)

$$X_1 \leq 5,400$$
$$X_2 \leq 9,504$$
$$X_3 \leq 8,760$$
$$X_4 \leq 10,680$$
$$X_5 \leq 11,016$$
$$X_6 \leq 5,472$$
$$X_7 \leq 9,072$$
$$X_8 \leq 3,300$$
$$X_9 \leq 8,160$$
$$X_{10} \leq 6,900$$
$$X_1 + X_2 + X_3 + X_4 + X_5 + X_6 + X_7 + X_8 + X_9 + X_{10} \leq 25,000$$

where X_1 is the amount to be invested in media 1 (*Consulting Engineer Magazine*), X_2 is the amount to be invested in media 2 (*Electrical Construction Magazine*), and so

---

[1] See Dennis Gensch, "Different Approaches to Advertising Media Selection," *Operational Research Quarterly*, vol. 21, no. 2 (June 1970), pp. 193–219; Philip Kotler, *Marketing Management* (Englewood Cliffs, N.J.: Prentice-Hall, Inc., 1967), Chapter 18.

[2] "Allocating Advertising Dollars by Linear Programming," *Journal of Advertising Research*, vol. 4, no. 3 (September 1964), pp. 42–48.

on. The coefficients 0 of X_1, 15.15 of X_2, 32.87 of X_3, and so on, in the linear function f represent the number of plant engineers reached by each magazine per advertising dollar invested, so that f represents the total number of plant engineers reached. The last constraint expresses the condition that no more than \$25,000 is to be spent on advertising and the other constraints are to prevent more dollars from being invested in any one monthly magazine than is necessary to buy 12 insertions.

Although linear programming models were satisfactory for crude versions of the media-selection problem, it soon became clear that the features exhibited by more sophisticated versions of the problem could not be modeled in linear programming terms. Frank Bass and Ronald Lonsdale[3] found linear-programming models to be

> crude devices to apply to the media-selection problem. The linearity assumption itself, is the source of much of the difficulty. Justifying an assumption of linear response to advertising exposures on theoretical grounds would be difficult. Assumptions about the nature of response to advertising cause most difficulties in models of the type examined in this article.

Philip Kotler[4] noted the following limitations:

> Linear programming assumes that repeat exposures have a constant marginal effect.
> It assumes constant media costs (no discounts).
> It cannot handle the problem of audience duplication.
> It says nothing about when ads should be scheduled.

Although later linear programming approaches to the media-selection problem sought to overcome the criticisms that had been voiced, the message was clear: although linearity, as a mathematical tool, is too good not to be true, a linear-programming model is not always a suitable fit for a media-selection problem; that is, the assumptions that must be made to force a fit are not always sufficiently realistic. When the model doesn't fit, don't use it.

3.8. SOMETIMES YOU CAN'T HAVE IT ALL

EXAMPLE 1   AN EQUIPMENT PURCHASE PROBLEM

The Inter-City Bus Company plans to invest up to \$3 million on new equipment. Two bus models, B4 and B9, are being considered for purchase. Bus B4 is expected to average 16 hours a day at 55 miles an hour with an average of 45 passengers. Bus B4 costs \$120,000. Bus B9, a double-decker, is expected to average 18 hours a day at 50 miles an hour with an average of 60 passengers. Bus B9 costs \$180,000. The company wishes to purchase at least 30 new buses.

---

[3]"An Exploration of Linear Programming in Media Selection," *Journal of Marketing Research,* vol. III, no. 2 (May 1966), pp. 179–188.
[4]*Marketing Management* (Englewood Cliffs, N.J.: Prentice-Hall, Inc.) p. 478.

How many vehicles of each model should be purchased so that capacity in passenger-miles per day is maximized?

Let x and y denote the number of B4 and B9 buses to be purchased, respectively. The problem is to

$$\text{Maximize capacity}$$

subject to

$$x \geq 0$$
$$y \geq 0$$
$$\text{(amount spent)} \leq 3{,}000{,}000$$
$$\text{(number of buses purchased)} \geq 30.$$

The function

$$C(x, y) = 45\,(55)\,(16)x + 60\,(50)\,(18)y = 39{,}600x + 54{,}000y$$

expresses capacity in passenger-miles per day. The constraint

$$120{,}000x + 180{,}000y \leq 3{,}000{,}000,$$

or equivalently,

$$2x + 3y \leq 50$$

expresses the condition that the amount spent must not exceed $3 million. The constraint

$$x + y \geq 30$$

expresses the condition that at least 30 buses are to be purchased.

From Section 3.3, Exercise 8 (p. 88) we see that this bus purchase linear program has no solution. The conditions are incompatible; sometimes you can't have it all.

3.9. PROBLEMS REQUIRING SOLUTIONS IN INTEGERS

EXAMPLE 1.   HOW MANY COATS AND DRESSES SHOULD BE MADE?

The Hoffman Clothing Manufacturers, Inc., has available 120 square yards of cotton and 100 square yards of wool for the manufacture of coats and dresses. Two square yards of cotton and 4 square yards of wool are used in making a coat while 4 square yards of cotton are used in making a dress. Cotton costs $5 per square yard and wool cost $20 per square yard. Four hours of labor are needed to make a coat and 2 hours of labor are needed to make a dress. The cost of labor is $25 per hour. At most 110 hours of labor are available for the manufacture of the coats and dresses.

If a coat sells for $300 and a dress sells for $140, how many of each should be made if net income is to be maximized?

Let x and y denote the number of coats and dresses to be made, respectively. The data given are summarized in Table 3.11.

Table 3.11

	Selling price	Cotton used per item	Wool used per item	Labor-hours per item	Number made
Coat	$300	2	4	4	x
Dress	$140	4	0	2	y
		$5 per sq yd; 120 sq yd available	$20 per sq yd; 100 sq yd available	$25 per labor-hour; 110 labor-hours available	

Since net income is to be maximized we turn our attention to expressing net income in terms of x and y.

Net income = Amt from sales − Production cost

$$I(x, y) = \underbrace{300x + 140y}_{\text{sales}} - \underbrace{[5(2)x + 20(4)x]}_{\text{cost of coat material}} - \underbrace{5(4)y}_{\text{cost of dress material}}$$

$$- \underbrace{[25(4)x + 25(2)y]}_{\text{labor cost}}$$

By multiplying and collecting terms we obtain:

$$I(x, y) = 110x + 70y$$

The constraint $4x \leq 100$ or, equivalently, $x \leq 25$, expresses the condition that the amount of wool used cannot exceed 100 square yards; $2x + 4y \leq 120$ expresses the condition that the amount of cotton used cannot exceed 120 square yards; $4x + 2y \leq 110$ expresses the condition that the number of labor-hours employed cannot exceed 110.

We thus emerge with the following linear program:

Maximize $I(x, y) = 110x + 70y$

subject to

$$x \geq 0, y \geq 0$$
$$x \leq 25$$
$$2x + 4y \leq 120$$
$$4x + 2y \leq 110$$

From Section 3.3, Exercise 7 (p. 88) we see that (50/3, 65/3) yields the maximum value 3350.

What is actually required is a feasible point expressed in integers which maximizes $I(x, y)$. As one would expect, such linear programs are called **integer programs**. Sometimes, as in the case of the Austin Company's linear programs, it turns out that solutions in integers are obtained when a linear program solution method is applied. Sometimes not, as we have just seen.

We know that a solution of a linear program can be found among its corner points. Thus if an integer solution is desired and some of the corner points are not integers, the idea of modifying the given linear program by appending to it new constraints with the property that no integer feasible points are lost and the resulting corner points involve only integers is naturally suggested. Upon implementation of this idea, linear programming methods can then be applied to the modified linear program to obtain a solution in integers which will also be a solution of the given integer program. We illustrate the implementation of this idea in the case of two-variable integer programs by returning to the Hoffman Clothing Manufacturer's integer program:

Find integer values for x and y which

$$\text{maximize } I(x, y) = 110x + 70y$$

subject to

$$x \geq 0, y \geq 0$$
$$x \leq 25$$
$$2x + 4y \leq 120$$
$$4x + 2y \leq 110.$$

The feasible points of this problem viewed as a linear program are shown in Figure 3.11. Figure 3.12 shows in detail the region surrounding (50/3, 65/3).

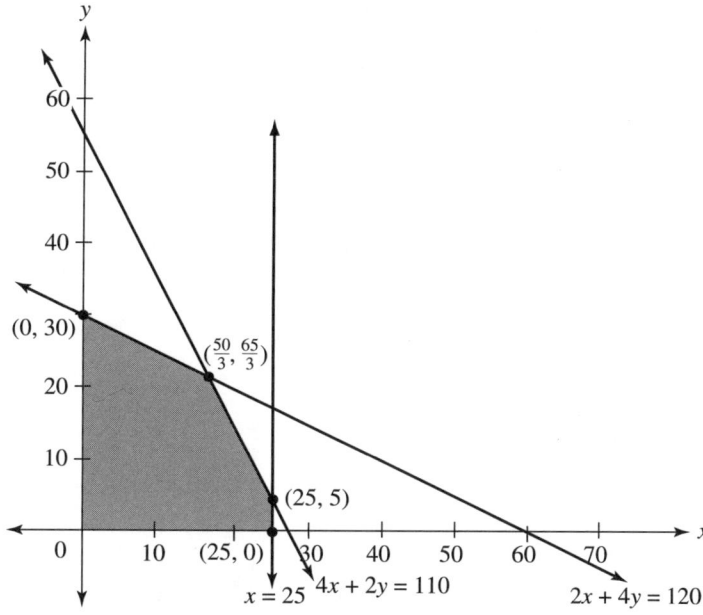

Figure 3.11

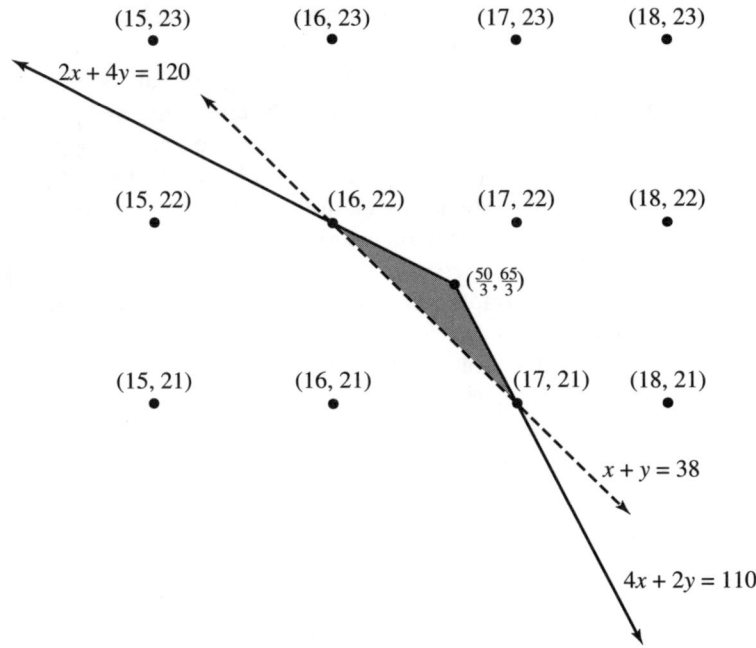

Figure 3.12

From Figure 3.12 we see that if the new constraints to be appended are to

1. eliminate $(\frac{50}{3}, \frac{65}{3})$ as a corner point,
2. not eliminate any integer feasible points,
3. yield corner points which are integers,

then we need only introduce the additional constraint $x + y \leq 38$, based on the boundary line ($x + y = 38$) passing through (16, 22) and (17, 21).

Thus the modified linear program is the following:

$$\text{Maximize } I(x, y) = 110x + 70y$$

subject to

$$x \geq 0, y \geq 0$$
$$x \leq 25$$
$$2x + 4y \leq 120$$
$$4x + 2y \leq 110$$
$$x + y \leq 38.$$

The feasible points of this modified program are shown in Figure 3.13. All of its corner points, (0, 0), (0, 30), (16, 22), (17, 21), (25, 5), and (25, 0), are expressed in

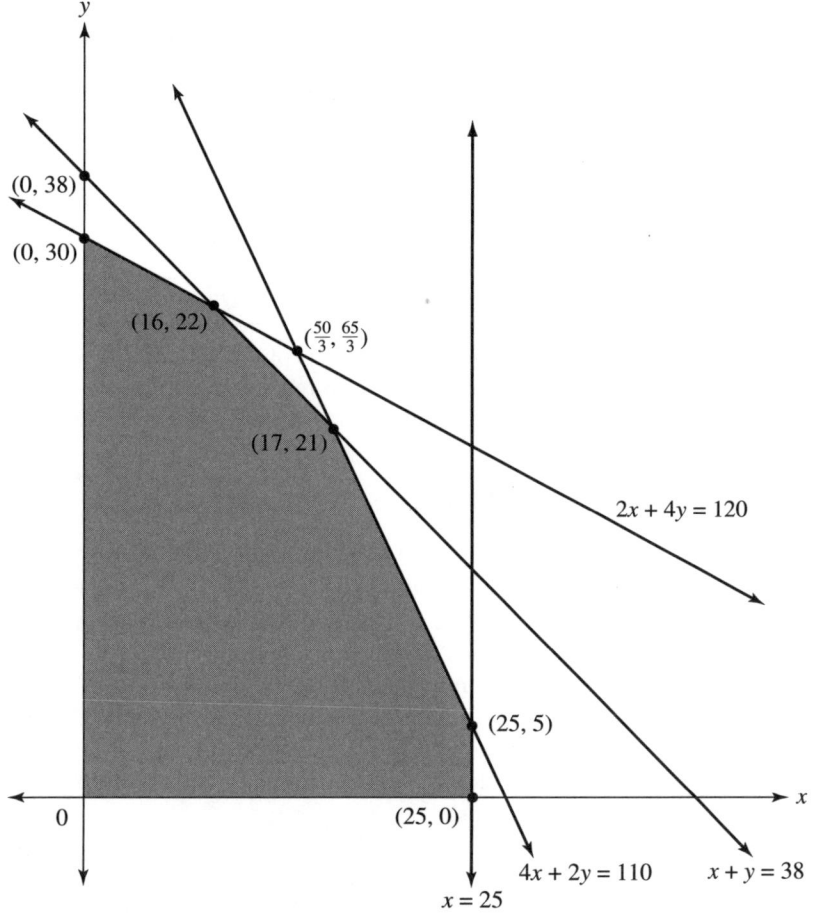

Figure 3.13

terms of integers. From Table 3.12 we see that (17, 21) is the solution and 3340 is the maximum value. Since this linear program differs from the original only in the loss non-integer feasible points (those in the triangular region with vertices (16, 22), (17, 21), (50/3, 65/3), above the line $x + y = 38$; see Figure 3.13), its solution (17, 21) is also a solution of the Hoffman clothing manufacturer's integer program.

When the number of variables is greater than two, graphical techniques like the one employed to determine the new constraints for obtaining a solution in integers cannot be used and other methods must be sought. Algebraic techniques for the determination of the required new constraints which are computationally effective in many situations have been developed. However, such techniques are beyond the scope of this book.

Table 3.12

Corner Point	$I(x, y) = 110x + 70y$
(0, 0)	0
(0, 30)	2100
(16, 22)	3300
(17, 21)	3340
(25, 5)	3100
(25, 0)	2750

The following situation illustrates problems which lead to integer programs which can be solved by linear programming methods. Due to certain structural features integer solutions are guaranteed.

EXAMPLE 2. THE VYTIS COMPANY'S SHIPPING PROBLEM

The Vytis Publishing Company has warehouses W1 in Williamstown and W2 in Jamesville. Copies of a newly published book which is in great demand are to be sent via air freight to distributors D1 in Chicago, D2 in New York, and D3 in Detroit. The Williamstown warehouse has 20,000 copies and the Jamesville warehouse has 15,000 copies. The Chicago distributor needs 8000 copies while the New York and Detroit distributors have each requested 10,000 copies. These distributors are able to accept larger supplies if it is economically advantageous to ship larger numbers. Table 3.13 is a shipping cost schedule which specifies the cost of shipping a book from each warehouse to each distributor.

Table 3.13

		To Distributor		
		D1	D2	D3
From Warehouse	W1	5¢	8¢	7¢
	W2	4¢	9¢	6¢

How should the shipments be made so as to minimize the total cost?

First, let us observe that we cannot analyze this problem in terms of two variables as we have other shipment problems since the total number of items needed (8000 + 10,000 + 10,000 = 28,000) is not equal to the number available (20,000 + 15,000 = 35,000). Our approach is to introduce a variable to stand for the number of books to be sent from each warehouse to each distributor. Let X_{11} denote the number of books to be sent from W1 to D1, X_{12} denote the number of books to be sent from W1 to D2, and so on. In general, X_{ij} is used to denote the number of items shipped from source i to destination j, and the number of variables is equal to the product of the number of sources multiplied by the number of destinations. Thus for the problem at hand, six variables are needed (see Table 3.14).

Table 3.14

		Received at		
		D1	D2	D3
Shipped from	W1	X_{11}	X_{12}	X_{13}
	W2	X_{21}	X_{22}	X_{23}

The shipping cost function to be minimized is

$$C = 5X_{11} + 8X_{12} + 7X_{13} + 4X_{21} + 9X_{22} + 6X_{23}.$$

In addition to the nonnegativity of the variables, there are two conditions which must be satisfied.

> First, the number of items that can be shipped from a source cannot exceed the number available at the source.

Thus we have

$$X_{11} + X_{12} + X_{13} \leq 20{,}000$$
$$X_{21} + X_{22} + X_{23} \leq 15{,}000.$$

> Second, at least the number of items required at a destination shall be sent to the destination.

Thus we have

$$X_{11} + X_{21} \geq 8000$$
$$X_{12} + X_{22} \geq 10{,}000$$
$$X_{13} + X_{23} \geq 10{,}000.$$

The rows of Table 3.14 yield the first two constraints whereas the columns yield the last three.

The integer program that emerges is the following:
Find nonnegative integer values for X_{ij} which

$$\text{minimize } C = 5X_{11} + 8X_{12} + 7X_{13} + 4X_{21} + 9X_{22} + 6X_{23}$$

subject to

$$X_{11} + X_{12} + X_{13} \leq 20{,}000$$
$$X_{21} + X_{22} + X_{23} \leq 15{,}000$$
$$X_{11} + X_{21} \geq 8000$$
$$X_{12} + X_{22} \geq 10{,}000$$
$$X_{13} + X_{23} \geq 10{,}000.$$

EXERCISES

1. Martins Plant Foods, Inc., plans to introduce Martins Miracle, hailed as a "blossom booster" for outdoors and indoors flowering plants. Martins Miracle is to be a mixture of bonemeal and processed vegetable matter. Each can of Martins Miracle is to contain at least 12 units of nitrogen and 9 units of phosphorus. Each pound of bonemeal contains 3 units of nitrogen and 1 unit of phosphorus; each pound of vegetable matter contains 2 units of nitrogen and 2 units of phosphorus.

    The problem is to determine the number of pounds of bonemeal and vegetable matter that should be used to make up a can of Martins Miracle so that the nutrient requirements are met at minimal cost.

    (a) Set up and solve the linear program model for this situation. What is the minimal cost in terms of this model and what mix would achieve it?
    (b) Management has been advised to obtain the best solution in integers to its component mixture problem. What additional constraint(s) must be added to the linear program obtained from (a) to guarantee a solution in integers?
    (c) Determine the solution to the integer program model that emerges. What does it mean to Martins Plant Foods, Inc.?

For the integer programs given in Exercises 2–4, determine the additional constraints that must be imposed to guarantee a solution in integers.

2. Find nonnegative integers which minimize $F(x, y) = 3x + 2y$ subject to

$$3x + y \geq 6$$
$$x + y \geq 5.$$

3. Find nonnegative integers which maximize $P(x, y) = 16x + 6y$ subject to

$$3x + y \leq 6$$
$$x + y \leq 5.$$

4. Find nonnegative integers which minimize $C(x, y) = 4x + 5y$ subject to
$$3x + y \geq 6$$
$$x + 3y \geq 6.$$

5. The Hudson Furniture Manufacturing Company, Inc., has available 300 square yards of pine veneer and 228 square yards of walnut veneer for the manufacture of model D-1 desks and model B-14 bookcases. 8 square yards of pine veneer and 6 square yards of walnut veneer are used in making a desk; 10 square yards of pine veneer and 7 square yards of walnut veneer are used in making a bookcase. Pine veneer costs $2 per square yard and walnut veneer costs $4 per square yard. 5 labor-hours are needed to manufacture a desk and 4 labor-hours are needed to manufacture a bookcase. The cost of labor is $4 per labor-hour. At most 150 labor-hours are available for the manufacture of the desks and bookcases.

A desk sells at $100 and a bookcase sells at $108.

The problem is to determine the number of each that should be made and sold to maximize net income.

(a) Determine the additional constraint(s) that must be imposed to guarantee a solution in integers.

(b) Solve the integer program that emerges.

(c) What advice would you give to the Hudson Company?

6. Formulate the following problem as an integer program but do not solve.

Cartons of citrus fruits are to be shipped from Lakeland and Orlando to distributors in Portland, New York, and Baltimore. Lakeland has available 10,000 cartons and Orlando has available 18,000 cartons. The Portland distributor needs 1000 cartons, the New York distributor needs 14,000 cartons, and the Baltimore distributor needs 9000 cartons. The shipping cost Table 3.15 specifies the cost of shipping a carton from each source to each destination. How should the shipments be made so as to minimize cost?

Table 3.15

	Portland	New York	Baltimore
Lakeland	7¢	6¢	5¢
Orlando	5¢	5¢	6¢

7. Formulate the following problem as a linear program but do not solve.

Bauxite ore is to be shipped from the Turin and Johnston mines of the Alexander Aluminum Company to refineries in Baltimore, Cincinnati, and Pittsburgh. The Turin mine has 15,000 tons and the Johnston mine has 20,000 tons. The Baltimore refinery requires at least 8000 tons, the Cincinnati refinery at least 10,000 tons, and the Pittsburgh refinery at least 12,000 tons. Table 3.16 specifies the cost (in dollars) of shipping 1 ton of bauxite from each source to each destination.

126 Finite Mathematics, Models, and Structure

Table 3.16

	Baltimore	Cincinnati	Pittsburgh
Turin	2	3	2
Johnston	3	4	2

The problem is to determine how the shipments should be made so as to satisfy the requirements of the refineries at minimum total cost.

Structurally speaking, this problem is the same as problem 6. In this case the shipping units (tons) admit noninteger solutions. This would not be the case had we been shipping cartons, automobiles, appliances, machinery, etc. which require integer values.

Problems Leading to 0–1 Integer Programs

A number of problems lead to integer program models which, more specifically, require that the integers be 0's and 1's. The assignment situation considered in Case 6 of Section 3.6 (p. 104) is one such problem. A value of 1 for an assignment variable means that the person or item or task is to be included in the assignment; a value of 0 means that it is not to be included.

As another example of a 0–1 integer program situation consider the following selection problem.

EXAMPLE 3. THE MARSDEN COMPANY'S SELECTION PROBLEM

The Marsden Company has seven items $I_1, I_2, I_3, I_4, I_5, I_6, I_7$, that are to be shipped from New York to Detroit as quickly as possible. A company airplane with a freight capacity of 1500 pounds is available for this purpose. Table 3.17 gives the weight and

Table 3.17

	Weight (lb)	Value ($)
I_1	300	600
I_2	350	610
I_3	400	650
I_4	250	400
I_5	260	405
I_6	280	410
I_7	500	660

value of each of the seven items. Since the sum of the weights of these items is 2340 pounds and the freight capacity of the plane is 1500 pounds, not all of the items can be taken.

How should the plane be loaded so that the value of its contents is maximized and its freight capacity is not exceeded?

Corresponding to items $I_1, \ldots, I_7$, we introduce variables $X_1, \ldots, X_7$, respectively. X_1 can take on one of two values, 0 if item I_1 is not taken and 1 if item I_1 is taken; X_2 can take on one of two values, 0 if item I_2 is not taken and 1 if item I_2 is taken; and so on.

The function

$$V = 600X_1 + 610X_2 + 650X_3 + 400X_4 + 405X_5 + 410X_6 + 660X_7$$

is formed by multiplying X_1 by the value of item I_1 (600), multiplying X_2 by the value of item I_2 (610), and so on, then adding. V is the value function to be maximized. Whenever $X_1, \ldots, X_7$ are given values (0's and 1's), V becomes a sum of item values. For example, when $X_1 = X_2 = X_3 = 1$ and $X_4 = X_5 = X_6 = X_7 = 0$, then

$$V = 600 + 610 + 650 = 1860$$

the sum of the values of I_1, I_2, and I_3.

The problem of loading the plane so that the value of its contents is maximized and its freight capacity is not exceeded is expressed by the following integer program.

Find nonnegative integers which maximize

$$V = 600X_1 + 610X_2 + 650X_3 + 400X_4 + 405X_5 + 410X_6 + 660X_7$$

subject to

$$X_1 \leq 1$$
$$X_2 \leq 1$$
$$X_3 \leq 1$$
$$X_4 \leq 1$$
$$X_5 \leq 1$$
$$X_6 \leq 1$$
$$X_7 \leq 1$$
$$300X_1 + 350X_2 + 400X_3 + 250X_4 + 260X_5 + 280X_6 + 500X_7 \leq 1500.$$

The nonnegativity statement and the first seven constraints express the requirement that the integer values of the variables be 0's or 1's; the last one expresses the condition that the freight capacity of the plane (1500 pounds) not be exceeded.

Example 3 illustrates problems with the following general structure. A container of some sort (truck, car, plane) is to be loaded with items of various values and weights. For the items involved there is a limitation on the weight that can be loaded in the container but not on the volume. The problem is to load the container in such a way that its weight limit is not exceeded and the value of the items loaded is the largest possible.

EXERCISES

Formulate the following problems as integer programs.

8. The Vroman Institute has two jobs to fill, physiologist (job 1) and biochemist (job 2), and is considering three candidates, Ann (candidate 1), Gena (candidate 2), and Marty (candidate 3), for the jobs. Each candidate's qualifications for each of the jobs have been assigned a numerical rating (see Table 3.18).

Table 3.18

		Job	
		Physiologist (job 1)	Biochemist (job 2)
Candidate	Ann (candidate 1)	3	$\frac{5}{2}$
	Gena (candidate 2)	2	$\frac{5}{2}$
	Marty (candidate 3)	2	$\frac{3}{2}$

These ratings are interpreted as a measure of a candidate's potential for a particular job. The Institute's problem is to fill the jobs in such a way that potential is maximized.

9. A legal advisory group is to make recommendations on two positions, State Supreme Court Judge and Civil Court Judge, and is considering three candidates, M. Jones, R. Johnson, and A. Marks. Table 3.19 describes potential ratings that have been assigned by the advisory group as a quantitative measure of each person's qualifications for each position.

Table 3.19

Candidate	Job	
	Supreme Court Judge	Civil Court Judge
Jones	9	8
Johnson	8	9
Marks	10	8

The advisory group wishes to make its recommendations on how these positions should be filled on the basis of maximization of potential.

10. The Onute Land Development Company has identified five sites in Albuquerque, Dallas, Miami, Phoenix, and San Diego for the construction of condominiums. The anticipated cost of construction on these sites (in millions of dollars) and the expected profit to be realized from each development (in millions of dollars) are described in Table 3.20. The company can commit at most $32 million to these developments, which is insufficient to undertake them all. The problem is to determine the selection of sites that yields the largest total expected profit, but for which the total cost does not exceed the amount available.

Table 3.20

Sites		Cost	Profit
(S_1)	Albuquerque	10	0.19
(S_2)	Dallas	12	0.23
(S_3)	Miami	11	0.20
(S_4)	Phoenix	15	0.30
(S_5)	San Diego	9	0.16

11. Ecap University has two jobs to fill, Dean of the Graduate School (job 1) and Dean of the School of Arts and Science (job 2), and is considering three candidates, J. Frank (candidate 1), M. Smith (candidate 2), and T. James (candidate 3). The search committee of the university, which is to make recommendations to the president, has assigned a numerical rating to each candidate's qualifications for each job (see Table 3.21). These ratings are interpreted as a quantitative measure of each candidate's potential for each job. The search committee wishes to make its recommendations in such a way that potential is maximized.

Table 3.21

	Job	
Candidate	Dean, Graduate School (job 1)	Dean, Arts and Science School (job 2)
Frank (candidate 1)	9	8
Smith (candidate 2)	7	7
James (candidate 3)	6	8

12. The Birute Investment Company has $50,000 available for investment. Four stocks, S_1 (oil), S_2 (computers), S_3 (airlines), and S_4 (steel), are being considered. Table 3.22 specifies the current price per share of each stock and the

expected net profit per share of stock. At most 200 shares of S_1 stock, 300 shares of S_2 stock, 400 shares of S_3 stock, and 300 shares of S_4 stock are to be purchased.

Table 3.22

Stock		Price	Profit
S_1	(oil)	$75	$6.00
S_2	(computers)	$80	$5.60
S_3	(airlines)	$50	$4.00
S_4	(steel)	$60	$4.50

The problem is to determine the number of shares of each stock that should be purchased so as to maximize the total return.

3.10. MATHEMATICAL DUALITY

The application of the corner point method presupposes that the linear porgram it is applied to has a solution to begin with. That this is not always the case is clear from the equipment purchase linear program developed in Example 1 of Section 3.8 (p. 114). Here the problem is with inconsistent constraints, but, as illustrated by the following example, a linear program may have an abundance of feasible points and still not have a solution.

$$\text{Maximize } F(x, y) = 3x + 2y$$

subject to

$$x \geq 0$$
$$y \geq 0$$
$$x + 2y \geq 4$$
$$x - 2y \geq 2$$

The graph of the feasible points of this linear program, shown in Figure 3.14, is unbounded. This linear program does not have a solution because $F(x, y) = 3x + 2y$ does not have a largest value. No matter what value is given, we can make $F(x, y) = 3x + 2y$ exceed that value by choosing a suitable feasible point. For example, if 100 were prescribed, choose (100, 0) as your feasible point; $F(100, 0) = 300$ is greater than 100. If 1000 were prescribed, choose (1000, 0) as your feasible point; $F(1000, 0) = 3000$ is greater than 1000; and so on. If we were to blindly apply the corner-point method to this linear program, we would emerge with $(3, \frac{1}{2})$ as the solution and 10 as the maximum value, which is nonsense.

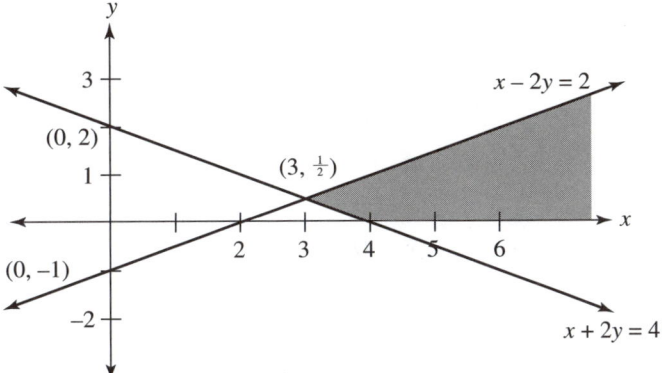

Figure 3.14

Some general approach to determining when a linear program has a solution is highly desirable. An important duality theorem, due to John von Neumann, provides us with such an approach. Its importance, however, goes far beyond this consideration. In Chapter 4 it will be seen to provide us with a bridge for applying the simplex method to minimize-type linear programs, and in Chapter 10 it will be seen to be pivotal in the application of linear-programming methods to game theory. Before stating this duality theorem some preliminaries are needed.

A **maximum linear program** with nonnegativity constraints is said to be in **standard form** if no nonzero constant appears in the objective function, and the constraints other than the nonnegativity conditions are in the less than or equal to ($\leq$) form. A **minimum linear program** with nonnegativity constraints is said to be in **standard form** if no nonzero constant appears in the objective function, and the constraints are in the greater than or equal to ($\geq$) form.

EXAMPLE 1

The Ramuné Company's linear program (from Case 1, Sec. 3.3, p. 95)

$$\text{Maximize } P(x, y) = 10x + 15y$$

subject to

$$x \geq 0$$
$$y \geq 0$$
$$x + 3y \leq 90$$
$$2x + y \leq 80$$

is in standard form.

EXAMPLE 2

The linear program

$$\text{Minimize } F(x, y) = 5x + 4y + 10$$

subject to

$$x \geq 0$$
$$y \geq 0$$
$$5x + 2y \geq 30$$
$$3x + y \geq 20$$

is not in standard form because of the nonzero constant 10 in the objective function $F(x, y) = 5x + 4y + 10$.

EXAMPLE 3

The maximize linear program

$$\text{Maximize } F(x, y) = 3x + 2y$$

subject to

$$x \geq 0 \qquad (3.1)$$
$$y \geq 0 \qquad (3.2)$$
$$x + 2y \geq 4 \qquad (3.3)$$
$$x - 2y \geq 2 \qquad (3.4)$$

considered at the beginning of our discussion is not in standard form because constraints (3.3) and (3.4) are not in the less than or equal to ($\leq$) form. Standard form can be obtained by multiplying (3.3) and (3.4) by -1 to reverse their sense. Doing so yields

$$\text{Maximize } F(x, y) = 3x + 2y$$

subject to

$$x \geq 0$$
$$y \geq 0$$
$$-x - 2y \leq -4$$
$$-x + 2y \leq -2$$

Now that standard form has been considered we can turn our attention to the central duality relationship that underlies von Neumann's duality theorem. Consider the Ramunė Company's linear program along with the minimize linear program that was stated in Section 3.3, Exercise 12, p. 88):

Maximize $P(x, y) = 10x + 15y$
subject to
$$x \geq 0$$
$$y \geq 0$$
$$x + 3y \leq 90$$
$$2x + y \leq 80$$

Minimize $G(s, t) = 90s + 80t$
subject to
$$s \geq 0$$
$$t \geq 0$$
$$s + 2t \geq 10$$
$$3s + t \geq 15$$

The coefficients of these linear programs are stated in tabular form in Figures 3.15(a) and (b).

Let us note the following structural properties:

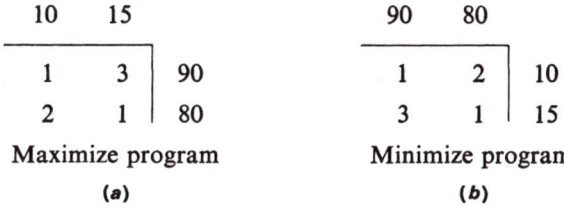

Maximize program
(a)

Minimize program
(b)

Figure 3.15

1. The row of objective function constants of either linear program appears as the column of constraint constants of the other program.
2. The rest of either tableau can be obtained from the tableau of the other by interchanging corresponding rows and columns. Row one, [1 3], of Figure 3.15(a) is column one, $\begin{bmatrix} 1 \\ 3 \end{bmatrix}$, of Figure 3.15(b); row two, [2 1], of Figure 3.15(a) is column two, $\begin{bmatrix} 2 \\ 1 \end{bmatrix}$, of Figure 3.15(b).

Because of these structural features, the two linear programs are said to be **mathematical duals** (or **duals**) of each other. Each linear program is said to be the mathematical dual (or dual) of the other. More generally, a maximum linear program in standard form and a minimum linear program in standard form are said to be **mathematical duals** (or **duals**) of each other if their tableaus have properties 1 and 2. Each is said to be the **dual** of the other.

EXAMPLE 4

Find the dual of the linear program

$$\text{Minimize } C(x, y) = 5x + 4y$$

subject to

$$x \geq 0$$
$$y \geq 0$$
$$2x + y \geq 23$$
$$4x + 5y \geq 75$$

Our first concern is standard form. Let us note that this linear program is in standard form, so that no adjustment is required. From its tabular array, shown in Figure 3.16(a), we obtain the tabular array of its maximum dual. See Figure 3.16(b).

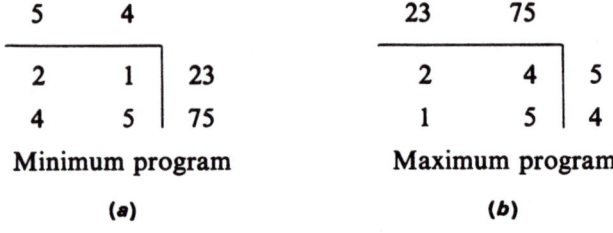

Figure 3.16

The row of objective function constants [5 4] and column of constraint constants $\begin{bmatrix} 23 \\ 75 \end{bmatrix}$ are interchanged, and rows one and two, [2 1] and [4 5], are carried over as columns one and two, $\begin{bmatrix} 2 \\ 1 \end{bmatrix}$ and $\begin{bmatrix} 4 \\ 5 \end{bmatrix}$, respectively, to obtain the tabular array of the dual.

The maximum dual is obtained from the tabular array shown in Figure 3.16(b) by introducing variables, s and t, say, and following the requirements of standard form. This yields

$$\text{Maximize } F(s, t) = 23s + 75t$$

subject to

$$s \geq 0$$
$$t \geq 0$$
$$2s + 4t \leq 5$$
$$s + 5t \leq 4$$

EXAMPLE 5

Find the dual of the linear program

Maximize $F(x, y) = 3x + 2y$

subject to

$$x \geq 0$$
$$y \geq 0$$
$$x + 2y \geq 4 \tag{3.5}$$
$$x - 2y \geq 2 \tag{3.6}$$

We first put the program into standard form by multiplying constraints (3.5) and (3.6) by -1. This yields

Maximize $F(x, y) = 3x + 2y$

subject to

$$x \geq 0$$
$$y \geq 0$$
$$-x - 2y \leq -4$$
$$-x + 2y \leq -2$$

From the tabular array of this program [Figure 3.17(a)], we obtain the tabular array of its minimum dual [Figure 3.17(b)].

	3	2	
-1	-2	-4
-1	2	-2

Maximum program
(a)

	-4	-2	
-1	-1	3
-2	2	2

Minimum program
(b)

Figure 3.17

This tabular array corresponds to the following linear program:

Minimize $G(s, t) = -4s - 2t$

subject to

$$s \geq 0$$
$$t \geq 0$$
$$-s - t \geq 3$$
$$-2s + 2t \geq 2$$

We now turn to the fundamental duality theorem.

> **Duality Theorem of John von Neumann.** If a given linear program has a feasible point and its dual also has a feasible point, then both linear programs have a solution and the same optimal value for the objective function. If either the given linear program or its dual does not have a feasible point, then neither have solutions.

EXAMPLE 6

As an illustration of the first part of the duality theorem, consider the Ramuné Company's linear program and its dual.

Maximize $P(x, y) = 10x + 15y$
subject to
$$x \geq 0$$
$$y \geq 0$$
$$x + 3y \leq 90$$
$$2x + y \leq 80$$

Minimize $G(s, t) = 90s + 80t$
subject to
$$s \geq 0$$
$$t \geq 0$$
$$s + 2t \geq 10$$
$$3s + t \geq 15$$

By inspection we see that (0, 0) is a feasible point of the Ramuné Company's program, and (10, 5) is a feasible point of its minimum dual. It follows from von Neumann's duality theorem that both have solutions and the same value. The duality theorem does not tell us what the solutions and common value are, but from Sections 3.6, Case 1 (p. 95) and 3.3, Exercise 12 (p. 88) we have that (30, 20) is the solution of the Ramuné Company's program, (4, 3) is the solution of its minimum dual, and 600 is their common optimal value.

EXAMPLE 7

To illustrate the second part of the duality theorem, we turn to the maximum linear program shown not to have a solution at the beginning of this section and its dual (see Example 5).

Maximize $F(x, y) = 3x + 2y$
subject to
$$x \geq 0$$
$$y \geq 0$$
$$x + 2y \geq 4$$
$$x - 2y \geq 2$$

Minimize $G(s, t) = -4s - 2t$
subject to

$$s \geq 0 \quad (3.7)$$
$$t \geq 0 \quad (3.8)$$
$$-s - t \geq 3 \quad (3.9)$$
$$-2s + 2t \geq 2 \quad (3.10)$$

The minimum program has no feasible points; if nonnegative values for s and t [required by (3.7) and (3.8)] are substituted into (3.9), we obtain either zero or a negative number, not a value greater than or equal to 3. Thus (3.7) through (3.10) cannot be satisfied simultaneously, which means that this program has no feasible points. It follows from the duality theorem that its dual, our maximum program, has no solution.

EXERCISES

Find the duals of the following linear programs.

1. Max. $P(x, y) = 18x + 12y$
   subject to
   $x \geq 0, y \geq 0$
   $4x + 3y \leq 320$
   $5x + 2y \leq 330$

2. Min. $C(x, y) = 80x + 75y$
   subject to
   $x \geq 0, y \geq 0$
   $2x + y \geq 4$
   $x + y \geq 3$
   $x + 2y \geq 4$

3. Max. $P(x, y, z) = 22x + 30y + 20z$
   subject to
   $x \geq 0, y \geq 0, z \geq 0$
   $2x + 3y + z \leq 11$
   $x + y + 3z \leq 10$
   $2x + 2y + z \leq 10$

4. Max. $I(x, y) = 0.10x + 0.08y$
   subject to
   $x \geq 0, y \geq 0$
   $x + y \leq 60$
   $-x + 3y \geq 0$
   $x \geq 15$

5. Max. $F(x, y) = x + y$
   subject to
   $x \geq 0, y \geq 0$
   $x + y \geq 1$
   $x - y \leq 1$

6. Use von Neumann's duality theorem to determine which of the linear programs stated in Exercises 1–5 have solutions.

CHAPTER 4

The Simplex Method

4.1. SOLUTION OF MAXIMUM TYPE LINEAR PROGRAMS

Although satisfactory for two-variable linear programs, the corner point method is not computationally effective when extended beyond three-variable situations. Its usefulness is restricted to introducing the general idea of linear programming, and the need for more powerful solution techniques is obvious. The simplex method, devised by George Dantzig in the late 1940's, was central to the explosion in linear-programming applications that occurred in the 1950's and 1960's. Although more powerful techniques exist for solving special problems, the simplex method is still the most computationally effective general method available for solving the widest variety of linear-programming problems. Its proof is beyond the scope of this book, but its application is based on the conversion operations used in the tableau method to convert a column of numbers in a tableau to zero–one form. To illustrate the nature of the simplex method, we return to the Ramunė Company's linear program.

EXAMPLE 1

Consider the Ramunė Company's linear program:

$$\text{Maximize } P(x, y) = 10x + 15y$$

subject to

$$x \geq 0 \tag{4.1}$$
$$y \geq 0 \tag{4.2}$$
$$x + 3y \leq 90 \tag{4.3}$$
$$2x + y \leq 80 \tag{4.4}$$

Let us observe that this problem is a maximize-type linear program in standard form with nonnegative constraint constants (90 and 80). These conditions are basic to the version of the simplex method described in this chapter.

We begin by formulating the inequality conditions (4.3) and (4.4) in equation form. [The inequalities (4.1) and (4.2) tell us that x and y must be nonnegative, and do not enter the scene at this point.] To formulate an inequality in equation form, we must add the proper **nonnegative amount** to the smaller side to achieve equality. Let S_1 denote the amount that must be added to $x + 3y$ to obtain 90, and let S_2 denote the amount that must be added to $2x + y$ to obtain 80. Appropriately, S_1 and S_2 are called **slack variables,** since S_1 takes up the slack between $x + 3y$ and 90, and S_2 takes up the slack between $2x + y$ and 80. The introduction of $S_1 \geq 0$ and $S_2 \geq 0$ yields the following system of equations:

$$x + 3y + S_1 \quad\quad = 90 \tag{4.5}$$
$$2x + y \quad\quad + S_2 = 80 \tag{4.6}$$

This system has infinitely many solutions, one of which describes the solution of our linear program. The simplex method can be viewed as a sequence of procedures for locating this particular solution and reading off the answer when it has been reached.

Our next step is to set up the initial simplex tableau, ST_1 (shown in Figure 4.1), which consists of the tableau of Equations (4.5) and (4.6), together with an additional row $[-10 \quad -15 \quad 0 \quad 0 \quad 0]$ consisting of the negatives of the coefficients of the objective function $P(x, y) = 10x + 15y$ and zeros. The role of this last row in the simplex tableau is to indicate the column to be converted to zero–one form, and we term it the **indicator row.**

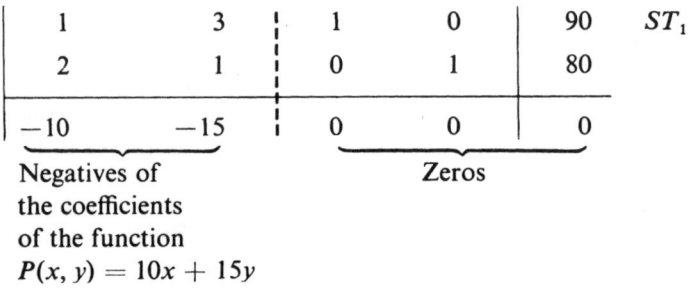

Figure 4.1

Column-Selection Principle. Locate the most negative number in the indicator row. The column containing this number is to be converted to zero–one form.

In our example, -15 is the most negative number in the indicator row of tableau ST_1. This singles out column 2 for conversion to zero–one form.

Pivot-Value-Selection Principle. Divide each positive number in the column selected into the corresponding value in the last column of the simplex tableau and choose as the pivot value that positive number which yields the smallest quotient.

In tableau ST_1, 3 and 1 are the positive numbers in column 2, the column selected for conversion to zero–one form, and 90 and 80, respectively, are the corresponding values in the last column. Thus, we form the quotients $\frac{90}{3} = 30$ and $\frac{80}{1} = 80$. Since 30 is less than 80, we must pivot on 3, even though it would be much more to our liking to pivot on 1. Although the same conversion operations are employed, a basic difference between the procedures of the tableau method and those of the simplex method is in the mechanism for choosing pivot values. The conversion of column 2 of tableau ST_1 into zero–one form with 3 as the pivot value leads to tableau ST_2, shown in Figure 4.2. Row ④ is obtained by multiplying row ① by $\frac{1}{3}$; row ⑤ = (-1) row ④ + row ②, and row ⑥ = (15) row ④ + row ③.

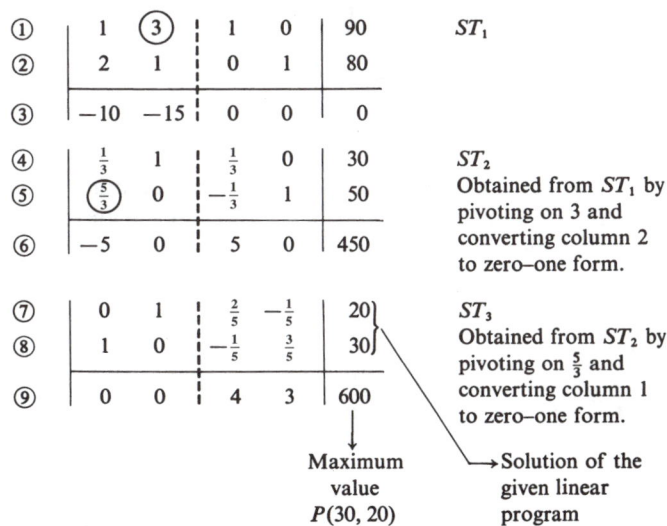

Figure 4.2

Since there is a negative number, -5, in the indicator row of tableau ST_2, our next step is to repeat the process. The process is repeated as long as the indicator row of the simplex tableau obtained contains negative numbers. The value -5 is the only negative number in row ⑥, and this singles out column 1 for conversion to zero–one form. The pivot-value-selection principle leads us to form the quotients $30 \div \frac{1}{3} = 90$ and $50 \div \frac{5}{3} = 30$. Since 30 is less than 90, we pivot on $\frac{5}{3}$ in row ⑤. The conversion of column 1 of tableau ST_2 into zero–one form with $\frac{5}{3}$ as the pivot value leads to tableau ST_3, shown in Figure 4.2. Row ⑧ is obtained by multiplying row ⑤ by $\frac{3}{5}$. Row ⑦ $= (-\frac{1}{3})$row ⑧ + row ④, and row ⑨ $=$ (5)row ⑧ + row ⑥.

Since the indicator row (row ⑨) of tableau ST_3 has no negative numbers, ST_3 is the final simplex tableau. The solution of our linear program appears as the x and y values of an "appropriate" solution of the system

$$x + 3y + S_1 = 90 \qquad (4.5)$$
$$2x + y + S_2 = 80 \qquad (4.6)$$

This "appropriate" solution is read off from the final simplex tableau ST_3 by means of the following reading principle.

> **Reading Principle.** If a column for a variable is in zero–one form and there are no identical columns in zero–one form (that is, ones and zeros in the same rows), then assign to the variable of the column the value in the row of the 1 that appears at the extreme right in the last column of the tableau. Assign 0 to the variable of a column not in zero–one form. The value of the objective function for the values of the linear program read off from the tableau is found in the last column of the indicator row.

Application of the reading principle to columns 1 and 2 (the columns of x and y, respectively) of tableau ST_3 yields the same solution obtained by the corner point method (see Section 3.6, Case 1, p. 95), $x = 30$, $y = 20$. The maximum value $P(30, 20) = 10(30) + 15(20) = 600$ is obtained from the value box in the last column of the indicator row. Just as a point of interest, further application of the reading principle to columns 3 and 4 (the columns of S_1 and S_2, respectively) of tableau ST_3 yields $S_1 = 0$, $S_2 = 0$ since these columns are not in zero–one form. Thus the "appropriate" solution of the system

$$x + 3y + S_1 = 90 \qquad (4.5)$$
$$2x + y + S_2 = 80 \qquad (4.6)$$

which contains the solution of our linear program is (30, 20; 0, 0).

Reading Principle: Part 2. If two or more columns of linear-program variables have the same zero–one form, then assign the usual value to the variable of one of these columns and 0 to the variables of the other columns in identical zero–one form.

Suppose, for example, that the simplex tableau displayed in Figure 4.3 arises in connection with solving a linear program whose variables are x and y. We are ready to read off a solution, but if we assign x the value 10, then we must assign y the value 0 since the columns of x and y have identical zero–one form. Also, we could take y equal to 10; we then take 0 for x.

The underlying linear program has solutions (10, 0) and (0, 10) with maximum value 120.

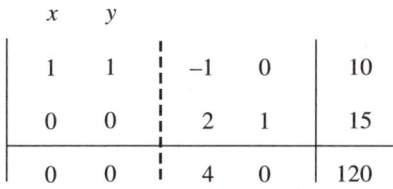

Figure 4.3

EXAMPLE 2

The Linnus Company's production problem (see Section 3.6, Exercise 11, p. 110) leads to the linear program

$$\text{Maximize } P(x, y, z) = 24x + 30y + 26z$$

subject to

$$x \geq 0$$
$$y \geq 0$$
$$z \geq 0$$
$$2x + 3y + z \leq 16$$
$$x + y + 3z \leq 17$$
$$2x + y + z \leq 12$$

where x, y, and z are the number of C15, C24, and C51 minicomputer units to be made, respectively. Solve this linear program by the simplex method.

Let us note that this linear program is a maximize linear program in standard form with nonnegative constraint constants (that is, 16, 17, and 12), so that the version of the simplex method discussed in Example 1 is applicable. By introducing the slack variables S_1, S_2, and S_3, we obtain the following system of three equations from the last three inequalities in the system of constraints.

$$2x + 3y + z + S_1 = 16 \tag{4.7}$$
$$x + y + 3z + S_2 = 17 \tag{4.8}$$
$$2x + y + z + S_3 = 12 \tag{4.9}$$

The tableau of this system of equations together with the indicator row [−24 −30 −26 0 0 0 0], consisting of the negatives of the objective function constants and zeros, is shown as tableau ST_1 in Figure 4.4.

①	2	③	1	1	0	0	16	ST_1
②	1	1	3	0	1	0	17	Since -30 is the most negative number in the indicator row, column 2 is chosen for conversion to zero–one form. We pivot on 3 in row ① since $\frac{16}{3}$ is the smallest quotient.
③	2	1	1	0	0	1	12	
④	−24	−30	−26	0	0	0	0	
⑤	$\frac{2}{3}$	1	$\frac{1}{3}$	$\frac{1}{3}$	0	0	$\frac{16}{3}$	ST_2
⑥	$\frac{1}{3}$	0	$\frac{8}{3}$	$-\frac{1}{3}$	1	0	$\frac{35}{3}$	Since -16 is the most negative number in the indicator row, column 3 is chosen for conversion to zero–one form. We pivot on $\frac{8}{3}$ since the quotient $\frac{35}{3} \div \frac{8}{3} = 4.38$ is smallest.
⑦	$\frac{4}{3}$	0	$\frac{2}{3}$	$-\frac{1}{3}$	0	1	$\frac{20}{3}$	
⑧	−4	0	−16	10	0	0	160	
⑨	$\frac{5}{8}$	1	0	$\frac{3}{8}$	$-\frac{1}{8}$	0	$\frac{31}{8}$	ST_3
⑩	$\frac{1}{8}$	0	1	$-\frac{1}{8}$	$\frac{3}{8}$	0	$\frac{35}{8}$	Since -2 is the most negative number in the indicator row, column 1 is chosen for conversion to zero–one form. We pivot on $\frac{5}{4}$ since the quotient $\frac{45}{12} \div \frac{5}{4} = 3$ is smallest.
⑪	$\frac{5}{4}$	0	0	$-\frac{1}{4}$	$-\frac{1}{4}$	1	$\frac{45}{12}$	
⑫	−2	0	0	8	6	0	230	
⑬	0	1	0	$\frac{1}{2}$	0	$-\frac{1}{2}$	2	ST_4
⑭	0	0	1	$-\frac{1}{10}$	$\frac{2}{5}$	$-\frac{1}{10}$	4	Since there are no negative numbers in the indicator row, this is the final simplex tableau. We have $x = 3, y = 2, z = 4$; maximum value 236.
⑮	1	0	0	$-\frac{1}{5}$	$-\frac{1}{5}$	$\frac{4}{5}$	3	
⑯	0	0	0	$\frac{38}{5}$	$\frac{28}{5}$	$\frac{8}{5}$	236	

Figure 4.4

Tableau ST_4 is the final simplex tableau since there are no negative numbers in the indicator row. Application of the reading rule yields the solution $x = 3$, $y = 2$, $z = 4$ and the maximum value 236. Implementation of this solution calls for a daily production schedule of 3 C15 units, 2 C24 units, and 4 C51 units with an anticipated daily profit of $236.

EXERCISES

Each of the following simplex tableaus arose from solving a maximum type linear program. Read off all solutions and the value of the objective function, or determine the pivot value to generate the next simplex tableau, as appropriate

1.

x	y	z				
1	0	0	-2	1	6	10
0	1	0	1	1	$\frac{1}{2}$	12
0	0	1	4	5	-1	15
0	0	2	3	2	2	50

2.

1	1	0	2	-1	4	14
0	0	1	-2	4	3	15
0	0	0	0	2	1	18
0	0	1	5	2	1	75

3.

$\frac{1}{2}$	2	$\frac{1}{3}$	4	-1	3	18
$\frac{1}{4}$	1	2	3	$\frac{1}{2}$	6	16
-1	4	$\frac{1}{2}$	2	8	2	3
-5	-1	2	2	3	0	98

4.

2	$\frac{1}{4}$	$\frac{1}{3}$	1	$\frac{1}{2}$	4	10
3	$\frac{1}{2}$	3	-1	$\frac{1}{3}$	1	9
2	3	-1	2	-1	4	4
3	0	0	2	-2	0	82

5.

3	1	$\frac{1}{4}$	1	$\frac{1}{4}$	$\frac{1}{2}$	12
1	-4	2	3	2	-1	1
2	$-\frac{1}{2}$	3	2	-1	-3	4
3	-6	4	0	2	1	56

6.

s	t		
1	0	5	-1	12
0	1	2	3	3
0	0	5	4	55

7.

$\frac{1}{3}$	$\frac{1}{2}$	1	0	$\frac{1}{9}$
$\frac{1}{4}$	1	-3	$\frac{1}{2}$	$\frac{1}{8}$
6	-2	0	2	46

8.

1	2	$\frac{1}{2}$	-4	$\frac{1}{4}$
$\frac{1}{2}$	4	$\frac{1}{3}$	5	$\frac{1}{6}$
0	2	8	-1	89

9.

3	0	-1	3	9
$-\frac{1}{2}$	1	1	$\frac{1}{2}$	4
-1	0	3	2	18

10.

1	0	-1	2	8
0	0	0	-1	12
0	1	1	4	15
0	0	5	6	27

11.

1	$\frac{1}{5}$	2	1	$\frac{1}{10}$
0	$\frac{1}{6}$	-3	0	$\frac{1}{3}$
0	1	-4	2	18

EXAMPLE 3

The linear program

$$\text{Maximize } F(x, y) = 20x + 14y$$

subject to

$$x \geq 0$$
$$y \geq 0$$
$$x \leq 25$$
$$2x + y \leq 75$$

yields the system of equations

$$x \quad\quad + S_1 \quad\quad = 25 \tag{4.10}$$
$$2x + y \quad\quad + S_2 = 75 \tag{4.11}$$

where S_1 and S_2 are slack variables. A sequence of simplex tableaus leading to the solution of this linear program is shown in Figure 4.5.

①	①	0	1	0	25	ST_1
②	2	1	0	1	75	Convert column 1 to zero–one form. Pivot on 1 since $\frac{25}{1}$ is less than $\frac{75}{2}$.
③	−20	−14	0	0	0	
④	1	0	1	0	25	ST_2
⑤	0	①	−2	1	25	Convert column 2 to zero–one form by pivoting on 1, the only eligible pivot value.
⑥	0	−14	20	0	500	
⑦	1	0	①	0	25	ST_3
⑧	0	1	−2	1	25	Convert column 3 to zero–one form. Since we cannot pivot on negative values, we must pivot on 1, the only eligible pivot value.
⑨	0	0	−8	14	850	
⑩	1	0	1	0	25	ST_4
⑪	2	1	0	1	75	Since there are no negative numbers in the indicator row, this is the final simplex tableau. We have $x = 0$, $y = 75$ and maximum value 1050.
⑫	8	0	0	14	1050	

Figure 4.5

EXAMPLE 4

The linear program

$$\text{Maximize } G(x, y) = 3x + 5y$$

subject to

$$x \geq 0$$
$$y \geq 0$$
$$-2x + y \leq 1$$
$$x - 3y \leq 6$$

yields the system of equations

$$-2x + y + S_1 = 1 \quad (4.12)$$
$$x - 3y + S_2 = 6 \quad (4.13)$$

where S_1 and S_2 are slack variables. From this linear program we obtain the simplex tableaus ST_1 and ST_2 shown in Figure 4.6. Tableau ST_2 is curious to say the least.

①	-2	①	1	0	1	ST_1 Convert column 2 to zero-one form. Since we cannot pivot on negative values, we must pivot on 1, the only eligible pivot candidate.
②	1	-3	0	1	6	
③	-3	-5	0	0	0	
④	-2	1	1	0	1	ST_2 Convert column 1 to zero-one form. But this cannot be done, because **there is no eligible pivot candidate. Implication: the linear program has no solution.**
⑤	-5	0	3	1	9	
⑥	-13	0	5	0	5	

Figure 4.6

Since the indicator row contains a negative value, -13, column 1 must be converted to zero–one form. But at the same time we cannot proceed, since both of the remaining values in column 1 are negative and cannot serve as pivot values. The implication of this dilemma is that the given linear program does not have a solution, which can be established independently by means of von Neumann's duality theorem or consideration of the program's region of feasible points (which is unbounded).

EXAMPLE 5

The linear program

Maximize $F(x, y, z) = 6x + 6y + 4z$

subject to

$$x \geq 0$$
$$y \geq 0$$
$$z \geq 0$$
$$x + y + z \leq 12$$
$$x + y + 3z \leq 9$$

yields the simplex tableaus ST_1 and ST_2 shown in Figure 4.7.

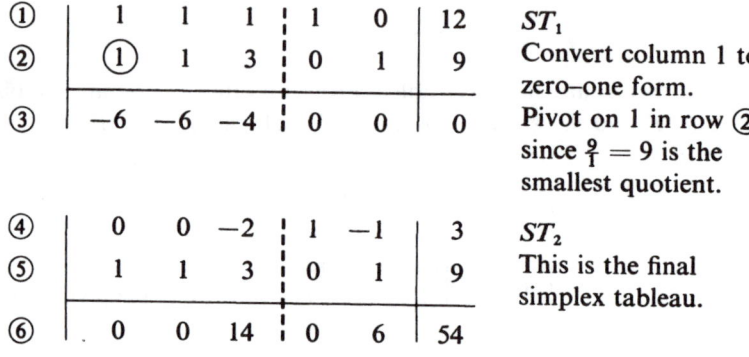

Figure 4.7

If we assign 9 to x (column 1), then we must assign 0 to y (column 2). Since column 3 is not in zero–one form, $z = 0$. This yields the solution (9, 0, 0). If we assign 9 to y, then we must assign 0 to x. Since $z = 0$, we obtain the solution (0, 9, 0). The maximum value is 54.

In connection with setting up the initial simplex tableau, it is useful to note that the part of the tableau corresponding to the placement of slack variables consists of a square block of zeros and ones, with ones running down the main diagonal and zeros elsewhere. With this in mind, we can actually omit writing down the system of equations from which the initial simplex tableau is obtained and set down the initial simplex tableau directly.

EXAMPLE 6

The linear program

Maximize $F(x, y, z, w) = 10x + 4y + 8z + 10w$

subject to

$x \geq 0, y \geq 0, z \geq 0, w \geq 0$

$x + 2y + w \leq 20$

$y + 2z + w \leq 18$

$x + 2y + z + 3w \leq 18$

yields the simplex tableaus ST_1 and ST_2 shown in Figure 4.8.

$$\begin{array}{|cccc:ccc|c|} 1 & 2 & 0 & 1 & 1 & 0 & 0 & 20 \\ 0 & 1 & 2 & 1 & 0 & 1 & 0 & 18 \\ 1 & 2 & 1 & 3 & 0 & 0 & 1 & 18 \\ \hline -10 & -4 & -8 & -10 & 0 & 0 & 0 & 0 \end{array}$$

ST_1
Convert column 1 to zero–one form by pivoting on 1 in row ③.

$$\begin{array}{|cccc:ccc|c|} 0 & 0 & -1 & -2 & 1 & 0 & -1 & 2 \\ 0 & 1 & 2 & 1 & 0 & 1 & 0 & 18 \\ 1 & 2 & 1 & 3 & 0 & 0 & 1 & 18 \\ \hline 0 & 16 & 2 & 20 & 0 & 0 & 10 & 180 \end{array}$$

ST_2
Final simplex tableau. (18, 0, 0, 0) is the solution and 180 is the maximum value.

Figure 4.8

Note that the size of the block of zeros and ones is equal to the number of major constraints of the linear program (in this case, 3) and not the number of variables (in this case, 4).

Summary

> The simplex method developed in this section is applicable to maximize type linear programs in standard form with nonnegative constraint constants.

If these conditions are not satisfied, a refinement on these procedures is needed. Such a refinement is discussed in: W. J. Adams *Fundamentals of Mathematics for Business, Social, and Life Sciences* (Prentice-Hall, 1979), Sec. 30.

Procedures

1. Be sure that the maximize linear program is in standard form with nonnegative constraint constants.
2. Set up the initial simplex tableau.
3. Column-Selection Principle. Locate the most negative number in the indicator row; the column containing this number is to be converted to zero–one form.*
4. Pivot-Value-Selection Principle. Divide each positive number in the column selected in step 3 into the corresponding value in the last column of the tableau, and choose as your pivot value that number which yields the smallest quotient.

If two or more values yield the same smallest quotient, then choose one at random.†
5. Convert the column selected in step 3 to zero–one form by pivoting on the value selected in step 4.
6. Repeat steps 3 through 5 until no pivot value can be obtained.

The simplex procedures will terminate if a tableau is obtained with a negative number in the indicator row with no eligible pivot candidates in the corresponding column, which means that the linear program has no solution (see Example 4), or if a tableau is reached that has nonnegative numbers in its indicator row. Such a tableau is the final simplex tableau from which a solution can be read off.

---

*It is worthy of note that it is not incorrect to choose a column determined by any negative number in the indicator row for conversion to zero–one form. It is, however, generally more efficient to work with the column determined by the most negative number in the indicator row in the sense that the least number of simplex tableaus is generated. There are exceptions; a more efficient column-selection principle exists [the method of steepest ascents; see, for example, G. Hadley, *Linear Programming* (Reading, Mass: Addison-Wesley Publishing Co., 1962), pp. 111–112], but its efficiency is offset by its cumbersomeness. Most computer codes for applying the simplex method employ the column-selection principle cited.

†Examples have been constructed which show that when different values in the same column yield the same smallest quotient, the choice of certain of these values as pivots can result in becoming entrapped in a loop so that the optimal feasible point is not reached. [See, for example, A. Charnes, "Optimality and Degeneracy in Linear Programming," *Econometrica*, vol. 20 (1952), pp. 160–170; or L. Cooper and D. Steinberg, *Methods and Applications of Linear Programming* (Philadelphia: W. B. Saunders Co., 1974), pp. 105–114.] Special rules have been constructed for avoiding such loops, but such rules have been virtually ignored in actual applications, since loops, while theoretically possible, have rarely, if ever, occurred in applications.

> **7. Reading Principle.** If a column of a linear-program variable is in zero–one form and no other column of a linear-program variable has exactly the same zero–one form, then assign the number, in the row of the 1, that appears in the last column of the tableau to the variable of this column. Assign 0 to the variable of a column not in zero–one form.
>
> If two or more columns of linear-program variables have the same zero–one form, then assign the value as described above to the variable of one of these columns and 0 to the variables of the other columns in identical zero–one form.
>
> The maximum value of the objective function for the solution obtained is found in the last column of the indicator row.

EXERCISES

Each of the tableaus in Exercises 12–23 arose from solving a maximum type linear program. Read off all solutions and the value of the objective function, determine pivot value options, or determine that the underlying linear program has no solution (with an explanation), as is appropriate.

12.

x	y	z				
2	3	-1	$\frac{1}{2}$	-4	2	4
1	4	2	2	-1	3	1
0	1	6	-3	$\frac{1}{3}$	4	6
6	-1	0	4	-2	6	20

13.

1	1	0	2	$\frac{1}{2}$	$\frac{1}{4}$	8
0	0	1	4	3	-2	12
0	1	0	6	0	4	59

14.

1	0	$\frac{1}{2}$	4	2	6	10
0	1	0	6	3	1	12
0	0	0	1	4	2	16
0	0	-2	0	-5	7	110

15.

1	0	0	2	3	2	10
0	1	0	1	1	0	15
0	0	1	$\frac{1}{2}$	$\frac{1}{3}$	1	18
1	0	0	0	1	5	90

16.
$$\begin{array}{ccc|cc|c} 1 & -1 & 1 & 2 & -2 & 6 \\ 0 & 2 & 0 & 4 & 1 & 8 \\ \hline 0 & -1 & 0 & 0 & 5 & 25 \end{array}$$

$\quad\quad x \quad\quad y \quad\quad z \quad\quad w$

17.
$$\begin{array}{cccc|cccc|c} 0 & 0 & 0 & 0 & 2 & 0 & 4 & 5 & 12 \\ 0 & 0 & 0 & 0 & 1 & 2 & 0 & 1 & 15 \\ 0 & 1 & 1 & 1 & 4 & 6 & 1 & 1 & 16 \\ 1 & 0 & 0 & 0 & 1 & -\frac{1}{2} & 2 & -2 & 18 \\ \hline 0 & 0 & 0 & 0 & 2 & 0 & 1 & 0 & 230 \end{array}$$

$\quad\quad u \quad\quad v$

18.
$$\begin{array}{cc|ccc|c} \frac{1}{2} & \frac{1}{3} & -1 & \frac{1}{2} & \frac{1}{12} & \\ \frac{2}{3} & \frac{1}{5} & 4 & 2 & \frac{1}{10} & \\ \hline -5 & -8 & 2 & 3 & 100 \end{array}$$

21.
$$\begin{array}{cc|cc|c} 1 & 1 & 4 & 2 & 10 \\ 0 & 0 & -1 & 3 & 12 \\ \hline 1 & 0 & 2 & 1 & 28 \end{array}$$

19.
$$\begin{array}{cc|cc|c} \frac{1}{3} & -2 & 1 & \frac{1}{2} & 8 \\ 4 & -\frac{1}{2} & -1 & 4 & 2 \\ \hline 0 & -5 & 5 & 7 & 14 \end{array}$$

22.
$$\begin{array}{cc|cc|c} 1 & 1 & 3 & \frac{1}{2} & 14 \\ 0 & 0 & -\frac{1}{4} & 2 & 18 \\ \hline 0 & 0 & 3 & 5 & 68 \end{array}$$

20.
$$\begin{array}{cc|cc|c} 1 & 3 & 1 & \frac{1}{7} & \frac{1}{14} \\ 0 & 1 & \frac{1}{2} & \frac{1}{4} & \frac{1}{8} \\ \hline -1 & 8 & 4 & -6 & 20 \end{array}$$

23.
$$\begin{array}{cc|cc|c} 1 & \frac{1}{6} & 4 & -2 & \frac{1}{18} \\ 0 & \frac{1}{5} & \frac{1}{2} & 1 & \frac{1}{15} \\ \hline -3 & -8 & 1 & 4 & 28 \end{array}$$

If possible, solve the following linear programs by the simplex method.

24. Max. $F(x, y) = 10x + 12y$
  subject to
  $x \geq 0, y \geq 0$
  $3x + y \leq 18$
  $x + y \leq 8$

25. Max. $F(x, y) = 6x + 5y$
  subject to
  $x \geq 0, y \geq 0$
  $x + 4y \leq 140$
  $x + 2y \leq 80$

26. Max. $G(x, y) = 5x + 10y$
  subject to
  $x \geq 0, y \geq 0$
  $6x + 2y \leq 90$
  $x + 2y \leq 40$

27. Max. $H(x, y) = 20x + 18y$
  subject to
  $x \geq 0, y \geq 0$
  $x + 5y \leq 280$
  $x + y \leq 80$

28. Max. $G(x, y) = 14x + 16y$
    subject to
    $x \geq 0, y \geq 0$
    $2x + y \leq 32$
    $x + 3y \leq 46$

29. Max. $F(x, y) = 8x + 10y$
    subject to
    $x \geq 0, y \geq 0$
    $-x + y \leq 6$
    $x - 3y \leq 9$

30. Max. $G(x, y, z) = 10x + 12y + 8z$
    subject to
    $x \geq 0, y \geq 0, z \geq 0$
    $x + y + 2z \leq 12$
    $y + 2z \leq 8$
    $2x + z \leq 10$

31. Max. $H(x, y, z) = 5x + 6y + 4z$
    subject to
    $x \geq 0, y \geq 0, z \geq 0$
    $x + 2y + z \leq 20$
    $x + y \leq 12$
    $y + 3z \leq 18$

32. "The precision of the simplex method is well known. No one would take issue with the conclusions obtained by use of the simplex method because the simplex method is the surest way of obtaining truth in a linear-programming context."

    Do you agree with this assertion? Answer yes or no, and explain in appropriate detail.

33. Set up a linear-programming model for the following program and solve it by use of the simplex method. Interpret the solution obtained in terms of the problem.

    One of the methods used by the Thorland Company to separate copper, lead, and zinc from ores is the flotation separation process. The flotation process consists of three steps: oiling, mixing, and separation, which must be applied for 2, 2, and 1 hours, respectively, to produce 1 unit of copper (1 unit is 100 pounds); 2, 3, and 1 hours, respectively, to produce 1 unit of lead; and 1, 1, and 3 hours, respectively, to produce 1 unit of zinc. The oiling and separation phases of the flotation process can each be in operation at most 10 hours a day; the mixing phase can be in operation at most 11 hours a day. The Thorland Company anticipates a profit of $45 on a unit of copper, $30 on a unit of lead, and $35 on a unit of zinc. The demand for these metals is unlimited. The problem is to determine how many units of each metal should be produced daily by means of the flotation process to maximize profit.

34. Solve the advertising media selection linear program formulated by James Engel and Martin Warshaw for the McGraw-Edison Company (p. 113). This is not a violation of the Constitution's ban on cruel and excessive punishment, but an illustration which shows off the power of the simplex method. Only three simplex tableaus are generated beyond ST_1.

4.2. AN INTERPRETATION OF THE SIMPLEX METHOD

We have seen that, by introducing slack variables, the constraints of a linear program other than the nonnegativity conditions can be expressed as a system of linear equations that has infinitely many solutions. One of these solutions describes the solution of the linear program, and the simplex method can be viewed as a sequence of procedures for locating this solution and reading off the solution and value of the linear program.

The simplex method can also be interpreted in terms of the corner point method. To illustrate, we return to the Ramuné Company's linear program:

$$\text{Maximize } P(x, y) = 10x + 15y$$

subject to

$$x \geq 0$$
$$y \geq 0$$
$$x + 3y \leq 90$$
$$2x + y \leq 80$$

The graph of the feasible points of this linear program is reproduced in Figure 4.9 and Table 4.1 shows the value of the objective function for each corner point. In Figure 4.10 the simplex-method solution is reproduced. Note that, if we apply the

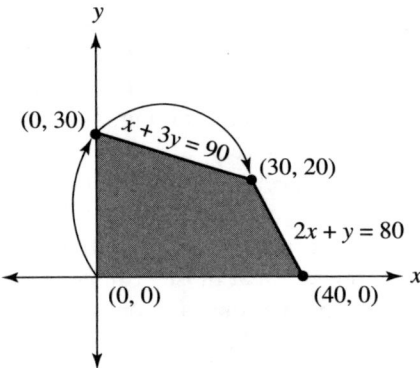

Figure 4.9

Table 4.1

Corner Point	$P(x, y) = 10x + 15y$
(0, 0)	0
(40, 0)	400
(0, 30)	450
(30, 20)	600

$$\begin{array}{cc|cc|c} 1 & ③ & 1 & 0 & 90 \\ 2 & 1 & 0 & 1 & 80 \\ \hline -10 & -15 & 0 & 0 & 0 \end{array} \quad ST_1 \;\; x=0, y=0 \\ \text{Value } 0$$

$$\begin{array}{cc|cc|c} \frac{1}{3} & 1 & \frac{1}{3} & 0 & 30 \\ ⑤/3 & 0 & -\frac{1}{3} & 1 & 50 \\ \hline -5 & 0 & 5 & 0 & 450 \end{array} \quad ST_2 \;\; x=0, y=30 \\ \text{Value } 450$$

$$\begin{array}{cc|cc|c} 0 & 1 & \frac{2}{5} & -\frac{1}{5} & 20 \\ 1 & 0 & -\frac{1}{5} & \frac{3}{5} & 30 \\ \hline 0 & 0 & 4 & 3 & 600 \end{array} \quad ST_3 \;\; x=30, y=20 \\ \text{Value } 600$$

Figure 4.10

reading rule to each of the simplex tableaus ST_1, ST_2, and ST_3, we obtain a corner point and the value of the objective function at the corner point. From ST_1 we have (0, 0) and $P(0, 0) = 0$; from ST_2 we have (0, 30) and $P(0, 30) = 450$; from ST_3 we have (30, 20) and $P(30, 20) = 600$. This example illustrates that for maximize linear programs the procedures of the simplex method can be viewed as taking us from corner point to corner point, until a corner point that yields the maximum value is reached.

4.3. SOLUTIONS OF MINIMUM TYPE LINEAR PROGRAMS

With a simple adjustment, the simplex method developed for maximum linear programs can also be used to solve minimum linear programs. The adjustment is based on the fact that when the simplex method is applied to a maximum linear program we obtain as a free extra bonus, so to speak, the solution of its minimum dual. The value of the minimum dual is, by von Neumann's theorem (see Section 3.10), equal to the value of the maximum program. To illustrate, we return once more to the Ramuné Company's linear program:

Maximize $P(x, y) = 10x + 15y$

subject to

$$x \geq 0$$
$$y \geq 0$$
$$x + 3y \leq 90$$
$$2x + y \leq 80$$

Its minimum dual,

$$\text{Minimize } G(s, t) = 90s + 80t$$

subject to

$$s \geq 0$$
$$t \geq 0$$
$$s + 2t \geq 10$$
$$3s + t \geq 15$$

has solution $s = 4$, $t = 3$ and minimum value 600 (see Section 3.3, Exercise 12, p. 88). The simplex-method solution of the Ramuné Company's linear program is reproduced in Figure 4.11. Let us note that the solution of the minimum dual (4, 3) is found in the indicator row of the final tableau ST_3 under what was originally a block of zeros and ones. The first value, 4, is the value of the first variable, s, and the second value, 3, is the value of the second variable, t. The minimum value 600 is the common value of the dual linear programs.

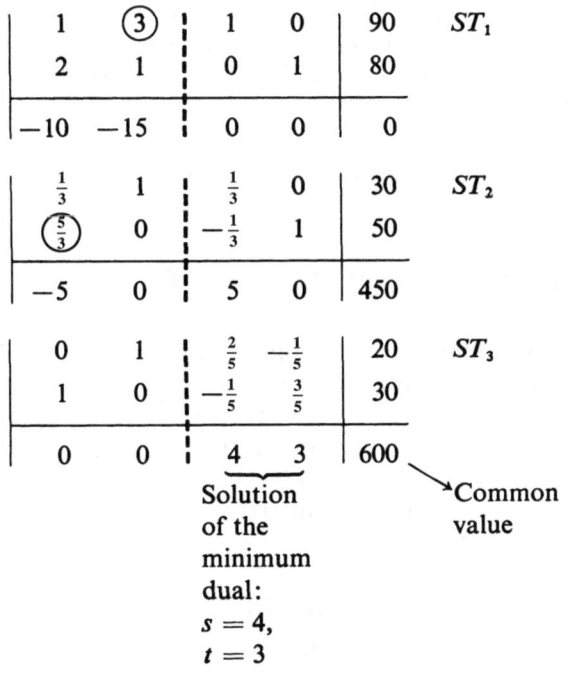

Figure 4.11

If only the maximum program were of interest to us, we would not pay attention to the solution of the minimum dual; it would be present as a free unneeded extra. On the other hand, if we needed to solve a minimum linear program, the simplex method could be employed in the following way:

Chapter 4 / The Simplex Method

1. Obtain the maximum dual program of the given minimum program.
2. Apply the simplex method to the maximum dual program.
3. Read off the solution and value of the minimum program from the indicator row of the final simplex tableau.

EXAMPLE 1

Use the simplex method to solve the linear program

$$\text{Minimize } F(x, y) = 180x + 210y$$

subject to

$$x \geq 0$$
$$y \geq 0$$
$$4x + y \geq 12$$
$$x + 3y \geq 15$$

We begin by determining the maximum dual. From the tabular array of this program [Figure 4.12(a)], we obtain the tabular array of its minimum dual [Figure 4.12(b)].

180	210	
4	1	12
1	3	15

Minimum program
(a)

12	15	
4	1	180
1	3	210

Maximum program
(b)

Figure 4.12

This tabular array corresponds to the following maximum program:

$$\text{Maximize } G(s, t) = 12s + 15t$$

subject to

$$s \geq 0$$
$$t \geq 0$$
$$4s + t \leq 180$$
$$s + 3t \leq 210$$

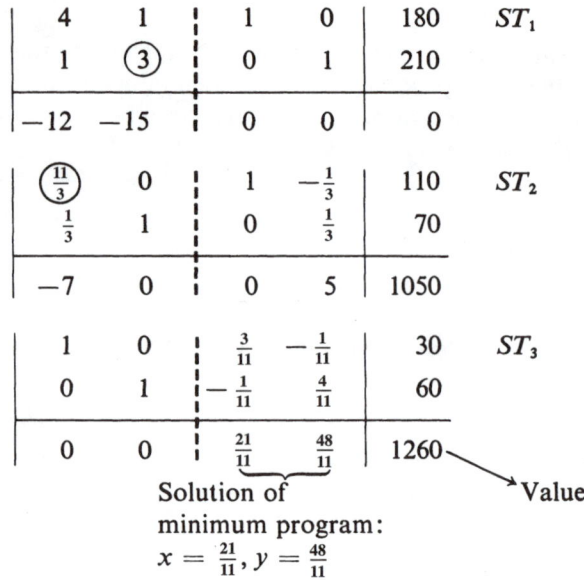

Figure 4.13

Applying the simplex method to this maximum dual program yields the sequence of simplex tableaus shown in Figure 4.13. From the indicator row of tableau ST_3, we obtain the solution $x = \frac{21}{11}$, $y = \frac{48}{11}$ and the minimum value 1260.

EXAMPLE 2

Solve the linear program

$$\text{Minimize } F(x, y, z) = 36x + 28y + 30z$$

subject to

$$x \geq 0$$
$$y \geq 0$$
$$z \geq 0$$
$$4x + 3z \geq 12$$
$$3x + 2y \geq 10$$
$$x + y + z \geq 14$$

From the tabular array of this program [see Figure 4.14(a)], we obtain the tabular array of its minimum dual [see Figure 4.14(b)].
This tabular array corresponds to the following maximum program:

$$\begin{array}{ccc|c}
36 & 28 & 30 & \\
\hline
4 & 0 & 3 & 12 \\
3 & 2 & 0 & 10 \\
1 & 1 & 1 & 14 \\
\end{array} \qquad \begin{array}{ccc|c}
12 & 10 & 14 & \\
\hline
4 & 3 & 1 & 36 \\
0 & 2 & 1 & 28 \\
3 & 0 & 1 & 30 \\
\end{array}$$

Minimum program   Maximum program

(a)   (b)

Figure 4.14

Maximize $G(s, t, u) = 12s + 10t + 14u$

subject to

$$s \geq 0, \ t \geq 0, \ u \geq 0$$
$$4s + 3t + u \leq 36$$
$$2t + u \leq 28$$
$$3s \ \ \ \ + u \leq 30$$

Applying the simplex method to this maximum dual yields the sequence of simplex tableaus shown in Figure 4.15. From the indicator row of tableau ST_3, we obtain the solution $x = 0$, $y = 10$, $z = 4$ and the maximum value 400.

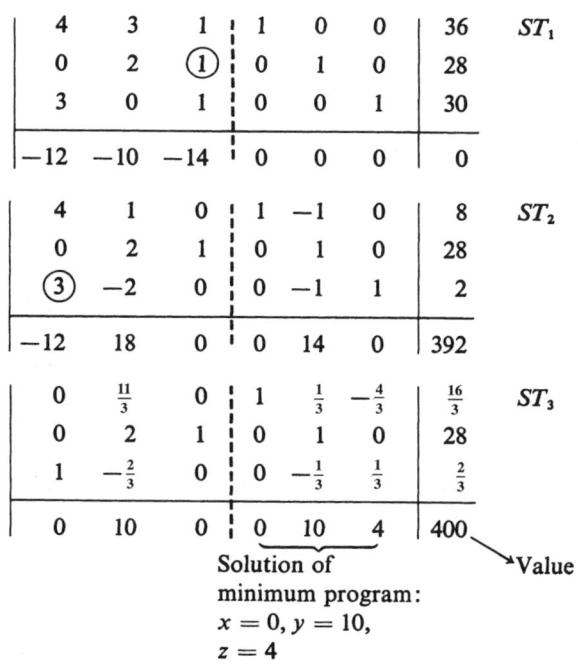

Figure 4.15

EXERCISES

Solve the following linear programs by the simplex method.

1. Min. $G(x, y) = 10x + 12y$
   subject to
   $x \geq 0, y \geq 0$
   $4x + y \geq 13$
   $2x + y \geq 9$

2. Min. $F(x, y) = 8x + 10y$
   subject to
   $x \geq 0, y \geq 0$
   $3x + y \geq 15$
   $4x + y \geq 18$

3. Min. $F(x, y) = 10x + 9y$
   subject to
   $x \geq 0, y \geq 0$
   $5x + 2y \geq 16$
   $3x + y \geq 9$

4. Min. $G(x, y) = 9x + 5y$
   subject to
   $x \geq 0, y \geq 0$
   $x + 4y \geq 22$
   $2x + y \geq 9$

5. Min. $F(x, y) = 18x + 20y$
   subject to
   $x \geq 0, y \geq 0$
   $3x + 2y \geq 28$
   $x + 4y \geq 26$

6. Min. $G(x, y, z) = 10x + 11y + 10z$
   subject to
   $x \geq 0, y \geq 0, z \geq 0$
   $2x + 2y + z \geq 45$
   $2x + 3y + z \geq 30$
   $x + y + 3z \geq 35$

7. Min. $F(x, y, z) = 10x + 12y + 20z$
   subject to
   $x \geq 0, y \geq 0, z \geq 0$
   $x + 2y + z \geq 8$
   $2x + y + 3z \geq 12$
   $x + y \geq 6$

8. Set up a linear programming model for the following problem and solve it by means of the simplex method. Interpret the solution obtained in terms of the problem.

   The Tuman Company is well known for its industrial transformers. Two models, W3 and W9, are produced at the Charlestown and Dubury plants of the company. At the Charlestown plant, 3 hours are required to complete a W3 unit and 2 hours are required to complete a W9 unit. Three hours are required to complete a unit of either model at the Dubury plant. Both plants can be in operation at most 18 hours a day. At the Charlestown plant it costs $3000 to produce a W3 unit and $3600 to produce a W9 unit; at the Dubury plant it costs $3300 to produce a unit of either model. To fulfill past orders the company must produce daily at least six W3 units and nine W9 units. The problem is to determine how production should be organized so that at least the required number of each model is produced at minimal cost.

CHAPTER 5

Life in the Land of Matrices

5.1. ALGEBRAIC OPERATIONS ON MATRICES

A matrix of numbers is a rectangular array of numbers, called entries, in which the locations of the entries are distinguished. Figure 5.1(a) illustrates a 2 by 1 matrix (2 rows and 1 column); Figure 5.1(b) a 1 by 3 matrix (1 row and 3 columns), and Figure 5.1(c) a 3 by 3 matrix. More generally, a matrix with m rows and n columns is called an **m by n matrix.** Two matrices with the same number of rows and same number of columns are said to be the **same size.** No two matrices in Figure 5.1 are the same size.

$$A = \begin{bmatrix} 1 \\ -1 \end{bmatrix} \qquad B = \begin{bmatrix} 4 & -\frac{1}{2} & 8 \end{bmatrix} \qquad C = \begin{bmatrix} 2 & -1 & 1 \\ 1 & 4 & 0 \\ 3 & 0 & -1 \end{bmatrix}$$

(a) (b) (c)

Figure 5.1

Matrices C and D shown in Figure 5.2 are the same size; both are 2 by 2 matrices. Two matrices are said to be **equal** if they are the same size and corresponding entries are equal. Matrices C and D in Figure 5.2 are not equal since the entry in row 2, column 2 of C is not equal to the entry in row 2, column 2 of D. If a matrix has the same number of rows as columns, it is said to be a **square matrix.** Matrices C and D in Figure 5.2 are square 2 by 2 matrices, and matrix C in Figure 5.1(c) is a square 3 by 3 matrix. To simplify language, we shall term the entry in the ith row and jth column of a matrix the i–j entry. Thus the 1–2 entry of a matrix is the term in row 1 and column 2; the 2–1 entry is the term in row 2 and column 1. The 1–2 entry of matrix C in Figure 5.2 is -1; the 2–1 entry of matrix C is 3.

$$C = \begin{bmatrix} 2 & -1 \\ 3 & -2 \end{bmatrix} \qquad D = \begin{bmatrix} 2 & -1 \\ 3 & 4 \end{bmatrix}$$

Figure 5.2

Matrices acquire mathematical life by virtue of the algebraic operations that are defined on them. It is to these algebraic operations that we now turn our attention.

Matrix Addition and Subtraction

If A and B are matrices of the same size, then

$$C = A + B$$

is the matrix obtained by adding corresponding entries of A and B.

$$D = A - B$$

is the matrix obtained by subtracting the entries of B from the corresponding entries of A. Thus if

$$A = \begin{bmatrix} 4 & 3 & 1 \\ 2 & -3 & -2 \end{bmatrix} \text{ and } B = \begin{bmatrix} 2 & -1 & 4 \\ -3 & 2 & 1 \end{bmatrix}$$

then

$$C = A + B = \begin{bmatrix} 4+2 & 3-1 & 1+4 \\ 2-3 & -3+2 & -2+1 \end{bmatrix} = \begin{bmatrix} 6 & 2 & 5 \\ -1 & -1 & -1 \end{bmatrix}$$

and

$$D = A - B = \begin{bmatrix} 4-2 & 3-(-1) & 1-4 \\ 2-(-3) & -3-2 & -2-1 \end{bmatrix} = \begin{bmatrix} 2 & 4 & -3 \\ 5 & -5 & -3 \end{bmatrix}$$

Inner Product of a Row and Column

In preparation for the definition of matrix multiplication, we introduce the following definition of inner product of a row and column. Consider the following two matrices A and B:

$$A = \begin{bmatrix} 2 & -1 \\ 3 & 2 \\ 0 & 4 \end{bmatrix}, \quad B = \begin{bmatrix} 2 & 0 & -1 \\ 4 & 3 & -2 \end{bmatrix}$$

The inner product of row 1 of matrix A, $[2\ -1]$, and column 1 of matrix B, $\begin{bmatrix} 2 \\ 4 \end{bmatrix}$, is the number $2(2) + (-1)4 = 0$ obtained by multiplying corresponding entries (first entry in row 1 of A and first entry in column 1 of B, second entry in row 1 of A and second entry in column 1 of B) and adding. The inner product of row 1 of B, $[2\ 0\ -1]$, and column 1 of A, $\begin{bmatrix} 2 \\ 3 \\ 0 \end{bmatrix}$, is $2(2) + 0(3) + (-1)0 = 4$. The inner product of row 1 of A and column 1 of A is not defined, since this row and column do not have the same number of entries. More generally, if A and B are matrices and the number of entries in a row of A equals the number of entries in a column of B, the **inner product of row i of matrix A and column j of matrix B** is the number obtained by multiplying corresponding entries in row i and column j and adding.

EXERCISES

1. For $A = \begin{bmatrix} 2 & -1 & 3 \\ 1 & 2 & 0 \end{bmatrix}$ and $B = \begin{bmatrix} 4 & 2 & -1 \\ 5 & 2 & 6 \end{bmatrix}$, find $A + B$, $A - B$, and $B - A$.

2. For $A = \begin{bmatrix} 2 & 1 \\ 5 & -2 \\ -4 & 3 \end{bmatrix}$ and $B = \begin{bmatrix} -1 & 2 \\ 3 & 1 \\ 2 & -5 \end{bmatrix}$, find $A + B$ and $A - B$.

3. For $A = \begin{bmatrix} 2 & -1 \\ 6 & -2 \end{bmatrix}$ and $B = \begin{bmatrix} 8 & -7 \\ 3 & -4 \end{bmatrix}$, find $A + B$ and $A - B$.

4. For $A = \begin{bmatrix} 2 & 3 \\ 4 & 5 \end{bmatrix}$ and $I_2 = \begin{bmatrix} 1 & 0 \\ 0 & 1 \end{bmatrix}$, find $I_2 - A$.

5. For $A = \begin{bmatrix} 3 & 2 & -1 \\ 4 & 6 & 2 \\ 3 & 0 & 1 \end{bmatrix}$ and $I_3 = \begin{bmatrix} 1 & 0 & 0 \\ 0 & 1 & 0 \\ 0 & 0 & 1 \end{bmatrix}$, find $I_3 - A$.

6. $A = \begin{bmatrix} 1 \\ 2 \\ 4 \end{bmatrix}$ and $B = \begin{bmatrix} -1 \\ 3 \\ -2 \end{bmatrix}$, find $A + B$ and $B - A$.

7. For $A = [3\ 2\ 4]$ and $B = [2\ -1\ -5]$, find $A + B$ and $A - B$.

8. For the matrices $A = \begin{bmatrix} 2 & -1 \\ 3 & 2 \\ 0 & 4 \end{bmatrix}$ and $B = \begin{bmatrix} 2 & 0 & -1 \\ 4 & 3 & -2 \end{bmatrix}$, find the inner product of (a) row 1 of A and column 2 of B, (b) row 1 of A and column 3 of B, (c) row 2 of A and column 1 of B, (d) row 2 of A and column 2 of B, (e) row 2 of A and column 3 of B, (f) row 3 of A and column 1 of B, (g) row 3 of A and column 2 of B, and (h) row 3 of A and column 3 of B.

9. For the matrices A and B defined in the preceding exercise, find the inner product of (a) row 1 of B and column 1 of A, (b) row 1 of B and column 2 of A, (c) row 2 of B and column 1 of A, (d) row 2 of B and column 2 of A.

Matrix Multiplication

Consider the 3 by 2 matrix A and 2 by 3 matrix B defined as follows:

$$A = \begin{bmatrix} 2 & -1 \\ 3 & 2 \\ 0 & 4 \end{bmatrix}, \quad B = \begin{bmatrix} 2 & 0 & -1 \\ 4 & 3 & -2 \end{bmatrix}$$

The matrix AB has as many rows as A and as many columns as B, and is thus a 3 by 3 matrix whose entries are determined in the following way:

$AB = \begin{bmatrix} 0 & — & — \\ — & — & — \\ — & — & — \end{bmatrix}$ The 1–1 entry of AB is the inner product of row 1 of A and column 1 of B; $2(2) + (-1)4 = 0$.

$AB = \begin{bmatrix} 0 & -3 & — \\ — & — & — \\ — & — & — \end{bmatrix}$ The 1–2 entry of AB is the inner product of row 1 of A and column 2 of B; $2(0) + (-1)3 = -3$.

$AB = \begin{bmatrix} 0 & -3 & 0 \\ — & — & — \\ — & — & — \end{bmatrix}$ The 1–3 entry of AB is the inner product of row 1 of A and column 3 of B; $2(-1) + (-1)(-2) = 0$.

$AB = \begin{bmatrix} 0 & -3 & 0 \\ 14 & — & — \\ — & — & — \end{bmatrix}$ The 2–1 entry of AB is the inner product of row 2 of A and column 1 of B; $3(2) + 2(4) = 14$.

$AB = \begin{bmatrix} 0 & -3 & 0 \\ 14 & 6 & — \\ — & — & — \end{bmatrix}$ The 2–2 entry of AB is the inner product of row 2 of A and column 2 of B; $3(0) + 2(3) = 6$.

$$AB = \begin{bmatrix} 0 & -3 & 0 \\ 14 & 6 & -7 \\ — & — & — \end{bmatrix}$$

The 2–3 entry of AB is the inner product of row 2 of A and column 3 of B; $3(-1) + 2(-2) = -7$.

$$AB = \begin{bmatrix} 0 & -3 & 0 \\ 14 & 6 & -7 \\ 16 & — & — \end{bmatrix}$$

The 3–1 entry of AB is the inner product of row 3 of A and column 1 of B; $0(2) + 4(4) = 16$.

$$AB = \begin{bmatrix} 0 & -3 & 0 \\ 14 & 6 & -7 \\ 16 & 12 & — \end{bmatrix}$$

The 3–2 entry of AB is the inner product of row 3 of A and column 2 of B; $0(0) + 4(3) = 12$.

$$AB = \begin{bmatrix} 0 & -3 & 0 \\ 14 & 6 & -7 \\ 16 & 12 & -8 \end{bmatrix}$$

The 3–3 entry of AB is the inner product of row 3 of A and column 3 of B; $0(-1) + 4(-2) = -8$.

More generally, if A is an n by m matrix and B is an m by k matrix (that is, the number of columns of A equals the number of rows of B), then the **matrix product AB** is defined and is an n by k matrix (same number of rows as A and same number of columns as B) whose i–j entry is the inner product of row i of A and column j of B.

EXAMPLE 1

For the following matrices A and B find, provided that it is defined, the matrix product BA.

$$B = \begin{bmatrix} 2 & 0 & -1 \\ 4 & -3 & -2 \end{bmatrix}, \quad A = \begin{bmatrix} 2 & -1 \\ 3 & 2 \\ 0 & 4 \end{bmatrix}$$

Since B is a 2 by 3 matrix and A is a 3 by 2 matrix, the matrix product BA is defined and is a 2 by 2 matrix. We have

$$BA = \begin{bmatrix} 2(2) + 0(3) + (-1)0 & 2(-1) + 0(2) + (-1)4 \\ 4(2) + 3(3) + (-2)0 & 4(-1) + 3(2) + (-2)4 \end{bmatrix} = \begin{bmatrix} 4 & -6 \\ 17 & -6 \end{bmatrix}$$

Let us note that the matrix products AB and BA are not the same:

$$AB = \begin{bmatrix} 0 & -3 & 0 \\ 14 & 6 & -7 \\ 16 & 12 & -8 \end{bmatrix}, \quad BA = \begin{bmatrix} 4 & -6 \\ 17 & -6 \end{bmatrix}$$

In multiplying matrices one must pay careful attention to the order in which they are multiplied.

EXAMPLE 2

Find, if they exist, the matrix products AB and BA, where

$$A = \begin{bmatrix} -2 \\ 4 \\ 3 \end{bmatrix}, \quad B = [-1 \ 3 \ 5]$$

Since A is a 3 by 1 matrix and B is a 1 by 3 matrix, the product AB is defined and is a 3 by 3 matrix with the following entries:

$$AB = \begin{bmatrix} -2(-1) & -2(3) & -2(5) \\ 4(-1) & 4(3) & 4(5) \\ 3(-1) & 3(3) & 3(5) \end{bmatrix} = \begin{bmatrix} 2 & -6 & -10 \\ -4 & 12 & 20 \\ -3 & 9 & 15 \end{bmatrix}$$

Since B is a 1 by 3 matrix and A is a 3 by 1 matrix, the product BA is a 1 by 1 matrix:

$$BA = [-1(-2) + 3(4) + 5(3)] = [29]$$

EXAMPLE 3

Find the matrix product PC, where P and C are defined as follows:

$$P = \begin{matrix} & \overbrace{\text{Acc.} \quad \text{Main.} \quad \text{Mar.} \quad \text{Pur.}}^{\text{Service Departments}} & \\ & \begin{bmatrix} 0.24 & 0.20 & 0.30 & 0.20 \\ 0.26 & 0.20 & 0.30 & 0.20 \\ 0.28 & 0.20 & 0.40 & 0.30 \end{bmatrix} & \begin{matrix} \text{Production dept. } P_1 \\ \text{Production dept. } P_2 \\ \text{Production dept. } P_3 \end{matrix} \end{matrix}$$

$$C = \begin{bmatrix} 23{,}423 \\ 19{,}902 \\ 87{,}756 \\ 14{,}333 \end{bmatrix} \begin{matrix} \text{Total cost: accounting} \\ \text{Total cost: maintenance} \\ \text{Total cost: marketing} \\ \text{Total cost: purchasing} \end{matrix}$$

Referring back to the service-charge-allocation problem considered in Example 1 of Section 1.5 (pp. 35–38), we see that the entries in the first row of matrix P express the fraction of the total costs of the service departments—accounting, maintenance, marketing, and purchasing, respectively—that are assigned to production department P_1 of the Arkin Company. Similarly, the entries in the second row relate to production department P_2, and the entries in the third row relate to P_3. Matrix C states the total costs of the service departments. The matrix product PC is given by

$$PC = \begin{bmatrix} 0.24(23,423) + 0.20(19,902) + 0.30(87,756) + 0.20(14,333) \\ 0.26(23,423) + 0.20(19,902) + 0.30(87,756) + 0.20(14,333) \\ 0.28(23,423) + 0.20(19,902) + 0.40(87,756) + 0.30(14,333) \end{bmatrix}$$

$$= \begin{bmatrix} 38,795 \\ 39,264 \\ 49,941 \end{bmatrix} \begin{matrix} \text{Service departments' costs allocated to } P_1 \\ \text{Service departments' costs allocated to } P_2 \\ \text{Service departments' costs allocated to } P_3 \end{matrix}$$

Matrix PC describes the allocation of service departments' costs to the production departments of the firm.

EXERCISES

10. For the matrices $G = \begin{bmatrix} 1 & 2 \\ 3 & -1 \end{bmatrix}$ and $H = \begin{bmatrix} -2 & 4 \\ 1 & 3 \end{bmatrix}$, find the matrix products GH, HG, G^2, H^2, G^2H^2.

11. For $A = \begin{bmatrix} 1 & 3 & 2 \\ -1 & 0 & 1 \\ -2 & 1 & 3 \end{bmatrix}$ and $B = \begin{bmatrix} 2 & 1 & -1 \\ 1 & 3 & 0 \\ 0 & 1 & -2 \end{bmatrix}$, find AB, BA, A^2, and B^2.

12. For $A = \begin{bmatrix} 1 \\ 2 \\ -1 \end{bmatrix}$ and $B = [2 \ 4 \ -7]$, find, if possible, AB and BA.

13. For $A = \begin{bmatrix} 1 & 1 \\ 1 & 1 \end{bmatrix}$ and $B = \begin{bmatrix} -1 & -1 \\ 1 & 1 \end{bmatrix}$, find AB.

14. For $A = \begin{bmatrix} 1 & 2 \\ 2 & 4 \end{bmatrix}$ and $B = \begin{bmatrix} 2 & 4 \\ -1 & -2 \end{bmatrix}$, find AB.

15. For $A = \begin{bmatrix} 2 & 4 & -1 \\ -4 & -8 & 2 \\ -2 & -4 & 1 \end{bmatrix}$ and $B = \begin{bmatrix} 4 & 1 & -1 \\ 1 & 1 & 1 \\ 12 & 6 & 2 \end{bmatrix}$, find AB.

16. For $A = \begin{bmatrix} a & b \\ c & d \end{bmatrix}$ and $I_2 = \begin{bmatrix} 1 & 0 \\ 0 & 1 \end{bmatrix}$, find AI_2 and I_2A.

17. For $A = \begin{bmatrix} a & b & c \\ d & e & f \\ g & h & i \end{bmatrix}$ and $I_3 = \begin{bmatrix} 1 & 0 & 0 \\ 0 & 1 & 0 \\ 0 & 0 & 1 \end{bmatrix}$, find AI_3 and I_3A.

18. For $A = \begin{bmatrix} 2 & -1 \\ 1 & -2 \end{bmatrix}$, $X = \begin{bmatrix} 0 & 1 \\ 3 & 2 \end{bmatrix}$, and $Y = \begin{bmatrix} -1 & -1 \\ 1 & -2 \end{bmatrix}$, find AX and AY and compare.

19. For $A = \begin{bmatrix} 1 & 3 \\ -1 & 2 \end{bmatrix}$, $B = \begin{bmatrix} -1 & 3 \\ 2 & -1 \end{bmatrix}$, and $C = \begin{bmatrix} 2 & -1 \\ 3 & -2 \end{bmatrix}$,

   (a) Find $A(BC)$ and $(AB)C$ and compare.
   (b) Find $A(B + C)$ and $AB + AC$ and compare.
   (c) Find $(B + C)A$ and $BA + CA$ and compare.

20. For $A = \begin{bmatrix} 1 & 2 & -1 \\ 2 & -1 & 3 \\ 1 & 0 & 2 \end{bmatrix}$, $B = \begin{bmatrix} 2 & 0 & 1 \\ 1 & 3 & 2 \\ 0 & 1 & -1 \end{bmatrix}$, and $C = \begin{bmatrix} -2 & 2 & -1 \\ 3 & -1 & 0 \\ 2 & 0 & 1 \end{bmatrix}$,

   (a) Find BC, AB, and AC.
   (b) Find $A(BC)$ and $(AB)C$ and compare.
   (c) Find $A(B + C)$ and $AB + AC$ and compare.
   (d) Find $(B + C)A$ and $BA + CA$ and compare.

21. For $A = \begin{bmatrix} a & b \\ c & d \end{bmatrix}$, $B = \begin{bmatrix} e & f \\ g & h \end{bmatrix}$, and $C = \begin{bmatrix} j & k \\ m & n \end{bmatrix}$,

   (a) Find BC, AB, and AC.
   (b) Find $A(BC)$ and $(AB)C$ and show that they are equal.
   (c) Find $A(B + C)$ and $AB + AC$ and show that they are equal.
   (d) Find $(B + C)A$ and $BA + CA$ and show that they are equal.

22. This concerns the Sonin Company's situation described in Section 1.5, Exercise 1, p. 40. Let

$$P = \begin{bmatrix} S_1 & S_2 & S_3 \\ 0.22 & 0.40 & 0.30 \\ 0.40 & 0.35 & 0.42 \end{bmatrix} \begin{matrix} P_1 \\ P_2 \end{matrix} \text{ and } C = \begin{bmatrix} 22{,}000 \\ 30{,}000 \\ 44{,}000 \end{bmatrix}$$

Matrix P describes the fractions of the total costs of the service departments S_1, S_2, and S_3 of a firm that are assigned to its production departments P_1 and P_2. Matrix C describes the total costs of the service departments. Find the matrix that describes the allocation of service departments' costs to the production departments. (See Example 3.)

23. The Karkin Steel Works operates out of two plants, P_1 and P_2, and produces two types of steel, a chromium–manganese–molybdenum alloy (CMM steel) and a chromium–silicon–tungsten alloy (CST steel). Of concern are three main types of pollutants—particulate matter, sulfur oxides, and hydrocarbons—which are by-products of the operation of the plants. Air-quality standards require that specified minimal amounts of pollutants be removed. These minimal amounts (in tons) for the two types of steel are shown in matrix B. The cost (in dollars) of removing 1 ton of each kind of pollutant at each plant is shown in matrix E. Determine matrix BE and explain what it describes.

$$B = \begin{bmatrix} \text{Particulates} & \text{Sulfur oxides} & \text{Hydrocarbons} \\ 20 & 30 & 25 \\ 30 & 25 & 40 \end{bmatrix} \begin{matrix} \text{CMM steel} \\ \text{CST steel} \end{matrix}$$

$$E = \begin{bmatrix} \text{Plant } P_1 & \text{Plant } P_2 \\ 200 & 220 \\ 160 & 150 \\ 180 & 200 \end{bmatrix} \begin{matrix} \text{Particulates} \\ \text{Sulfur oxides} \\ \text{Hydrocarbons} \end{matrix}$$

5.2. PROPERTIES OF MATRICES

It is interesting to compare the properties of matrix addition and multiplication with those of real-number addition and multiplication. There are many similarities, but also some striking differences. First, we recall some properties of number addition. All letters in the following list represent arbitrary real numbers.

1. *Closure property of number addition:* If a and b are any numbers in the real number system R, then the sum of a and b, denoted by $a + b$, is in R.
2. *Commutative property of number addition:* $a + b = b + a$.
3. *Associative property of number addition:* $(a + b) + c = a + (b + c)$.
4. *Zero element:* There is an unique real number, called *zero* and denoted by 0, with the property $a + 0 = 0 + a$, where a is any real number.
5. *Negative of a number:* If a is any real number, there is an unique real number, called the negative of a and denoted by $-a$, with the property $a + (-a) = (-a) + a = 0$.

From the way in which matrix addition is defined, it is easy to see that it also has these properties. All letters in the following list denote arbitrary matrices of the same size.

Closure property of matrix addition. If A and B are any matrices of the same size, then there is an unique matrix of that size, denoted by $A + B$, called the sum of A and B.
   As we have seen, $A + B$ is obtained by adding corresponding entries of A and B.

Commutative property of matrix addition. $A + B = B + A$.
   This follows from the definition of matrix addition and the commutativity of number addition.

Associative property of matrix addition. $(A + B) + C = A + (B + C)$.
   This follows from the definition of matrix addition and the associativity of number addition.

Zero matrix. There is a unique matrix, called the **zero matrix** and denoted by $\bar{0}$, with the property $A + \bar{0} = \bar{0} + A = A$.

$\bar{0}$ is the matrix that is the same size as A and whose entries are all zeros. Thus for the class of 2 by 2 matrices

$$\bar{0} = \begin{bmatrix} 0 & 0 \\ 0 & 0 \end{bmatrix}$$

For the class of 3 by 3 matrices

$$\bar{0} = \begin{bmatrix} 0 & 0 & 0 \\ 0 & 0 & 0 \\ 0 & 0 & 0 \end{bmatrix}$$

Negative of a matrix. If A is any matrix, then there is an unique matrix, called the **negative of A** and denoted by $-A$, with the property $A + (-A) = (-A) + A = \bar{0}$. For example, if

$$A = \begin{bmatrix} a & b \\ c & d \end{bmatrix}$$

then the negative of A is

$$-A = \begin{bmatrix} -a & -b \\ -c & -d \end{bmatrix}$$

Thus the negative of $A - [4 \ -2]$ is $-A = [-4 \ 2]$.

Also, we have the following properties of real-number multiplication.

1. *Closure property of number multiplication:* If a and b are any numbers in the real number system R, then the product of a and b, denoted by $a \cdot b$ or ab, is in R.
2. *Commutative property of number multiplication:* $ab = ba$.
3. *Associative property of number multiplication:* $(ab)c = a(bc)$.
4. *Identity element:* There is an unique number, called one and denoted by 1, with the property $1a = a1 = a$, where a is any real number.
5. *Inverse property:* If a is any nonzero real number, then there is an unique real number, called the multiplicative inverse of a or reciprocal of a, and denoted by a^{-1}, such that $aa^{-1} = a^{-1}a = 1$.
6. *Nonzero products of nonzero factors property:* The product of any two nonzero real numbers is nonzero; that is, if $a \neq 0$ and $b \neq 0$, then $ab \neq 0$.
7. *Cancellation property:* If $ax = ay$, where $a \neq 0$, then $x = y$.

In considering the behavior of matrix multiplication, we restrict ourselves to square matrices of the same size to ensure that all matrix products are at least defined. In the following discussion all letters denote arbitrary square matrices of the same size.

Closure property of matrix multiplication. If A and B are square matrices of the same size, then there is an unique square matrix of that size, denoted by AB, called the product of A and B. This follows from the way in which matrix multiplication is defined.

Commutativity of matrix multiplication? **No.** Matrix multiplication is not commutative since $AB = BA$ does *not* hold for all square matrices of the same size. For example, if

$$A = \begin{bmatrix} 1 & 3 \\ 2 & -1 \end{bmatrix} \text{ and } B = \begin{bmatrix} 1 & 4 \\ -2 & -3 \end{bmatrix}$$

then $AB \neq BA$ since

$$AB = \begin{bmatrix} -5 & -5 \\ -2 & -3 \end{bmatrix} \text{ whereas } BA = \begin{bmatrix} 9 & -1 \\ -8 & -3 \end{bmatrix}$$

Associative property of matrix multiplication. $(AB)C = A(BC)$. Associativity for the class of 2 by 2 matrices is established in Exercise 21(b) of Section 5.1.

Identity matrices. The analog for the class of 2 by 2 matrices, for example, of the identity element 1 of the real-number system is a matrix, call it I_2, with the property that $AI_2 = I_2A = A$, where A is any 2 by 2 matrix. For the class of 2 by 2 matrices, the identity matrix I_2 is defined by

$$I_2 = \begin{bmatrix} 1 & 0 \\ 0 & 1 \end{bmatrix}$$

For any 2 by 2 matrix $A = \begin{bmatrix} a & b \\ c & d \end{bmatrix}$, $AI_2 = I_2A = A$ (see Exercise 16 of Section 5.1).

For the class of 3 by 3 matrices, the identity matrix is

$$I_3 = \begin{bmatrix} 1 & 0 & 0 \\ 0 & 1 & 0 \\ 0 & 0 & 1 \end{bmatrix}$$

since for any 3 by 3 matrix A, $AI_3 = I_3A = A$ (see Exercise 17 of Section 5.1). More generally, for the class of n by n matrices, the identity matrix is the n by n matrix

$$I_n = \begin{bmatrix} 1 & 0 & 0 & \cdots & 0 \\ 0 & 1 & 0 & \cdots & 0 \\ 0 & 0 & 1 & \cdots & 1 \\ \cdot & \cdot & \cdot & & \cdot \\ \cdot & \cdot & \cdot & & \cdot \\ \cdot & \cdot & \cdot & & \cdot \\ 0 & 0 & 0 & \cdots & 1 \end{bmatrix}$$

whose entries are 0's except for 1's that run down the main diagonal.

Inverse of a matrix. If $A \neq \bar{0}$ is an n by n matrix, and there is an n by n matrix B with the property $AB = BA = I_n$, then B is denoted by A^{-1}, is called the **inverse of A,** and A is said to be **nonsingular.**

Thus, for example, the inverse of

$$A = \begin{bmatrix} 1 & 3 \\ -1 & 2 \end{bmatrix} \quad \text{is} \quad B = \begin{bmatrix} \frac{2}{5} & -\frac{3}{5} \\ \frac{1}{5} & \frac{1}{5} \end{bmatrix}$$

since $AB = BA = I_2$, the identity matrix for the class of 2 by 2 matrices.

If the inverse of a matrix A does not exist, then A is said to be a **singular matrix.** Many nonzero square matrices are singular. Consider, for example,

$$A = \begin{bmatrix} 1 & 0 \\ 0 & 0 \end{bmatrix}$$

There is no 2 by 2 matrix B such that $AB = BA = I_2$. To verify this, let $B = \begin{bmatrix} a & b \\ c & d \end{bmatrix}$ and consider the product AB:

$$AB = \begin{bmatrix} 1 & 0 \\ 0 & 0 \end{bmatrix} \begin{bmatrix} a & b \\ c & d \end{bmatrix} = \begin{bmatrix} a & b \\ 0 & 0 \end{bmatrix}$$

Clearly, $AB = \begin{bmatrix} a & b \\ 0 & 0 \end{bmatrix}$ cannot be made equal to $I_2 = \begin{bmatrix} 1 & 0 \\ 0 & 1 \end{bmatrix}$, no matter how a and b are chosen.

Distributive property. A link between addition and multiplication that holds for numbers as well as matrices is the distributive property. If a, b, and c are any real numbers, then $a(b + c) = ab + ac$. If A, B and C are any square matrices of the same size, then $A(B + C) = AB + AC$ and $(B + C) = BA + CA$.

The special case of the distributive property for the class of 2 by 2 matrices is established in Exercises 21(c) and 21(d) of Section 5.1.

Some Matrix Surprises

Products of nonzero matrices. If A and B are nonzero matrices, it does not follow that their product is a nonzero matrix. For example, for

$$A = \begin{bmatrix} 1 & 2 \\ 2 & 4 \end{bmatrix} \quad \text{and} \quad B = \begin{bmatrix} 2 & 4 \\ -1 & -2 \end{bmatrix}, \quad AB = \begin{bmatrix} 0 & 0 \\ 0 & 0 \end{bmatrix}$$

Matrix Cancellation? No. The cancellation property does not hold for matrices. If $AX = AY$, where $A \neq \bar{0}$, it does not follow that $X = Y$. For example, if A, X, and Y are defined by

$$A = \begin{bmatrix} 2 & -1 \\ 4 & -2 \end{bmatrix}, \quad X = \begin{bmatrix} 0 & 1 \\ 3 & 2 \end{bmatrix}, \quad Y = \begin{bmatrix} -1 & -1 \\ 1 & -2 \end{bmatrix}$$

then $AX = AY$, but $X \neq Y$.

EXERCISES

1. What is the zero matrix for the class of 4 by 2 matrices? Explain.

2. What is the negative of $\begin{bmatrix} 3 & 1 \\ -2 & 2 \\ 1 & -4 \\ -3 & 6 \end{bmatrix}$? Explain.

3. What is the identity matrix for the class of 4 by 4 matrices? Explain.

4. Is $\begin{bmatrix} -\frac{1}{4} & \frac{3}{8} \\ \frac{1}{2} & -\frac{1}{4} \end{bmatrix}$ the inverse of $\begin{bmatrix} 2 & 3 \\ 4 & 2 \end{bmatrix}$? Explain.

5. $\begin{bmatrix} 2 & \frac{1}{2} \\ \frac{1}{3} & -1 \end{bmatrix}$ the inverse of $\begin{bmatrix} 4 & 1 \\ 2 & 3 \end{bmatrix}$? Explain.

6. Is $\begin{bmatrix} \frac{1}{2} & \frac{1}{2} & -\frac{1}{2} \\ -\frac{1}{2} & \frac{1}{6} & \frac{1}{6} \\ \frac{1}{2} & -\frac{1}{2} & \frac{1}{2} \end{bmatrix}$ the inverse of $\begin{bmatrix} 1 & 0 & 1 \\ 2 & 3 & 1 \\ 1 & 3 & 2 \end{bmatrix}$? Explain.

5.3. MATRIX INVERSION

A matrix A is **nonsingular** if it has an inverse A^{-1}. As we have noted, $A = \begin{bmatrix} 1 & 3 \\ -1 & 2 \end{bmatrix}$ is nonsingular with inverse $A^{-1} = \begin{bmatrix} \frac{2}{5} & -\frac{3}{5} \\ \frac{1}{5} & \frac{1}{5} \end{bmatrix}$. But how is A^{-1} determined? This is the problem we now take up. To begin, we again turn to the 2 by 2 matrix

$$A = \begin{bmatrix} 1 & 3 \\ -1 & 2 \end{bmatrix}$$

174 Finite Mathematics, Models, and Structure

By placing the identity matrix $I_2 = \begin{bmatrix} 1 & 0 \\ 0 & 1 \end{bmatrix}$ next to A, we obtain the tableau of numbers

$$[A \mid I_2] = \begin{bmatrix} 1 & 3 & 1 & 0 \\ -1 & 2 & 0 & 1 \end{bmatrix}$$

To obtain A^{-1} we employ the following row operations, familiar from the tableau method for solving systems of linear equations (Section 1.4), on the tableau $[A \mid I_2]$ to convert the A part of $[A \mid I_2]$ to I_2; in the course of doing so, the I_2 part of $[A \mid I_2]$ will be converted to A^{-1}.

> **Pivoting.** To convert a chosen number c to 1, multiply the row containing c by $1/c$.
>
> **Conversion to Zero.** To convert a number n to 0, multiply the one row (that is, the row containing 1 from the pivoting operation) by n and add the resulting row to the row containing n.
>
> **Interchanging Rows.** Any two rows in a tableau may be interchanged.

The sequence of inversion tableaus that leads to the inverse of

$$A = \begin{bmatrix} 1 & 3 \\ -1 & 2 \end{bmatrix}$$

is shown in Figure 5.3.

$$[A \mid I_2] \quad \begin{matrix} \text{①} \\ \text{②} \end{matrix} \begin{bmatrix} ① & 3 & \vdots & 1 & 0 \\ -1 & 2 & \vdots & 0 & 1 \end{bmatrix}$$

$$T_2 \quad \begin{matrix} \text{③} \\ \text{④} \end{matrix} \begin{bmatrix} 1 & 3 & \vdots & 1 & 0 \\ 0 & ⑤ & \vdots & 1 & 1 \end{bmatrix} \quad \begin{matrix} \text{row ①} \\ \text{row ① + row ②} \end{matrix}$$

$$T_3 \quad \begin{matrix} \text{⑤} \\ \text{⑥} \end{matrix} \begin{bmatrix} 1 & 0 & \vdots & \frac{2}{5} & -\frac{3}{5} \\ 0 & 1 & \vdots & \frac{1}{5} & \frac{1}{5} \end{bmatrix} \quad \begin{matrix} (-3)\text{row ⑥ + row ③} \\ (\frac{1}{5})\text{row ④} \end{matrix}$$

Figure 5.3

From tableau T_3 we see that the conversion of A to I_2 has been completed, and that I_2 in turn has been converted to

$$A^{-1} = \begin{bmatrix} \frac{2}{5} & -\frac{3}{5} \\ \frac{1}{5} & \frac{1}{5} \end{bmatrix}$$

EXAMPLE 1

Find, provided that it exists, the inverse of

$$A = \begin{bmatrix} 2 & 4 \\ 1 & -1 \end{bmatrix}$$

By placing I_2 next to A we obtain the initial inversion tableau $[A\,|\,I_2]$.

$$[A\,|\,I_2] = \begin{bmatrix} 2 & 4 & | & 1 & 0 \\ 1 & -1 & | & 0 & 1 \end{bmatrix}$$

The sequence of inversion tableaus in which the A part of $[A\,|\,I_2]$ is converted to I_2 and the I_2 part of $[A\,|\,I_2]$ is converted to A^{-1} is displayed in Figure 5.4.

$$[A\,|\,I_2] \quad \begin{array}{c} ① \\ ② \end{array} \begin{bmatrix} 2 & 4 & | & 1 & 0 \\ 1 & -1 & | & 0 & 1 \end{bmatrix}$$

$$T_2 \quad \begin{array}{c} ③ \\ ④ \end{array} \begin{bmatrix} ① & -1 & | & 0 & 1 \\ 2 & 4 & | & 1 & 0 \end{bmatrix} \begin{array}{l} \text{row } ② \\ \text{row } ① \end{array}$$

$$T_3 \quad \begin{array}{c} ⑤ \\ ⑥ \end{array} \begin{bmatrix} 1 & -1 & | & 0 & 1 \\ 0 & ⑥ & | & 1 & -2 \end{bmatrix} \begin{array}{l} \text{row } ③ \\ (-2)\text{row } ⑤ + \text{row } ④ \end{array}$$

$$T_4 \quad \begin{array}{c} ⑦ \\ ⑧ \end{array} \begin{bmatrix} 1 & 0 & | & \frac{1}{6} & \frac{4}{6} \\ 0 & 1 & | & \frac{1}{6} & -\frac{2}{6} \end{bmatrix} \begin{array}{l} \text{row } ⑧ + \text{row } ⑤ \\ (\frac{1}{6})\text{row } ⑥ \end{array}$$

Figure 5.4

The simplest way to begin the conversion of the A part of $[A\,|\,I_2]$ to I_2 is to interchange rows ① and ② in tableau $[A\,|\,I_2]$; this interchange yields tableau T_2 in Figure 5.4. From tableau T_4 we see that the conversion of the A part of $[A\,|\,I_2]$ to I_2 has been completed, and that I_2 has been converted to

$$A^{-1} = \begin{bmatrix} \frac{1}{6} & \frac{4}{6} \\ \frac{1}{6} & -\frac{2}{6} \end{bmatrix}$$

As a check against error, we verify that $AA^{-1} = A^{-1}A = I_2$.

$$\begin{bmatrix} 2 & 4 \\ 1 & -1 \end{bmatrix} \begin{bmatrix} \frac{1}{6} & \frac{4}{6} \\ \frac{1}{6} & -\frac{2}{6} \end{bmatrix} = \begin{bmatrix} 1 & 0 \\ 0 & 1 \end{bmatrix}$$

$$\begin{bmatrix} \frac{1}{6} & \frac{4}{6} \\ \frac{1}{6} & -\frac{2}{6} \end{bmatrix} \begin{bmatrix} 2 & 4 \\ 1 & -1 \end{bmatrix} = \begin{bmatrix} 1 & 0 \\ 0 & 1 \end{bmatrix}$$

EXAMPLE 2

Find, provided that it exists, the inverse of

$$E = \begin{bmatrix} 1 & 0 & 1 \\ 2 & 3 & 1 \\ 1 & 3 & 2 \end{bmatrix}$$

By placing I_3, the identity matrix for the class of 3 by 3 matrices, next to E, we obtain the initial inversion tableau $[E|I_3]$ shown in Figure 5.5.

$$[E|I_3] \quad \begin{array}{c} ① \\ ② \\ ③ \end{array} \begin{bmatrix} ① & 0 & 1 & | & 1 & 0 & 0 \\ 2 & 3 & 1 & | & 0 & 1 & 0 \\ 1 & 3 & 2 & | & 0 & 0 & 1 \end{bmatrix}$$

$$T_2 \quad \begin{array}{c} ④ \\ ⑤ \\ ⑥ \end{array} \begin{bmatrix} 1 & 0 & 1 & | & 1 & 0 & 0 \\ 0 & 3 & -1 & | & -2 & 1 & 0 \\ 0 & 3 & ① & | & -1 & 0 & 1 \end{bmatrix} \begin{array}{l} \text{row } ① \\ (-2)\text{row } ④ + \text{row } ② \\ (-1)\text{row } ④ + \text{row } ③ \end{array}$$

$$T_3 \quad \begin{array}{c} ⑦ \\ ⑧ \\ ⑨ \end{array} \begin{bmatrix} 1 & -3 & 0 & | & 2 & 0 & -1 \\ 0 & ⑥ & 0 & | & -3 & 1 & 1 \\ 0 & 3 & 1 & | & -1 & 0 & 1 \end{bmatrix} \begin{array}{l} (-1)\text{row } ⑨ + \text{row } ④ \\ \text{row } ⑨ + \text{row } ⑤ \\ \text{row } ⑥ \end{array}$$

$$T_4 \quad \begin{array}{c} ⑩ \\ ⑪ \\ ⑫ \end{array} \begin{bmatrix} 1 & 0 & 0 & | & \tfrac{1}{2} & \tfrac{1}{2} & -\tfrac{1}{2} \\ 0 & 1 & 0 & | & -\tfrac{1}{2} & \tfrac{1}{6} & \tfrac{1}{6} \\ 0 & 0 & 1 & | & \tfrac{1}{2} & -\tfrac{1}{2} & \tfrac{1}{2} \end{bmatrix} \begin{array}{l} (3)\text{row } ⑪ + \text{row } ⑦ \\ (\tfrac{1}{6})\text{row } ⑧ \\ (-3)\text{row } ⑪ + \text{row } ⑨ \end{array}$$

Figure 5.5

The not unexpected modification in our procedure is that the E part of $[E|I_3]$ is to be converted to I_3; in the course of doing so, the I_3 part is converted to E^{-1}. In going from tableau T_2 to T_3, we pivot on 1 in row ⑥ and column (3), rather than on 3 in row ⑤ and column (2). This is for the sake of convenience; to obtain I_3, we need a 1 in that position, and it makes sense to take advantage of good fortune when it comes our way. Although 1's must be placed in the main diagonal, they need not be placed there in order. From tableau T_4 in Figure 5.5, we see that the conversion of the E part of $[E|I_3]$ has been completed, and that I_3 has been converted to E^{-1} given by

$$E^{-1} = \begin{bmatrix} \tfrac{1}{2} & \tfrac{1}{2} & -\tfrac{1}{2} \\ -\tfrac{1}{2} & \tfrac{1}{6} & \tfrac{1}{6} \\ \tfrac{1}{2} & -\tfrac{1}{2} & \tfrac{1}{2} \end{bmatrix}$$

As a check against error, this result can be verified by showing that $EE^{-1} = E^{-1}E = I_3$.

EXAMPLE 3

Find, provided that it exists, the inverse of

$$A = \begin{bmatrix} 1 & 2 \\ 2 & 4 \end{bmatrix}$$

The sequence of inversion tableaus is displayed in Figure 5.6. The entry in row ④, column (2) is 0, and thus cannot be converted to 1 by pivoting. Thus the A part of $[A|I_2]$ cannot be converted to I_2, which means that matrix A is singular and does not have an inverse.

Such is the case in general; if a matrix A cannot be reduced to the identity matrix by means of the row operations described, then A does not have an inverse.

$$[A|I_2] \quad \begin{matrix} ① \\ ② \end{matrix} \begin{bmatrix} ① & 2 & | & 1 & 0 \\ 2 & 4 & | & 0 & 1 \end{bmatrix}$$

$$T_2 \quad \begin{matrix} ③ \\ ④ \end{matrix} \begin{bmatrix} 1 & 2 & | & 1 & 0 \\ 0 & 0 & | & -2 & 1 \end{bmatrix} \begin{matrix} \text{row ①} \\ (-2)\text{row ③ + row ②} \end{matrix}$$

Figure 5.6

EXERCISES

Find, provided that it exists, the inverse of each of the following matrices.

1. $\begin{bmatrix} 1 & 3 \\ 4 & -1 \end{bmatrix}$

2. $\begin{bmatrix} 1 & -2 \\ 2 & 3 \end{bmatrix}$

3. $\begin{bmatrix} 2 & 3 \\ 4 & 1 \end{bmatrix}$

4. $\begin{bmatrix} 2 & 4 \\ 1 & 2 \end{bmatrix}$

5. $\begin{bmatrix} 1 & 0 & 0 \\ 2 & 2 & 0 \\ 5 & 5 & 5 \end{bmatrix}$

6. $\begin{bmatrix} 2 & -1 & 0 \\ 1 & 0 & 1 \\ 1 & -2 & 0 \end{bmatrix}$

7. $\begin{bmatrix} 2 & 0 & 1 \\ 1 & 1 & 3 \\ 1 & -1 & -2 \end{bmatrix}$

8. $\begin{bmatrix} 1 & 1 & 2 \\ 2 & 0 & 2 \\ 3 & 0 & 1 \end{bmatrix}$

9. $\begin{bmatrix} 1 & 2 & 3 \\ 3 & 2 & 1 \\ 5 & 4 & 5 \end{bmatrix}$

10. $\begin{bmatrix} 2 & -1 & 3 \\ 1 & -2 & 1 \\ -1 & 1 & 2 \end{bmatrix}$

11. $\begin{bmatrix} 2 & 3 & -1 \\ 1 & -2 & 3 \\ 3 & -1 & -2 \end{bmatrix}$

12. $\begin{bmatrix} 1 & -3 & 2 \\ 3 & -4 & 1 \\ 2 & -1 & -1 \end{bmatrix}$

5.4. MATRIX SOLUTIONS TO SYSTEMS OF LINEAR EQUATIONS

There are many methods for solving systems of linear equations, with each one possessing its own advantages and disadvantages. One of these methods is based on matrix inversion. To illustrate, consider the problem of determining x_1, x_2, and x_3 so as to satisfy the following system:

$$x_1 + x_3 = 4 \tag{5.1}$$
$$2x_1 + 3x_2 + x_3 = 6 \tag{5.2}$$
$$x_1 + 3x_2 + 2x_3 = 12 \tag{5.3}$$

If we define matrices E, X, and B by

$$E = \begin{bmatrix} 1 & 0 & 1 \\ 2 & 3 & 1 \\ 1 & 3 & 2 \end{bmatrix}, \quad X = \begin{bmatrix} x_1 \\ x_2 \\ x_3 \end{bmatrix}, \quad B = \begin{bmatrix} 4 \\ 6 \\ 12 \end{bmatrix}$$

then

$$EX = \begin{bmatrix} x_1 + x_3 \\ 2x_1 + 3x_2 + x_3 \\ x_1 + 3x_2 + 2x_3 \end{bmatrix}$$

and the problem, in matrix terms, is to find X such that

$$EX = B$$

If E^{-1} exists, then by multiplying both sides of $EX = B$ on the left by E^{-1}, we obtain

$$E^{-1}EX = E^{-1}B$$
$$I_3 X = E^{-1}B$$
$$X = E^{-1}B$$

Thus, if E, the matrix of coefficients of the system, is nonsingular, then the system has exactly one solution and that solution is given by $E^{-1}B$. From Example 2 of Section 5.3, we have

$$E^{-1} = \begin{bmatrix} \frac{1}{2} & \frac{1}{2} & -\frac{1}{2} \\ -\frac{1}{2} & \frac{1}{6} & \frac{1}{6} \\ \frac{1}{2} & -\frac{1}{2} & \frac{1}{2} \end{bmatrix}$$

Thus

$$X = E^{-1}B = \begin{bmatrix} \frac{1}{2} & \frac{1}{2} & -\frac{1}{2} \\ -\frac{1}{2} & \frac{1}{6} & \frac{1}{6} \\ \frac{1}{2} & -\frac{1}{2} & \frac{1}{2} \end{bmatrix} \begin{bmatrix} 4 \\ 6 \\ 12 \end{bmatrix} = \begin{bmatrix} -1 \\ 1 \\ 5 \end{bmatrix}$$

and we obtain $x_1 = -1$, $x_2 = 1$, $x_3 = 5$. It is easily verified, as a check against error, that $(-1, 1, 5)$ is the solution of system (5.1) through (5.3). More generally, consider the n by n system (n equations and n unknowns) of linear equations.

$$c_{11}x_1 + c_{12}x_2 + \cdots + c_{1n}x_n = b_1$$
$$c_{21}x_1 + c_{22}x_2 + \cdots + c_{2n}x_n = b_2$$
$$\vdots$$
$$c_{n1}x_1 + c_{n2}x_2 + \cdots + c_{nn}x_n = b_n$$

and define matrices

$$E = \begin{bmatrix} c_{11} & c_{12} & \cdots & c_{1n} \\ c_{21} & c_{22} & \cdots & c_{2n} \\ \vdots & \vdots & & \vdots \\ c_{n1} & c_{n2} & \cdots & c_{nn} \end{bmatrix}, \quad X = \begin{bmatrix} x_1 \\ x_2 \\ \vdots \\ x_n \end{bmatrix}, \quad B = \begin{bmatrix} b_1 \\ b_2 \\ \vdots \\ b_n \end{bmatrix}$$

E is the coefficient matrix of the system, X is the matrix of unknowns, and B is the matrix of constants. In terms of matrices, the problem is to find X such that

$$EX = B$$

If E^{-1} exists, then the system has a unique solution, which is given by

$$X = E^{-1}B$$

If E^{-1} does not exist, then the system either has no solution or infinitely many solutions. Let us observe that the existence of a unique solution depends only on the existence of E^{-1} and thus on the nature of the coefficient matrix E, and not on the nature of the matrix of constants B.

The matrix-inversion approach to solving n by n systems with a unique solution is advantageous in situations in which the coefficient matrix E does not change, but the matrix of constants B is to be varied. As long as E does not change, only one matrix inversion is required, and through the solution equation $X = E^{-1}B$ solutions corresponding to a number of possible B matrices can easily be obtained. We examine such situations in the next two sections.

EXERCISES

Solve each of the following systems of linear equations by means of matrix inversion.

1. $-x_1 + 2x_2 = 4$
   $3x_1 + x_2 = 10$

2. $x_1 - 2x_2 = 6$
   $2x_1 + 3x_2 = 9$

3. $3x + 5y = 8$
   $2x + 3y = 6$

4. $\begin{aligned} x_1 + x_2 + 2x_3 &= 3 \\ 2x_1 \qquad\quad + 2x_3 &= 4 \\ 3x_1 \qquad\quad + x_3 &= 6 \end{aligned}$

5. $\begin{aligned} 2x_1 + 3x_2 - x_3 &= 11 \\ x_1 - 2x_2 + 3x_3 &= 2 \\ 3x_1 - x_2 - 2x_3 &= 5 \end{aligned}$

6. $\begin{aligned} x_1 - x_2 - x_3 &= -4 \\ 3x_1 - 5x_2 + 5x_3 &= 6 \\ 2x_1 + 3x_2 + 4x_3 &= 7 \end{aligned}$

5.5. RETURN TO SERVICE CHARGE ALLOCATION AND INCOME CONSOLIDATION PROBLEMS

Service Charge Allocation

In Example 1 of Section 1.5 (p. 35), we saw that the problem of determining the total costs of the service departments of the Arkin Company leads to the system of equations

$$\begin{aligned} x - 0.1y \qquad\quad - 0.1w &= 20{,}000 \\ -0.02x + y \qquad\quad - 0.1w &= 18{,}000 \\ -0.1x - 0.2y + z - 0.1w &= 80{,}000 \\ -0.1x - 0.1y \qquad + w &= 10{,}000 \end{aligned}$$

where x, y, z, and w denote the total costs of the accounting, maintenance, marketing, and purchasing departments, respectively, and \$20,000, \$18,000, \$80,000 and \$10,000 are the respective overhead costs of these departments for the month of January.

In matrix terms we have

$$EX = B$$

where

$$E = \begin{bmatrix} 1 & -0.1 & 0 & -0.1 \\ -0.02 & 1 & 0 & -0.1 \\ -0.1 & -0.2 & 1 & -0.1 \\ -0.1 & -0.1 & 0 & 1 \end{bmatrix}, \quad X = \begin{bmatrix} x \\ y \\ z \\ w \end{bmatrix}, \quad B = \begin{bmatrix} 20{,}000 \\ 18{,}000 \\ 80{,}000 \\ 10{,}000 \end{bmatrix}$$

For E^{-1} we obtain

$$E^{-1} = \begin{bmatrix} 1.0135134 & 0.1126125 & 0 & 0.1126125 \\ 0.0307124 & 1.0135134 & 0 & 0.1044225 \\ 0.1179360 & 0.2252251 & 1 & 0.1343160 \\ 0.1044225 & 0.1126125 & 0 & 1.0217034 \end{bmatrix}$$

Thus

$$X = E^{-1}B = \begin{bmatrix} 23{,}423.43 \\ 19{,}901.72 \\ 87{,}755.93 \\ 14{,}332.51 \end{bmatrix}$$

describes the total costs of the service departments.

If for each service department there is no change in the percentage of its total cost assigned to the other service departments, so that matrix E, which reflects these percentages, is unchanged, then with one matrix inversion the total costs of the service departments can be obtained for each financial period as the matrix B, expressing overhead costs, changes.

For further discussion of the cost-allocation problem and the use of matrix methods, consult the following literature:

[1] N. CHURCHILL. "Linear Algebra and Cost Allocations: Some Examples." *Accounting Review*, vol. 39, no. 4 (October 1964), pp. 894–904.
[2] R. S. KAPLAN. "Variable and Self-Service Costs in Reciprocal Allocation Models." *Accounting Review*, vol. 48, no. 4 (October 1973), pp. 738–748.
[3] R. P. MANES. "Comment on Matrix Theory and Cost Allocation." *Accounting Review*, vol. 40, no. 3 (July 1965), pp. 640–643.
[4] R. MINCH and E. PETRI. "Matrix Models of Reciprocal Service Cost Allocation." *Accounting Review*, vol. 47, no. 3 (July 1972), pp. 576–580.
[5] T. H. WILLIAMS and C. H. GRIFFIN. "Matrix Theory and Cost Allocation." *Accounting Review*, vol. 39, no. 3 (July 1964), pp. 671–678.

Income Consolidation

In Example 2 of Section 1.5 (p. 38) we saw that the problem of determining the net incomes of the Russel, Ferrara, and Thomas Companies on a consolidated basis leads to the system of equations

$$x - 0.8y - 0.6z = 100{,}000$$
$$y - 0.2z = 80{,}000$$
$$-0.2x - 0.1y + z = 60{,}000$$

where x, y, and z denote the net incomes of the Russel, Ferrara, and Thomas Companies, respectively, on a consolidated basis, and \$100,000, \$80,000 and \$60,000 are the respective net incomes of these companies from their own operations.

In matrix terms this system is expressed by

$$EX = B$$

where

$$E = \begin{bmatrix} 1 & -0.8 & -0.6 \\ 0 & 1 & -0.2 \\ -0.2 & -0.1 & 1 \end{bmatrix}, \quad X = \begin{bmatrix} x \\ y \\ z \end{bmatrix}, \quad B = \begin{bmatrix} 100{,}000 \\ 80{,}000 \\ 60{,}000 \end{bmatrix}$$

Calculating E^{-1} yields

Finite Mathematics, Models, and Structure

$$E^{-1} = \begin{bmatrix} 1.183575 & 1.038647 & 0.917874 \\ 0.048309 & 1.062802 & 0.241546 \\ 0.241546 & 0.314009 & 1.207729 \end{bmatrix}$$

We thus obtain

$$X = E^{-1}B = \begin{bmatrix} 1.183575 & 1.038647 & 0.917874 \\ 0.048309 & 1.062802 & 0.241546 \\ 0.241546 & 0.314009 & 1.207729 \end{bmatrix} \begin{bmatrix} 100{,}000 \\ 80{,}000 \\ 60{,}000 \end{bmatrix} = \begin{bmatrix} 256{,}521.7 \\ 104{,}347.8 \\ 121{,}739.1 \end{bmatrix}$$

As is not surprising, this result agrees with the one obtained in Example 2 of Section 1.5. But the advantage of this solution procedure over the other is that if there are no changes in the intercorporate shareholdings, so that matrix E is not changed, then with one matrix inversion the consolidated basis net incomes can easily be obtained for each financial period (month, quarter, etc.) as the matrix B, expressing the net incomes of these affiliates from their own operations, changes.

Thus, for example, if in the next financial period the net incomes of these affiliates from their own operations were \$110,000, \$90,000 and \$50,000, respectively, then the consolidated basis net incomes would simply be obtained as follows:

$$X = E^{-1}B = \begin{bmatrix} 1.183575 & 1.038647 & 0.917874 \\ 0.048309 & 1.062802 & 0.241546 \\ 0.241546 & 0.314009 & 1.207729 \end{bmatrix} \begin{bmatrix} 110{,}000 \\ 90{,}000 \\ 50{,}000 \end{bmatrix} = \begin{bmatrix} 269{,}565.2 \\ 113{,}043.5 \\ 115{,}217.3 \end{bmatrix}$$

EXERCISES

1. In connection with the service-charge-allocation problem, determine the costs of the accounting, maintenance, marketing and purchasing departments (a) if the overhead costs of these departments are \$22,000, \$20,000, \$76,000, and \$12,000, respectively; (b) if the overhead costs of these departments are \$24,000, \$22,000, \$90,000, and \$14,000, respectively.

2. In Exercise 1 of Section 1.5 (p. 40), the fraction of the total cost of each service department of the Sonin Company that is assigned to the service and production departments of the firm is specified. The overhead of the service departments for the month of March is also given.

    (a) State, in matrix terms, the system of linear equations that determines the total costs of the service departments.

    (b) Solve this system by matrix inversion to determine these total costs.

    (c) If the overhead of S_1, S_2, and S_3 for April is \$42,000, \$36,000, and \$24,000, respectively, and the assignment of costs to the service and production departments remains the same, determine the total costs of the service departments.

(d) State, in terms of a suitable matrix product, the allocation of the total costs of the service departments to the production departments.

3. In Exercise 2 of Section 1.5 (p. 40), the interdependency structure between the Ramunė, Algis, and Charles companies is shown.
   (a) State in matrix terms the system of linear equations that determines the net incomes of these affiliate companies on a consolidated basis.
   (b) Solve this system by matrix inversion.
   (c) If the net incomes of these companies are $90,000, $70,000, and $70,000, respectively, find the net incomes of these affiliate companies on a consolidated basis.

5.6. LEONTIEF INPUT–OUTPUT MODELS

Input–output models for economic systems were pioneered by the economist Wassily Leontief, a recipient of the 1973 Nobel Prize in Economics. In input–output analysis an economic system is viewed as a collection of interacting industries in which each industry produces an output that serves as raw materials, or input, for the industries of the system and requires input from the industries of the system. Let a_{ij} denote the amount of input (dollar's worth) of commodity i needed to produce $1 worth of commodity j; the first subscript refers to input, the second to output. Thus, for example, the equation $a_{21} = 0.20$ asserts that 20¢ worth of commodity 2 is needed to produce $1 worth of commodity 1. For an n-industry economy, the matrix

$$A = \begin{bmatrix} a_{11} & a_{12} & \cdots & a_{1n} \\ a_{21} & a_{22} & \cdots & a_{2n} \\ \vdots & \vdots & & \vdots \\ a_{n1} & a_{n2} & \cdots & a_{nn} \end{bmatrix}$$

called the **input-coefficient matrix** of the system, specifies the amount of each commodity that is needed to produce $1 worth of each commodity. The entries in the first column, for example, specify the inputs required from each of the n industries to produce $1 worth of the commodity produced by industry 1. The entries in the first row specify the amount of the commodity provided by industry 1 needed to produce $1 worth of the commodities produced by the n industries of the system.

We also assume that there is an **open sector** in the economy (consisting of households, for example) that absorbs a noninput demand for the product of each industry and supplies the primary input, labor. Let $d_1, d_2, \ldots, d_n$ denote the demand of the open sector for the commodities produced by industries $1, 2, \ldots, n$ and let $x_1, x_2, \ldots, x_n$ denote the total output (dollar's worth) of industries $1, 2, \ldots, n$. The product $a_{ij}x_j$ is (the amount of commodity i needed to produce $1 worth of commodity

j) × (total dollar's worth of commodity j produced), and thus expresses the input requirement of industry j for commodity i. For example, if $a_{ij} = 0.20$ (20¢ worth of commodity i is needed to produce \$1 worth of commodity j) and $x_j = 5000$ (\$5000 worth of commodity j is produced), then $0.20(5000) = \$1000$ worth of commodity i is needed to produce commodity j. Thus $a_{11}x_1$ is the input requirement of industry 1 for commodity 1, $a_{12}x_2$ is the input requirement of industry 2 for commodity 1, $a_{13}x_3$ is the input requirement of industry 3 for commodity 1, and so on. The sum

$$a_{11}x_1 + a_{12}x_2 + \cdots + a_{1n}x_n + d_1$$

is the sum of the input requirements of the n industries and the open sector for commodity 1. For x_1, the total output of industry 1, to satisfy this demand, we must have

$$x_1 = a_{11}x_1 + a_{12}x_2 + \cdots + a_{1n}x_n + d_1$$

Similarly, for x_2, the total output of industry 2, to satisfy the demand for commodity 2, we must have

$$x_2 = a_{21}x_1 + a_{22}x_2 + \cdots + a_{2n}x_n + d_2$$

More generally, for x_n, the total output of industry n, to satisfy the demand for commodity n, we must have

$$x_n = a_{n1}x_1 + a_{n2}x_2 + \cdots + a_{nn}x_n + d_n$$

Thus the conditions that must be satisfied by output levels $x_1, x_2, \ldots, x_n$ of the n industries in the economy to satisfy the demands of the open sector and the industries themselves are expressed by the following system of n equations:

$$x_1 = a_{11}x_1 + a_{12}x_2 + \cdots + a_{1n}x_n + d_1$$
$$x_2 = a_{21}x_1 + a_{22}x_2 + \cdots + a_{2n}x_n + d_2$$
$$\vdots$$
$$x_n = a_{n1}x_1 + a_{n2}x_2 + \cdots + a_{nn}x_n + d_n$$

Rewriting this system so that terms involving $x_1, x_2, \ldots, x_n$ appear on one side and the constants $d_1, d_2, \ldots, d_n$ appear on the other side yields

$$(1 - a_{11})x_1 - a_{12}x_2 - \cdots - a_{1n}x_n = d_1$$
$$-a_{21}x_1 + (1 - a_{22})x_2 - \cdots - a_{2n}x_n = d_2$$
$$\vdots$$
$$-a_{n1}x_1 - a_{n2}x_2 - \cdots + (1 - a_{nn})x_n = d_n$$

(5.4)

It is advantageous, as we shall see, to express this system in terms of a matrix equation involving a matrix product. To do so we introduce matrices I_n, X, and D, as follows:

$$I_n = \begin{bmatrix} 1 & 0 & 0 & \cdots & 0 \\ 0 & 1 & 0 & \cdots & 0 \\ 0 & 0 & 1 & \cdots & 0 \\ \cdot & \cdot & \cdot & & \cdot \\ \cdot & \cdot & \cdot & & \cdot \\ \cdot & \cdot & \cdot & & \cdot \\ 0 & 0 & 0 & \cdots & 1 \end{bmatrix}, \quad X = \begin{bmatrix} x_1 \\ x_2 \\ \cdot \\ \cdot \\ \cdot \\ x_n \end{bmatrix}, \quad D = \begin{bmatrix} d_1 \\ d_2 \\ \cdot \\ \cdot \\ \cdot \\ d_n \end{bmatrix}$$

Matrix I_n is an n by n matrix with 1's in the main diagonal and 0's elsewhere; X is called the **output matrix** of the system and D is called the **final-demand matrix** of the system. Subtracting A, the input-coefficient matrix, from I_n yields

$$I_n - A = \begin{bmatrix} (1 - a_{11}) & -a_{12} & \cdots & -a_{1n} \\ -a_{21} & (1 - a_{22}) & \cdots & -a_{2n} \\ \cdot & \cdot & & \cdot \\ \cdot & \cdot & & \cdot \\ \cdot & \cdot & & \cdot \\ -a_{n1} & -a_{n2} & \cdots & (1 - a_{nn}) \end{bmatrix}$$

Taking the product of $(I_n - A)$ and X, $(I_n - A)X$, yields the left side of system (5.4); the right side of system (5.4) is expressed by matrix D. Thus, in matrix terms, system (5.4) is expressed by the following matrix equation:

$$(I_n - A)X = D$$

In summary, then, the problem of satisfying the needs of the n industries of the economy, expressed by the input-coefficient matrix A, and the needs of the open sector of the system, expressed by the final-demand matrix D, reduces to the matrix problem of determining output matrix X such that the product $(I_n - A)X$ equals D. As long as the input-coefficient matrix A does not change, $(I_n - A)^{-1}$ does not change, and with one matrix inversion a variety of possible final-demand situations can be studied.

To illustrate, consider a two-industry economy governed by the input-coefficient matrix

$$A = \begin{bmatrix} 0.4 & 0.3 \\ 0.3 & 0.2 \end{bmatrix}$$

and having final-demand matrix $D = \begin{bmatrix} d_1 \\ d_2 \end{bmatrix}$. The problem is to find an output matrix $X = \begin{bmatrix} x_1 \\ x_2 \end{bmatrix}$ such that $(I_2 - A)X = D$. If $(I_2 - A)^{-1}$ exists, then $X = (I_2 - A)^{-1} D$.

$$I_2 - A = \begin{bmatrix} 0.6 & -0.3 \\ -0.3 & 0.8 \end{bmatrix}, \quad (I_2 - A)^{-1} = \begin{bmatrix} \frac{80}{39} & \frac{10}{13} \\ \frac{10}{13} & \frac{20}{13} \end{bmatrix}$$

Thus

$$X = (I_2 - A)^{-1} D = \begin{bmatrix} \frac{80}{39} & \frac{10}{13} \\ \frac{10}{13} & \frac{20}{13} \end{bmatrix} \begin{bmatrix} d_1 \\ d_2 \end{bmatrix} = \begin{bmatrix} \frac{80}{39} d_1 + \frac{10}{13} d_2 \\ \frac{10}{13} d_1 + \frac{20}{13} d_2 \end{bmatrix}$$

and we have

$$x_1 = \tfrac{80}{39} d_1 + \tfrac{10}{13} d_2$$
$$x_2 = \tfrac{10}{13} d_1 + \tfrac{20}{13} d_2$$

If $d_1 = \$39{,}000$ and $d_2 = \$61{,}000$ ($\$39{,}000$ worth of commodity 1 and $\$61{,}100$ worth of commodity 2 are required by the open sector), then $x_1 = \$127{,}000$ and $x_2 = \$124{,}000$. $\$127{,}000$ worth of commodity 1 and $\$124{,}000$ worth of commodity 2 must be produced to satisfy the input needs of industries 1 and 2 and the requirements of the open sector.

If the projected requirements of the open sector should change to $d_1 = \$46{,}800$ and $d_2 = \$65{,}000$, then $x_1 = \$146{,}000$ and $x_2 = \$136{,}000$. $\$146{,}000$ worth of commodity 1 and $\$136{,}000$ worth of commodity 2 must be produced to satisfy the input needs of industries 1 and 2 and the requirements of the open sector.

The successful application of the input–output model to an economy requires that a realistic input-coefficient matrix A and final demand matrix D be developed for the economy. Assuming that this could be done, the second major problem is to suitably refine these matrices so that they are realistic over time. Leontief's pioneering studies of the American economy from an input–output analysis point of view are discussed in [5] and [6] in the references noted.

For further discussion of input–output analysis, see the following works:

References

[1] W. J. BAUMOL. *Economic Theory and Operations Analysis*, 2d ed. Englewood Cliffs, N.J.: Prentice-Hall, Inc., 1965, Chapter 20.
[2] H. B. CHENERY and P. G. CLARK. *Interindustry Economics*. New York: John Wiley & Sons, Inc., 1959.

[3] Conference on Research in Income and Wealth, National Bureau of Economic Research. *Input–Output Analysis: An Appraisal.* Princeton, N.J.: Princeton University Press, 1955.

[4] R. DORFMAN, P. A. SAMUELSON, and R. M. SOLOW. *Linear Programming and Economic Analysis.* New York: McGraw-Hill Book Company, 1958, Chapters 9–12.

[5] W. W. LEONTIEF. *The Structure of American Economy, 1919–1939,* 2d ed. New York: Oxford University Press, 1951.

[6] W. W. LEONTIEF, ed. *Studies in the Structure of the American Economy.* New York: Oxford University Press, 1953.

EXERCISES

1. For a two-industry economy, let us suppose that $0.20 and $0.30 worth of the first industry's commodity is needed by the first and second industries to produce $1 worth of their respective commodities and that $0.30 and $0.10 worth of the second industry's commodity is needed by the first and second industries to produce $1 worth of their respective commodities.

   (a) Set up the input-coefficient matrix of the economy.

   (b) If the open sector of the economy requires $2520 worth of commodity 1 and $3150 worth of commodity 2, what output levels will satisfy the input needs of the industries and the requirements of the open sector? How much of these outputs will be consumed by industries 1 and 2?

(c) If the open sector of the economy requires $3465 worth of commodity 1 and $3780 worth of commodity 2, what output levels will satisfy the input needs of the industries and the requirements of the open sector? How much of these outputs will be consumed by industries 1 and 2?

SELF-TESTS FOR CHAPTERS 1–5

Allow 90 or so minutes for each self-test. Go over each one before proceeding to the next.

Self-Test 1

1. Write an equation for the line passing through (2, 5) with slope $\frac{1}{2}$.
2. Write an equation for the line passing through $(-1, 2)$ and $(3, 1)$.
3. Solve the system
$$x + 2y = 40$$
$$4x + 3y = 100$$
4. Solve the system
$$5s + 7t = 18$$
$$3s + 2t = 2$$
5. The following table shows employee age (x) in years and income (y) in thousands of dollars for a random sample of six employees of a very large corporation.

x	30	35	42	46	50	58
y	12	15	14	22	25	22

(a) Find the estimated linear regression equation specifying the average value of y in terms of x.

(b) Find the predicted income corresponding to age 45 and interpret the result obtained.

(c) Plot the original data as well as the regression line on one diagram.

6. Solve the system
$$2x + y - 4z = 8$$
$$3x + 2y - 7z = 12$$
$$x - y + 3z = -4$$

7. Solve the system
$$x + 6y + 4z = -1$$
$$x - y - 3z = -3$$
$$3x + 4y - 2z = -7$$

8. Is $f(x, y, z) = 3x + 2y + 2z$ a linear function? Explain.

9. Sketch the graph of the following system:
$$x \geq 0, y \geq 0$$
$$4x + 5y \leq 70$$
$$x + y \leq 16$$

10. Answer TRUE or FALSE. Explain the basis for your answer.
    (a) Linear programming could be used to maximize $F(x, y) = 3x + y^2$.
    (b) A feasible point of a linear program satisfies all of the constraints.
    (c) A constraint is a mathematical statement guaranteed to precisely reflect reality.
    (d) A company should implement a linear program's solution if they are certain that it is valid.
    (e) In real life it will always be the case that the solution of a linear programming model will be the optimum one for the company.

Self-Test 2

1. Set up the following problem as a linear program and solve.
   The Last National Bank has assets in the form of loans and securities that, it is assumed, bring returns of 10 and 8 percent, respectively, in a certain time period. The bank has a total of $60 million to allocate between loans and securities. Two major guidelines are imposed by the bank on its lending activity: (1) a securities balance equal to or greater than 25 percent of total assets must be maintained, and (2) at least $15 million must be available for loans.
   The bank is interested in determining, under these conditions, how funds should be allocated between loans and securities so that investment income is maximized.

2. Solve the system
$$x + 3y - z = -2$$
$$-x - y + 2z = 7$$
$$3x - 2y + z = -5$$

3. Solve: Maximize $F(x, y) = 3x + 2y$
   subject to
$$x \geq 0, y \geq 0$$
$$2x + 3y \leq 5$$
$$3x - y \leq 2$$

4. Set up the following problem as a linear program and solve.
   Audio Acoustics, Inc., plans to spend up to $12,000 in buying and stocking two speaker system models that are in great demand. The K17 model costs $50 per

unit and occupies 3 cubic feet of storage space; the K24 model costs $60 per unit and occupies 4 cubic feet of storage space. The store has 760 cubic feet of storage space available. The management expects to make a profit of $2 on each K17 unit and $3 on each K24 unit. The problem is to determine, with respect to the given conditions, how many units of each model should be bought and stocked so that profit is maximized.

5. The Atlantic Company, a lamp manufacturer, began production of two models, A14 and A51. To determine the best production schedule, two consulting firms, the Marks Company and Andrius Consultants, were hired to analyze the company's operations and make recommendations. The Marks Company set up a linear program for the production process, which when solved by a method called the simplex method yielded the solution (500, 280) with a maximum value of 3000. The Marks Company recommended that 500 A14 units and 280 A51 units be made per week, for an anticipated maximum profit of $3000 per week.

Andrius Consultants set up a different linear program for the production process, which when solved by the corner-point method yielded (450, 300) as the solution with a maximum value of 2500. Andrius Consultants recommended that 450 A14 units and 300 A51 units be made weekly to obtain a maximum profit of $2500 per week.

The management of the Atlantic Company found these developments puzzling and raised the following questions. Answer these questions in appropriate detail.

(a) How is it possible for different solutions to be obtained? After all, isn't mathematics a precise subject?

(b) Which solution is correct and in what sense is it correct?

(c) Which solution should be implemented and why?

6. Consider the integer program: find nonnegative integers which minimize $C(x, y) = 4x + 5y$ subject to

$$3x + y \geq 6$$
$$x + y \geq 5$$
$$x + 3y \geq 6.$$

(a) Determine the additional constraint(s) that must be imposed to guarantee a solution in integers.

(b) Solve the integer program that emerges by the corner point method.

7. Is the following a correct statement of von Neumann's duality theorem? Explain. If the statement must be corrected, then make the necessary correction.

If a linear program in standard form has a feasible point, then so does its dual, and both have the same value.

Self-Test 3

1. Consider the following linear programs.

   (i) Max. $F = 5x + 6y + 8z$

   subject to

   $x \geq 0, y \geq 0, z \geq 0$

   $3x + 2y + z \leq 13$

   $2x + y + 2z \leq 8$

   $3x + 2z \leq 12$

   (ii) Min. $H = 10x + 4y + 8z$

   subject to

   $x \geq 0, y \geq 0, z \geq 0$

   $-2x + y - 3z \geq 4$

   $2x - 4y + 8z \geq 3$

   $4x + y + 3z \geq 2$

   $3x + y + 4z \geq 3$

   (a) Find the dual of each L.P.

   (b) Use von Neumann's duality theorem to determine whether (i) and (ii) have solutions. Explain the basis for your conclusions.

   (c) In any case where an L.P. has a solution determine it, the solution of its dual, and the common value of their objective functions.

2. For $A = \begin{bmatrix} 4 & -2 \\ 1 & 3 \end{bmatrix}$ and $B = \begin{bmatrix} -1 & 2 \\ 3 & 5 \end{bmatrix}$, find AB, BA, A^2, and B^2.

3. Is $B = \begin{bmatrix} 1 & -2 \\ 2 & -3 \end{bmatrix}$ the multiplicative inverse of $A = \begin{bmatrix} -3 & 2 \\ -2 & 1 \end{bmatrix}$? Explain.

4. Camera World, which makes VHS video cameras among other products, has three service departments—accounting, shipping and marketing, and two production departments. Each service department's total cost must be distributed to the service departments and to the production departments based on their respective usages of the services provided. For each service department listed in the left most column of Table 1 the fraction of its total cost assigned to the service departments of the firm is given. The October overhead of the service departments is also given.

Table 1

Service Dept.	Service Department			October Overhead
	Accounting	Shipping	Marketing	
Accounting	0.01	0.08	0.02	$30,000
Shipping	0.02	0	0.15	$20,000
Marketing	0.02	0.01	0	$95,000

Set up the system of equations that describes the conditions to be satisfied by the total costs of the service departments. Be sure to clearly state what your variables represent.

Write True or False next to each of the following statements. If a statement is false, give an example which shows that it's false.

5. For square matrices A and B of the same size, $AB = BA$.

6. If the matrix product $AB = \bar{0}$, then $A = \bar{0}$ or $B = \bar{0}$.

7. If the matrix product $AX = AY$, where $A \neq \bar{0}$, then $X = Y$.

8. Multiplication of square matrices of the same size is associative.

9. Every nonzero square matrix has an additive inverse.

10. Every nonzero square matrix has a multiplicative inverse.

11. Coal is to be sent from distribution centers in Brooks, Scranton, and Kearn to Smithtown and Warren. 4000 tons are in Brooks, 5000 tons are in Scranton, and 3500 tons are in Kearn. 5000 tons are to be sent to Smithtown and 7500 tons are to be sent to Warren. The cost (in dollars) of shipping 1 ton of coal from each distribution center to each destination is given in Table 2.

The problem is to determine the number of tons that should be shipped from each distribution center to each destination so that the total shipping cost is minimized.

Table 2

	Smithtown	Warren
Brooks	0.8	0.75
Scranton	0.9	1.0
Kearn	0.75	0.8

Self-Test 4

Find, provided that it exists, the multiplicative inverse of each of the following matrices.

1. $B = \begin{bmatrix} 1 & 4 \\ -2 & -8 \end{bmatrix}$

2. $F = \begin{bmatrix} 1 & -2 & 3 \\ 2 & 1 & -3 \\ -1 & 1 & 2 \end{bmatrix}$

Solve each of the following systems of equations by matrix inversion.

3. $4x_1 + 3x_2 = 10$
   $2x_1 + 5x_2 = 12$

4. $x_1 + x_2 + 3x_3 = 2$
   $x_1 + 2x_2 - x_3 = -4$
   $x_1 + x_2 + x_3 = 8$

5. Consider the linear program:

$$\text{Max. } F = 9s + 10t + 16u$$

subject to

$$s \geq 0, t \geq 0, u \geq 0$$
$$3s + 5t + u \leq 30$$
$$4s + t + 2u \leq 28$$

   (a) Find the dual program.
   (b) Use von Neumann's theorem to determine whether or not this linear program has a solution. Explain the basis for your conclusion.
   (c) If your answer to (b) is yes, then determine its solution, the solution of its dual, and the common value of their objective functions.

6. The Russo River Freight Company plans to invest up to $3.5 million in the purchase of at least 20 new barges. Two models, BM23 and BM30, are being considered. A BM23 barge costs $200,000 and is expected to average 18 hours per day at 8 miles per hour with an average 500 tons of freight per hour. A BM30 barge costs $250,000 and is expected to average 20 hours per day at 10 miles per hour with an average of 400 tones of freight per hour.

   The Russo Company would like to determine the number of barges of each model that should be purchased so that capacity in freight-miles-per-day is maximized.

7. For a two-industry economy, let us suppose that 50¢ and 20¢ worth of the first industry's product is needed by the first and second industries to produce $1 worth of their respective commodities, and that 20¢ and 50¢ worth of the second industry's product is needed by the first and second industries to produce $1 worth of their respective commodities.
   (a) Set up the input-coefficient matrix of the economy.
   (b) If the open sector of the economy requires $1575 worth of commodity 1 and $1890 worth of commodity 2, what output levels will satisfy the input needs of the industries and the requirements of the open sector? How much of these outputs will be consumed by industries 1 and 2?
   (c) If the open sector of the economy requires $2940 worth of commodity 1 and $3675 worth of commodity 2, what output levels will satisfy the input needs of the industries and the requirements of the open sector? How much of these outputs will be consumed by industries 1 and 2?

8. Do you agree or disagree with the following point of view. So state and explain in appropriate detail.

   "Mathematical methods have the advantage of certitude. No qualified person can resist the truth of a mathematical conclusion properly communicated. The job of communication may be difficult if the solution is complex, but when the communication is competent, agreement is inevitable. If anyone doubts a solution, he can recalculate the equations and check the steps in the derivation. Then he must either demonstrate that there has been an error or acknowledge the truth of the solution."

CHAPTER 6

Basic Concepts of Probability

6.1. SET NOTATION AND LANGUAGE

For discussion of basic probabilistic ideas, the language and operations that are part of elementary set theory are useful. The following is a brief survey of basic set theory ideas.

A **set** is said to be defined whenever a rule is given that allows us to distinguish those objects belonging to the set from those not belonging to the set. The objects that satisfy the requirements of the rule defining the set are called **members** or **elements** of the set. The symbol $x \in A$ is used to signify that x is a member of set A; if x is not a member of A, we write $x \notin A$. Some examples of sets are the set consisting of Chicago and New York, the set of letters of the English alphabet, the set of natural numbers, and the set consisting of the number 1.

Two sets A and B are said to be **equal** if they have the same members, that is, if every element of A is also an element of B, and vice versa. Thus the property of having the same elements is the important feature in set equality and not the description used to define the set. The set consisting of the number 1, the set of integers between 0 and 2, and the set of solutions of the equation $X - 1 = 0$ are all the same set.

A common way of designating a set is by displaying its elements between braces. For example, the set consisting of Chicago and New York can be denoted by {Chicago, New York} or {New York, Chicago}. When a set is described in this way, no ordering of the elements is recognized. In other words, we are not thinking of the object first listed as occupying first place; no placings are understood. When sets are described in this way, each element is listed only once. The set consisting of the numbers 1 and 2 can be denoted by $\{1, 2\}$ or $\{2, 1\}$; the set consisting of the number 0 can be denoted by $\{0\}$.

We should also note that logically speaking there is a major difference between a set with one element and the element itself. The number 0 is not the same as $\{0\}$, for example. The number 0 belongs to a number system, and it is subject to arithmetic operations; the set $\{0\}$ is not part of a number system and it makes no sense to perform arithmetic operations with it. (This fact has unfortunately not always deterred people from doing so.) To avoid being tied up in linguistic knots, the distinction between a one-element set and the element itself is often passed over in discussion.

As another example, consider the set of all living people in Chicago who are over 200 years old. This is a well-defined set, because the rule given allows us to distinguish members of the set from nonmembers. The curious feature of this set is that it has no members. A set defined by such a rule is called the **empty set,** or **null set,** and is denoted by the symbol $\emptyset$. Here, too, we should note that $\emptyset$ is not the same as the number 0; $\emptyset$ is not part of a number system and it makes no sense to perform arithmetic operations with it. Although they look similar, 0, $\{0\}$, and $\emptyset$ are different things. As an analogy, it might be helpful to picture a set as a box containing certain objects, its elements; the empty set, then, corresponds to an empty box.

Set A is said to be a **subset** of set B if $A = \emptyset$ or if every member of A is also a member of B. Thus, by definition, the empty set $\emptyset$ is a subset of every set. For example, the subsets of the set $\{a, b, c\}$ are

$$\emptyset, \quad \{a\}, \quad \{b\}, \quad \{c\},$$
$$\{a, b\}, \quad \{a, c\}, \quad \{b, c\}, \quad \{a, b, c\}$$

To avoid logical contradictions when working with sets, it is necessary to identify a set M, called the **universal** or **master set,** with the property that all sets which arise in the discussion are subsets of M. The specification of the master set depends on the nature of the discussion. If sets of numbers are being discussed, it might be appropriate to let M be the set of all real numbers; if sets of people are under discussion, it might be appropriate to let M be the set of all people in some community, or the set of all living and past persons.

Set Operations

If A is a subset of the universal set M, then the **complement of A,** denoted by A^c, is the set of objects in M that are not in A. If, for example, $M = \{a, b, c, d\}$ and $A = \{a, b\}$, then $A^c = \{c, d\}$, $M^c = \emptyset$, and $\emptyset^c = M$. If M and A are as shown in Figure 6.1, then A^c is shown as the shaded remainder of M.

If A and B are subsets of the universal set M, then the **union of A and B,** denoted by $A \cup B$, is the set of objects that belong to at least one of the sets A and B

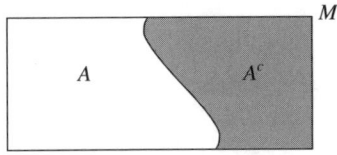

Figure 6.1

(equivalently, the set of objects that belong to either A or B). If $A = \{a, b\}$ and $B = \{b, c\}$, then $A \cup B = \{a, b, c\}$. If A, B, and C are sets, then the **union of A, B, and C,** denoted by $A \cup B \cup C$, is the set of objects that belong to at least one of the sets A, B, and C. The union of any number of sets is defined in an analogous manner. If A and B consist of points interior to the circles shown in Figure 6.2, then $A \cup B$ is represented by the total shaded region shown in Figure 6.2.

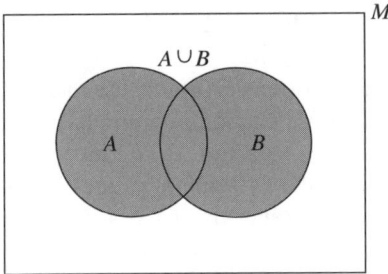

Figure 6.2

If A and B are subsets of the universal set M, then the **intersection of A and B,** denoted by $A \cap B$, is the set of objects that A and B have in common. More generally, the intersection of any number of sets is the set of objects that are common to all. If $A = \{a, b, c, d\}$, $B = \{c, d, f\}$, and $C = \{c, f\}$, then $A \cap B = \{c, d\}$ and $A \cap B \cap C = \{c\}$. If A and B consist of points interior to the circles shown in Figure 6.3, then $A \cap B$ is represented by the overlap shown as the shaded region.

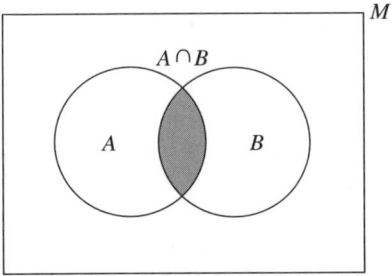

Figure 6.3

EXERCISES

1. If $A = \{1, 3, 5\}$, $B = \{3, 5, 6\}$, $C = \{4, 5\}$, and $M = \{1, 2, 3, 4, 5, 6\}$, find **(a)** $A \cup B$, **(b)** $A \cup C$, **(c)** $B \cup C$, **(d)** $B \cup M$, **(e)** A^c, **(f)** B^c, **(g)** C^c, **(h)** $A \cap B$, **(i)** $A \cap C$, **(j)** $B \cap C$, **(k)** $A^c \cap B$, **(l)** $A \cap B^c$, **(m)** $A \cup B^c$, **(n)** $A^c \cap B^c$, **(o)** $A \cup B \cup C$, **(p)** $A \cap B \cap C$, **(q)** $A^c \cup B^c \cup C$, **(r)** $A^c \cap B^c \cap C^c$.

2. If $M = \{HHH, HHT, HTH, HTT, THH, THT, TTH, TTT\}$, $A = \{HHT, HTH\}$, and $B = \{HTT, THT, TTH\}$, find **(a)** $A \cup B$, **(b)** $A \cap B$, **(c)** A^c, **(d)** B^c, **(e)** $A^c \cup B^c$, **(f)** $A^c \cup B$, **(g)** $A^c \cap B$, **(h)** $A \cup B^c$, **(i)** $A^c \cap B^c$.

3. If M is the set of all presidents of the United States, past and present, and A is the set of Republican presidents, what is A^c?

6.2. PREFACE TO PROBABILITY

By a **random process,** or **experiment,** we mean a process which gives rise to an outcome, but which outcome cannot be predicted with any certainty in advance. The processes of tossing a coin, tossing a die, and choosing a sample from a shipment of items are simple examples of random processes. Although the outcome of any repetition or occurrence of a random process cannot be predicted with any certainty in advance, if we focus on an event of interest over a long series of repetitions of the random process we find stability in the relative frequency with which the event occurs. This stability can serve as a basis for understanding and predicting the behavior of the random process.

To develop this point of view we introduce the following definition.

Let A denote an event connected with a random process. The **relative frequency of A,** denoted by $R(A)$, is the ratio

$$R(A) = \frac{\text{number of times } A \text{ occurs}}{\text{number of times the process is repeated}}.$$

Thus, if a coin is tossed 5 times and head shows on 2 of the 5 tosses, the relative frequency of heads for this series of 5 tosses is $\frac{2}{5}$ or 0.40. We say that head showed 40 percent of the time in the 5 tosses.

Over the short run, that is, for a small number of repetitions of a random process, the behavior of $R(A)$ for a single toss is greatly influenced by a unit increase in its denominator while the numerator remains the same or increases by one. As a result $R(A)$ is unstable and fluctuates considerably. In Figure 6.4 the relative frequency of the event head shows is plotted (vertical axis) against the number of tosses of a certain nickel for the first 15 tosses (horizontal axis). The relative frequency of this event is rather erratic and varies from 0 to 0.533. In Figure 6.5 the relative frequency of head shows is shown for the last 15 of a series of 200 tosses of the nickel. The relative frequency of this event is rather stable at this point, varying from 0.536 to 0.546, since the denominator of the relative frequency ratio is large and the change brought by an additional toss of the nickel is small.

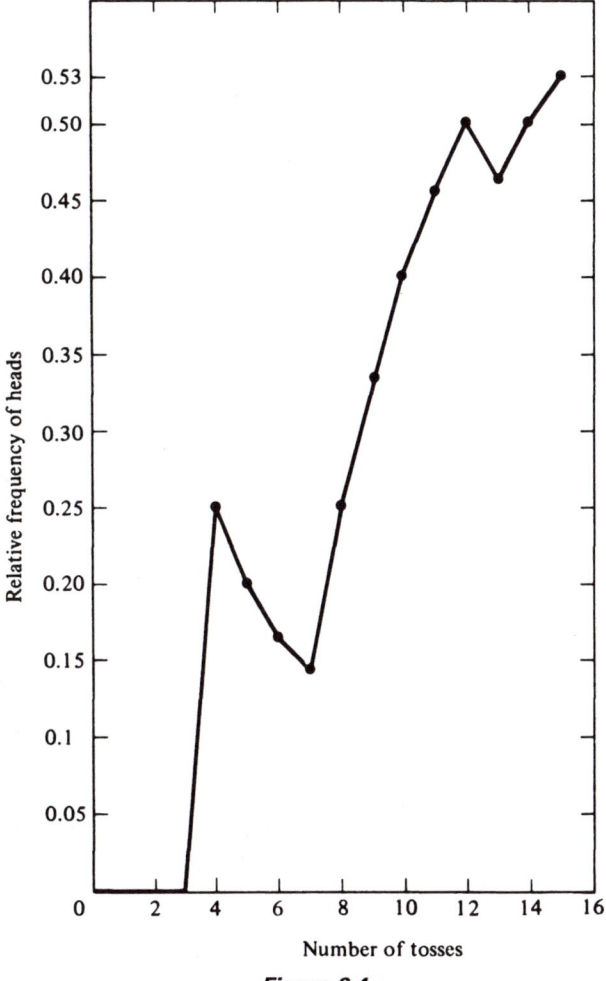

Figure 6.4

The desire to develop a mathematical structure that would help us understand and predict the long term relative frequency behavior of events connected with random processes played a leading role in the development of probability as this mathematical instrument.

Sample Space for a Random Process

Since our study focuses on the behavior of events, it is desirable to express the nature of the many events that arise in the study of a random process in terms of a set

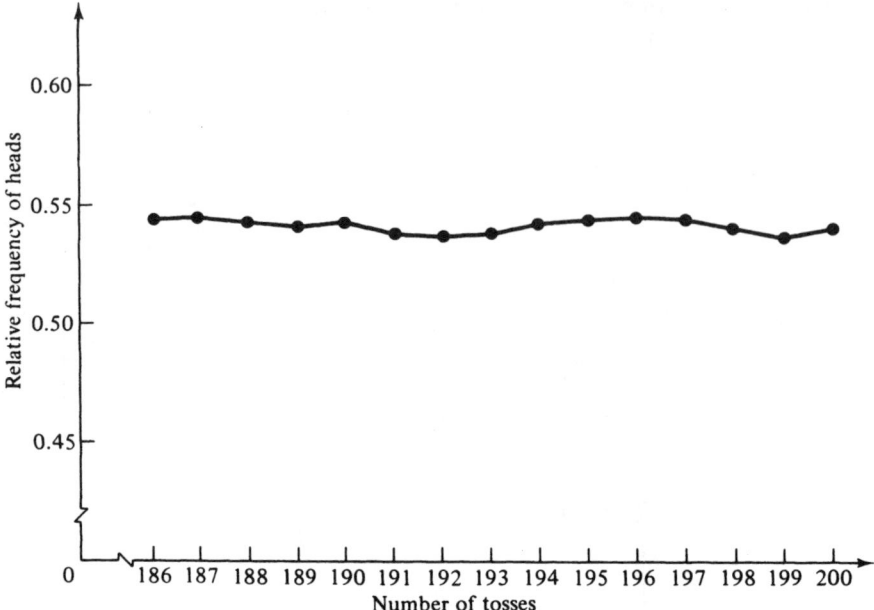

Figure 6.5

of comparatively few events which is adopted as a foundation. This leads us to the concept of sample space for a random process. To develop this concept consider the process of tossing a die. Some of the events connected with this process are 1 shows, 2 shows, an even number shows, a number between 2 and 5 shows, a number less than 4 shows. To describe events of interest concerning the number which shows when a die is tossed, consider the set of events

$$S = \{1 \text{ shows}, 2 \text{ shows}, 3 \text{ shows}, 4 \text{ shows}, 5 \text{ shows}, 6 \text{ shows}\},$$

which we further abbreviate by writing

$$S = \{1, 2, 3, 4, 5, 6\}.$$

This set of events has the property that whenever a die is tossed a unique event in S is determined; S is complete in the sense that some event in S occurs when a die is tossed, and unambiguous in the sense that only one event in S occurs when a die is tossed. Other events involving the character of the number showing when the die is tossed can be described in terms of subsets of S. For example, an even number shows is described by $\{2,4,6\}$; a number between 2 and 5 shows is described by $\{3,4\}$; a number greater than 4 but less than 3 shows is described by the humble, but important, empty set $\emptyset$. $S = \{1,2,3,4,5,6\}$ is said to be a sample space for the die tossing process because of its fundamental property that whenever a die is tossed, one and only one of the events in S occurs. S is further said to be a finite sample space because the number of events in it, 6, is a positive integer.

More generally, a collection of events

$$S = \{s_1, s_2, \ldots, s_n\},$$

is said to be a **finite sample space** for a random process if whenever the process is repeated, one and only one of the events in S occurs. The n events $s_1, s_2, \ldots, s_n$ which make up S are called **sample points.** The subsets of S describe events which can be analyzed in terms of S.

In general there are many possible sample spaces which can be given for a random process since all we need for a sample space is a collection of events with the fundamental defining property cited. However, not all sample spaces that may be given for a process are useful in helping us to further understand the process. $S_1 = \{O, E\}$, where O is the event an odd number shows and E is the event an even number shows, is also a sample space for the die tossing process since exactly one of these events occur when a die is tossed; S_1 is not a very useful sample space because there is very little which can be described in terms of it.

EXAMPLE 1

Specify two sample spaces for the experiment of tossing a pair of dice.

To easily distinguish the dice, we shall assume that one of them is red and the other is green. One sample space, which we shall call S_1, is the set of events listed in Table 6.1. The first number in each ordered pair specifies the number that shows on the red die and the second number specifies the number that shows on the green die. (4, 2), for example, is the event that red shows 4 and green shows 2; (2, 4) is the event that red shows 2 and green shows 4. S_1 is a sample space because it has the property that whenever the dice are tossed exactly one of the events in S_1 occurs.

Table 6.1

$$S_1 = \begin{cases} (1,1) & (1,2) & (1,3) & (1,4) & (1,5) & (1,6) \\ (2,1) & (2,2) & (2,3) & (2,4) & (2,5) & (2,6) \\ (3,1) & (3,2) & (3,3) & (3,4) & (3,5) & (3,6) \\ (4,1) & (4,2) & (4,3) & (4,4) & (4,5) & (4,6) \\ (5,1) & (5,2) & (5,3) & (5,4) & (5,5) & (5,6) \\ (6,1) & (6,2) & (6,3) & (6,4) & (6,5) & (6,6) \end{cases}$$

Another sample space S_2 is

$$S_2 = \{2, 3, 4, 5, 6, 7, 8, 9, 10, 11, 12\}$$

where 2 denotes the event that the sum of the numbers showing is 2, 3 denotes the event that the sum of the numbers showing is 3, and so on. S_2 is also a sample space because it too has the property that whenever the dice are tossed exactly one of the events in S_2 occurs.

Properties of the Relative Frequency Function R(A)

In developing a mathematical structure, called probability model, that would help us understand and predict the behavior of random processes it would make sense to look at properties of $R(A)$ to obtain a sense of direction. Let us recall that for an event A connected with a random process,

$$R(A) = \frac{\text{number of times } A \text{ occurs}}{\text{number of times the process is repeated}}.$$

1. $R(A) \geq 0$. Both numerator and denominator of $R(A)$ are nonnegative.
2. $R(A) \leq 1$. The numerator of $R(A)$ cannot exceed its denominator.
3. $R(A) = 0$. If A is an event whose occurrence is not possible (which happens when A is defined by incompatible conditions). Such an event A is identified with $\emptyset$.
4. $R(S) = 1$, where $S = \{s_1, s_2, \ldots, s_n\}$ is a sample space for the random process. Whenever the process is repeated one of the sample points in S occurs, and thus S occurs. Therefore, the numerator and denominator of $R(S)$ are equal.

In connection with the die tossing experiment, consider the event an even number shows. Let us observe that since an even number shows as often as 2, 4, and 6 show, the relative frequency of occurrence of this event is the sum of the relative frequencies of the events 2 shows, 4 shows and 6 shows. If, for example, 2, 4 and 6 are observed to show with relative frequencies 19/100, 27/100, and 14/100, respectively, then the relative frequency of an even number's occurrence is 19/100 + 27/100 + 14/100 = 60/100 = 60 percent. Such is the case in general, and we state this observation as follows:

5. $R(A) = $ sum of the relative frequencies of the sample points that describe A.
6. $R(S) = R(s_1) + R(s_2) + \cdots + R(s_n)$, where $S = \{s_1, s_2 \cdots, s_n\}$ is a sample space for the process. This property is obtained by applying property 5 to S.
7. $R(s_1) + R(s_2) + \cdots + R(s_n) = 1$; the sum of the relative frequencies of all sample points is 1. This follows from properties 4 and 6.

EXERCISES

1. Toss a coin 200 times, record the occurrence of the event tail shows, and determine the relative frequency of tail shows throughout the sequence of 200 tosses. Draw graphs to show the fluctuation in relative frequency for the first 15 tosses, tosses 86 through 100, and tosses 186 through 200.

2. Consider the process of tossing a die.
   (a) Let L denote the event that a number less than 3 shows, 4 denote the event that 4 shows, and B denote the event that a number greater than 4 shows. Is $S = \{L, 4, B\}$ a sample space for this process? Explain.
   (b) Let E denote the event that an even number shows, and A denote the event that a number greater than 3 shows. Is $S_1 = \{E, A\}$ a sample space for this process? Explain.
   (c) Let A denote the event that a number less than 3 shows, B denote the event that a number between 3 and 5 shows, and C denote the event that a number greater than 4 shows. Is $S_2 = \{A, B, C\}$ a sample space for the process? Explain.

3. Consider the process of tossing a pair of dice. Let A denote the event that the sum of the numbers showing is less than 5, B denote the event that the sum of the numbers showing equals 5, and C denote the event that the sum of the numbers showing is greater than 5. Is $S = \{A, B, C\}$ a sample space for this process. Explain.

4. Set up two sample spaces for the process of tossing a coin twice in succession.

5. Set up three sample spaces for the process of dealing a card from a standard deck of 52 cards.

6. Formulate two sample spaces for the process of tossing a coin three times in succession.

7. Formulate four sample spaces for the process of dealing a hand of five cards from a standard deck of 52 cards.

8. The Twolow Company makes light bulbs. Two plants, P1 and P2, carry out the production process. The daily output is 8000 bulbs with P1 producing 5000 bulbs of which 1 percent are defective, and P2 producing 3000 bulbs of which 0.5 percent are defective: A bulb is selected from the day's output. Set up three sample spaces for the selection process.

9. A college residence provides housing for 4 students, 3 of which are majoring in history. A sample of 2 students is chosen from the residence.
   (a) Set up two sample spaces for the selection process.
   (b) Is $S = \{s_1, s_2, s_3, s_4\}$, where s_1 is the event that student 1 is chosen, etc., a sample space for the selection process? Explain.

6.3. STRUCTURE OF A PROBABILITY MODEL

Since the concept of probability model is intended to be applicable to the study of the long term relative frequency behavior of events connected with random processes, the manner in which we define this concept is guided by properties of relative frequency.

> A **(finite) Probability model** for a random process consists of two components:
>
> (i) A sample space $S = \{s_1, s_2, \ldots, s_n\}$.
> (ii) A function P, called a probability function, which assigns to each subset A of S a value, denoted by $P(A)$ and called the probability of A, subject to the conditions listed below. We state these conditions on the sample points and then extend them to subsets of S.
>
> 1. $P(s_1) \geq 0, P(s_2) \geq 0, \ldots, P(s_n) \geq 0$
> 2. $P(s_1) \leq 1, P(s_2) \leq 1, \ldots, P(s_n) \leq 1$
> 3. $P(s_1) + P(s_2) + \ldots + P(s_n) = 1$
> 4. If A is a subset of S, $P(A)$ = sum of the probabilities of the sample points describing A; if $A = \emptyset$, $P(A) = 0$. (For example, if $A = \{s_1, s_5\}$, $P(A) = P(s_1) + P(s_5)$.)

The structural requirements of a probability model for a random process are somewhat analogous to a community's building code requirements for building a house. A building code does not tell us how to build a house. It tells us that however we build our house, for it to be legitimate in terms of the building code it must satisfy such and such conditions which are spelled out in the code. The concept of probability model does not tell us how to define a sample space and probability function for a random process; it tells us that in building a specific probability model for a random process, which can be done in many ways, we must satisfy the conditions stated in the definition of probability model (the building code in this case) in order for the model to be mathematically legitimate.

To take an example, Janet James was presented with the following structure which was claimed to be a probability model for the process of tossing a die.

$$S = \{O, E\}, P(O) = 1/3, P(E) = 2/3,$$

where O is the event that an odd number shows and E is the event that an even number shows on a throw of the die. Is this structure a probability model?

S is a sample space for the die tossing process since it satisfies the requirement that exactly one of its events occur when the die is tossed. The assignment by P of the numerical values to O and E satisfies the requirements of a probability function since

1/3 and 2/3 are non-negative, less than 1, and sum to 1. This establishes the mathematical legitimacy of $\{S, P\}$ as a probability model for the die tossing process; whether or not this model realistically describes the behavior of any die in Janet's possession is another question entirely. We turn to the issue of formulating a specific probability model for a specific random process in the next two sections.

At first sight the concept of probability model seems no different from the properties of relative frequency that were listed, except for the notation used. This is not the case. To be sure, the conditions required of a probability function closely mirror properties of relative frequency, but relative frequency is defined in a very specific way while probabilities can be assigned to events in a wide variety of ways so long as the conditions cited are satisfied. Probability assignments reflect properties of relative frequency, but go beyond them in much the same way that a son may reflect properties of his father, but goes beyond them.

There are a number of valid consequences which come out of the probability model structure, but the two theorems considered here are of particular interest.

Basic Results for Equally Likely Outcome Models

Theorem 1. Let $S = \{s_1, s_2, \ldots, s_n\}$ denote a sample space with n sample points. Let us suppose that for one reason or another we are led to the probability function P which assigns the same value x to all of the sample points in S; then this value is $x = 1/n$.

Proof: From the definition of probability model, we have
$$P(s_1) + (Ps_2) + \cdots + P(s_n) = 1$$
Since each of $P(s_1), P(s_2), \ldots, P(s_n)$ equals x, we obtain
$$\underbrace{x + x + \cdots + x}_{n \text{ terms}} = 1$$
$$nx = 1$$
$$x = 1/n$$

A probability model in which all sample points are assigned the same probability value is called an **equally likely outcome** model.

Theorem 2. If $S = \{s_1, s_2, \ldots, s_n\}$, $P(s_1) = \cdots = P(s_n) = 1/n$, and A is an event that is described by k sample points, then

$$P(A) = k/n.$$

Proof: For the sake of simplifying our discussion, let us suppose that A is described by the first k sample points $s_1, s_2, \ldots, s_k$. Then we have

$$P(A) = P(s_1) + P(s_2) + \cdots + P(s_k)$$

$$= \underbrace{1/n + 1/n + \cdots + 1/n}_{k \text{ terms}} = k/n$$

In the **special case of equally likely outcomes** probability questions reduce to counting questions. To determine the probability of A in this framework, count the number of sample points that describe A, count the number of sample points, and take their ratio. If S contains 10,000 sample points and A is described by 1000 of them, then $P(A) = 1000/10{,}000 = 1/10$.

6.4. A TALE OF THREE PROBABILITY MODELS

The following story might at first sight seem rather outrageous, but when we consider the rather weird doings which are reported every day in the press and on television, it's not at all outrageous and is in fact perfectly possible.

Rasa Adams's friend Indre was on the eve of her twenty-first birthday for which a big family celebration was planned. Since Indre was interested in the laws of chance, her mother decided to obtain a golden die as a birthday present for her. On learning of this in a conversation with Indre's mother, Rasa concluded that the best present she could get Indre to go with the golden die would be a probability model; after all, what good is a golden die without an accompanying probability model? But what's the best place to shop for a probability model, thought Rasa. She finally decided to try Trump Models, run by the noted model builder Herman J. Trump. Rasa was not disappointed at Trump Models. There was a large selection with many attractive models, and Herman Trump was willing to custom design a model to your specifications if you so desired. Rasa finally narrowed her search to three models, $R4$ (red), $B7$ (blue) and $Y3$ (yellow). These models had the following specifications:

*R*4 (red): $S = \{1, 2, 3, 4, 5, 6\}$; $P(1) = \cdots = P(6) = 1/6$

*B*7 (blue): $S = \{1, 2, 3, 4, 5, 6\}$; $P(1) = P(3) = P(5) = 1/9$
$P(2) = P(4) = P(6) = 2/9$

*Y*3 (yellow): $S = \{1, 2, 3, 4, 5, 6\}$; $P(1) = P(3) = P(5) = 1/12$
$P(2) = P(4) = P(6) = 3/12$

All three of these models are mathematically legitimate in that they all satisfy the probability model requirements. This is as far as Rasa can go in selecting a probability model for Indre's die. To determine which of these models, if any, is a realistic fit to Indre's die, we need to know more about the nature of this die. Rasa called Indre's mother and asked this question. "I had to pay extra for this," said Indre's mother, "but this is an unusual die in that platinum weights have been inserted inside

it so that the even numbers are favored to show over the odd ones by 2 to 1." This led Rasa to purchase the blue model B7 as the most realistic model for Indre's die.

To examine a valid consequence of the blue model, consider the event E that an even number shows. $E = \{2, 4, 6\}$.

$$P(E) = P(2) + P(4) + P(6) = 2/9 + 2/9 + 2/9 = 0.67$$

The relative frequency interpretation of this result is that if a die whose behavior is described by the blue model is tossed a large number of times, an even number will show approximately 67% of the time.

After Indre's birthday it came to pass that her die was tossed 500 times. Records kept of the outcomes of the tossings reveal that an even number showed on 246 of the 500 tosses, so that $R(E) = 0.492$. Since the actual relative frequency of an even number showing, 0.492, is very much at variance with the projected value of approximately 0.67, this shows that a valid consequence of Indre's model, interpreted in relative frequency terms, is false. Indre's model is not a realistic one for her die, and her mother paid for an unusual die which was not delivered. The result observed suggests that the red model, based on the assumption that the die in question is well balanced, of uniform construction, is a more realistic one for the die that Indre actually obtained.

EXERCISES

1. Consider the yellow model Y3 for the process of tossing a certain die. $S = \{1, 2, 3, 4, 5, 6\}$,

    $$P(1) = P(3) = P(5) = 1/12; P(2) = P(4) = P(6) = 3/12.$$

    Let E denote the event that an even number shows.
    (a) Find $P(E)$.
    (b) State the relative frequency interpretation of the result obtained in (a).
    (c) In tossing the die in question 1000 times, an even number was observed to show 665 times. Does this show that the conclusion obtained in (a) is not valid? Explain.
    (d) Is the conclusion reached in (a), interpreted in relative frequency terms, true? Explain.
    (e) Is the yellow model Y3 realistic for the die in question? Explain.

2. Consider the model G4 (green) for the process of tossing a certain die. $S = \{1, 2, 3, 4, 5, 6\}$,

    $$P(1) = P(3) = P(5) = 1/10, P(4) = 3/10, P(2) = P(6) = 2/10.$$

    Let A denote the event that an odd number shows.
    (a) Find $P(A)$.
    (b) State the relative frequency interpretation of the result obtained in (a).

(c) In tossing the die in question 1000 times an odd number was observed to show 302 times. Does this evidence establish that the conclusion obtained in (a) is valid? Explain.

(d) Is the conclusion obtained in (a), interpreted in relative frequency terms, true? Explain.

(e) Is the green model $G4$ realistic for the die in question? Explain.

6.5. PROBABILITY MODELS FOR RANDOM PROCESSES

EXAMPLE 1. IS DEALING A CARD AS STRAIGHTFORWARD AS IT SOUNDS?

Consider the process of dealing a card from a standard deck of 52 cards and let us address the problem of setting up a probability model for this process and determining the probability that a picture card is dealt.

For convenience in referring to the cards, let us set up a translation system so that we can refer to the cards as 1, 2, ..., 52; in this translation system 1 might denote the ace of spades, 2 the king of spaces, etc. We take as our sample space,

$$S = \{C_1, C_2, \cdots, C_{52}\},$$

where C_1 is the event that card 1 is dealt, C_2 is the event that card 2 is dealt, etc.

Our next task is to define a probability function P on S. This can be done in many ways, and the function P that emerges depends on the assumption we make about how the card will be dealt from the deck.

Suppose we assume what is usually assumed in such situations, but not always made explicit, that the card is dealt from a well shuffled deck in an unbiased way—at random, as we say. The probability function P which best reflects this assumption assigns the same value, 1/52, to each sample point in S. We thus emerge with the following probability model:

$$\text{Model 1:} \quad S = \{C_1, C_2, \cdots, C_{52}\}$$
$$P(C_1) = \cdots = P(C_{52}) = 1/52$$

From Model 1 it follows that the probability that a picture card is dealt is $\frac{12}{52} = 0.23$.

As is well known, some card dealers are less than honest. Suppose we assume, based on past experience, that the dealer intends to "arrange things" so that the card we are dealt is neither a picture card nor an ace, but that any of the other cards may be dealt without bias. For notational convenience suppose that the picture cards and aces are in the cards we numbered 1, 2, ... 16, and that the cards 17, 18, ... 52 correspond to the remaining cards. This leads to Model 2.

$$\text{Model 2:} \quad S = \{C_1, \ldots, C_{16}, C_{17}, \ldots, C_{52}\},$$
$$P(C_1) = \ldots = P(C_{16}) = 0, \ P(C_{17}) = \ldots = \ldots = P(C_{52}) = 1/36$$

From Model 2 it follows that the probability that a picture card is dealt is 0.

EXAMPLE 2. SUSAN'S PROBLEM

Susan Reti was interested in determining the probability that an even sum shows for the process of tossing a pair of well balanced dice of uniform construction (one red and one green). She asked two of her friends, Rachael and Laura, if they would help her set up probability models to determine the probability of this event. Both were glad to do so.

Rachael's Model: Rachael took as her sample space S_1 the set of events described by all ordered pairs of integers between 1 and 6, inclusive, as shown in Table 6.2. The first number in each ordered pair specifies the number which shows on the red die and the second number specifies the number which shows on the green die.

Table 6.2

$$S_1 = \begin{cases} (1,1) & (1,2) & (1,3) & (1,4) & (1,5) & (1,6) \\ (2,1) & (2,2) & (2,3) & (2,4) & (2,5) & (2,6) \\ (3,1) & (3,2) & (3,3) & (3,4) & (3,5) & (3,6) \\ (4,1) & (4,2) & (4,3) & (4,4) & (4,5) & (4,6) \\ (5,1) & (5,2) & (5,3) & (5,4) & (5,5) & (5,6) \\ (6,1) & (6,2) & (6,3) & (6,4) & (6,5) & (6,6) \end{cases}$$

The assumption that the dice are well balanced is best reflected by the probability function P which assigns the same value, 1/36, to each sample point in S_1.

The sample points with even sums are found in alternate diagnols of Table 6.2, beginning with the first, and are 18 in number. Thus, from Rachael's model it follows that the probability of E, that an even sum shows, is :

$$P(E) = 18/36 = 0.50$$

The relative frequency interpretation of this conclusion is that if a pair of well balanced dice are tossed a large number of times, an even sum will show approximately 50% of the time.

Laura's Model: Laura proceeded to analyze Susan's problem in a different way. She took as her sample space

$$S_2 = \{2, 3, 4, 5, 6, 7, 8, 9, 10, 11, 12\},$$

where 2 is the event that the sum of the numbers showing is 2, 3 is the event that the sum of the numbers showing is 3, and so on.

The assumption that the dice are well-balanced led Laura to the probability function P which assigns the same value, 1/11, to the eleven sample points in S_2.

$E = \{2, 4, 6, 8, 10, 12\}$, and Laura obtained

$$P(E) = \frac{6}{11} = 0.55$$

for the probability that an even sum shows.

The relative frequency interpretation of this result is that if a pair of well-balanced dice are tossed a large number of times, and even sum will show approximately 55% of the time.

Susan was more confused than ever and turned to her cousin Jack for help with her problem.

Jack's Model: Jack took as his sample space S_3 the collection of events shown in Table 6.3.

Table 6.3

$\{1,1\}, \{1,2\}, \{1,3\}, \{1,4\}, \{1,5\}, \{1,6\}$
$\{2,2\}, \{2,3\}, \{2,4\}, \{2,5\}, \{2,6\}$
$\{3,3\}, \{3,4\}, \{3,5\}, \{3,6\}$
$\{4,4\}, \{4,5\}, \{4,6\}$
$\{5,5\}, \{5,6\}$
$\{6,6\}$

212   Finite Mathematics, Models, and Structure

Here $\{1,1\}$ is the event that both dice show 1, $\{1,2\}$ is the event that one die shows 1 and the other shows 2, and so on. There are 21 sample points, Jack observed. Since the dice are assumed to be well balanced, Jack was led to take as his probability function P the one which assigns the same value, 1/21, to each sample point. There are 12 sample points with even sum and this led Jack to conclude that $P(E)$, the probability that an even sum shows, is 12/21 or 0.57.

The relative frequency interpretation of this result is that if a pair of well balanced dice are tossed a large number of times, an even sum will show approximately 57% of the time.

Susan's Dilemma

Does $P(E) = 0.50$ (Rachael's Model), 0.55 (Laura's Model), or 0.57 (Jack's Model)? If mathematics is such a precise subject, how could this happen? Which result should I accept, Susan pondered? All three conclusions are correct in the sense of being valid consequences of their respective probability models; from the point of view of validity there is no conflict since these conclusions arise from different sources.

As to the question of truth, these conclusions, interpreted in relative frequency terms, cannot all be true for well-balanced dice. To settle the question of truth, well-balanced dice would have to be tossed a large number of times and a record kept of how often an even sum shows. When this is done, it is found that an even sum shows in the neighborhood of 50 percent of the time. Such evidence, obtained by performing the process and observing, establishes the truth of $P(E) = 0.50$ and the falsity of $P(E) = 0.55$ and $P(E) = 0.57$ in terms of the relative frequency interpretation of probability, for well balanced dice. The appearance of a valid conclusion that is false alerts us to the unrealistic nature of Laura's assumption of equally likely outcomes for S_2. Indeed, it is not difficult to pinpoint the difficulty with Laura's assumption. It is unrealistic to assume, for example, that a sum of 2 is as likely to occur as a sum of 7 when well-balanced dice are tossed; a sum of 2 can only occur in one way—when (1,1) shows; a sum of 7 can occur in six ways—when (1,6) (6,1), (2,5), (5,2), (3,4) or (4,3) show.

EXAMPLE 3. A PRODUCTION PROCESS PROBLEM

The Bokson Company makes television tubes in two plants, B1 and B2. The weekly output is 3000 tubes, with 1800 tubes being produced in plant B1, of which 1 percent are defective, and 1200 tubes being produced in plant B2, of which 2 percent are defective. A tube is selected at random from the week's output. The problem is to set up a probability model for this selection process and find the probability that a defective tube is chosen.

We take as our sample space the set of 3000 events $S = \{t_1, t_2, \ldots, t_{3000}\}$, where t_1 is the event that tube 1 is chosen, t_2 is the event that tube 2 is chosen, and so on. To say that a tube is chosen at random means that a tube is chosen in an unbiased way so that certain tubes are not favored over others. In other words, all tubes have the same likelihood of being chosen. This assumption is best reflected by the probability function P defined by

$$P(t_1) = P(t_2) = \cdots = P(t_{3000}) = \tfrac{1}{3000}$$

To obtain the probability that a defective tube is chosen in terms of this probability model, we must find the number of ways of selecting a defective tube and divide by 3000, the total number of sample points. Since 1 percent of the output of plant B1 is defective, B1 contributes 0.01(1800) or 18 defectives to the total number of defectives. Plant B2 contributes 0.02(1200) or 24 defectives, and the total number of defectives is 42. Thus

$$P(\text{defective tube is chosen}) = \tfrac{42}{3000} = 0.014$$

The relative-frequency interpretation of this result is that, if the envisioned selection process is performed a large number of times, a defective tube will be chosen approximately 1.4 percent of the time.

> The basic assumption that led to this model is that the drawing of a tube is made at random; that is, in an unbiased way, no favoritism, deliberate or inadvertent. This is much more easily said than done. If the model constructed is to be realistic for the drawing, then we must do everything possible to ensure that the randomness assumed in theory is lived up to in practice.

EXERCISES

1. The appearance of a valid conclusion that is false also alerts us to the unrealistic nature of Jack's assumption of equally likely outcomes for S_3.
   (a) Pinpoint the difficulty with Jack's assumption.
   (b) How should Jack's probability function be modified to make it realistic for well-balanced dice?

2. How should Laura's probability function be modified to make it realistic for well-balanced dice?

3. The Twolow Company makes light bulbs. Two plants, P1 and P2, carry out the production process. The daily output is 8000 bulbs, with P1 producing 5000 bulbs of which 1 percent are defective, and P2 producing 3000 bulbs of which 0.5% are defective. A bulb is selected from the day's output.
   (a) When asked to determine the probability that a defective bulb is chosen, Mark Twolow set up the following probability model for the selection process, based on the assumption that the bulb is selected at random from the day's output. $S = \{GP1, GP2, DP1, DP2\}$, where $GP1$ is the event that a good bulb made by $P1$ is selected, and so on.

   $$P(GP1) = P(GP2) = P(DP1) = P(DP2) = 1/4$$

   Mark determined the probability that a defective bulb is chosen to be $P(DP1) + P(DP2) = 1/2$. (1) Is Mark's model satisfactory? How so? (2) Is Mark's conclusion correct? How so?

   (b) Mark's brother Bob suggested a simpler model, again based on the assumption that the bulb is chosen at random from the day's output. $S = \{G, D\}$, where G is the event that a good bulb is selected and D is the event that a defective bulb is chosen. $P(G) = P(D) = 1/2$. Bob pointed out that he and Mark had reached the same conclusion, so that each confirms the correctness of the other. (1) Would you agree or disagree with Bob's comment? Explain. (2) Is Bob's model satisfactory? Explain. (3) Is Bob's conclusion correct? (4) When asked for the probability that a defective bulb made in P1 is selected in terms of his model, Bob gave 1/4 as the answer. Would you agree or disagree? How so?

   (c) Formulate your own probability model for the light bulb selection process, assuming that the bulb is chosen at random from the day's output.

   (d) Should Mark's probability function be modified? How so? If modification is called for, how would you carry it out?

   (e) Should Bob's probability function be modified? Explain. If modification is called for, how would you carry it out?

4. Jason took a well-balanced coin from his pocket and asked his friend Andrew to help him determine the probability of throwing one head and one tail on two

successive tosses of the coin. Andrew took $S = \{0,1,2\}$, where 0 is the event that no heads show in the two tosses, 1 is the event that one head shows in the two tosses, etc., as his sample space. He defined a probability function P by $P(0) = P(1) = P(2) = 1/3$, so that the probability of throwing one head and one tail is 1/3. The relative frequency interpretation of this conclusion is that if a well-balanced coin is tossed twice in succession a large number of times, a head and tail will show approximately 33.3% of the time. When Jason's well-balanced coin was tossed twice in succession 500 times (which involves 1000 tosses), a head and tail were observed to show 246 times.

(a) Does this mean that Andrew's conclusion is not valid? How so?

(b) Is Andrew's conclusion, interpreted in relative frequency terms, true? How so?

(c) How is the discrepancy between the predicted relative frequency and the observed relative frequency to be explained?

(d) Set up your own probability model for tossing a coin, assumed to be well balanced, twice in succession.

(e) Should Andrew's probability function be modified? Explain. If modification is called for, how would you carry it out?

5. Bill Albert succeeded in finding a rare pair of loaded dice. With reference to the set S_1 of 36 outcomes described in Table 6.1 (p. 201), these dice have the property that outcomes with an even sum [such as (1,1), (1,3), and so on] are twice as likely to occur as outcomes with and odd sum [such as (2,1), (3,2), and so on].

(a) Define a probability function on S_1 that best reflects the nature of Bill's dice.

(b) Find the probability that (i) an even sum shows; (ii) a sum of 7 shows; (iii) a sum less than 6 shows.

6. Consider the statement: the event A, connected with a random process, has probability 0.96. Does the following correctly state the relative-frequency interpretation of this statement? Explain. "If the process is repeated over and over, in the long run the probability of A will be 0.96."

7. Consider the process of dealing a card from a standard deck of 52 cards.

(a) Construct a probability model for this process.

(b) What assumption underlies the probability model defined in answer to the preceding question?

(c) Find the probability that a club is dealt. What is the relative-frequency interpretation of your result?

(d) Find the probability that a picture card is dealt. What is the relative-frequency interpretation of your result?

8. The Jacini Company makes razor blades. Two machines, M1 and M2, are used in the production process. The daily output is 5000 blades, with machine M1 producing 3000 blades of which 0.5 percent are defective, and machine M2 pro-

ducing 2000 blades of which 1 percent are defective. A blade is selected at random from the day's output.

(a) Define three sample spaces for the selection process.
(b) Define a probability function on one of the sample spaces defined in answer to the preceding question.
(c) What assumption underlies the probability function defined?
(d) Find the probability that (1) a blade made by machine M1 is selected; (2) a defective blade made by machine M2 is selected; (3) a defective blade is selected.

9. Jack Jones was interested in the number of times head shows in three successive tosses of a well-balanced coin. He took as his sample space $S = \{0, 1, 2, 3\}$, where 0 is the event that no head shows in the three tosses, 1 is the event that one head shows in the three tosses, and so on. Jack assigned equal probabilities of $\frac{1}{4}$ to these sample points and determined the probability of head showing twice in the three tosses to be $\frac{1}{4}$. The relative-frequency interpretation of this conclusion is that, if the coin is tossed three times in succession a large number of times, head will show exactly twice approximately 25 percent of the time. Yet when Jack tossed a well-balanced coin three times in succession a large number of times, he observed that head showed exactly twice 38 percent of the time. In light of these developments, consider the following questions.

(a) Isn't Jack's conclusion valid? Explain.
(b) Isn't Jack's conclusion true? Explain.
(c) How is the discrepancy between the predicted relative frequency and the observed relative frequency to be explained?

10. For his birthday, Herman received a pair of dice that, he was told, were well-balanced. On the basis of the probability model that assigns the same value, $\frac{1}{36}$, to each of the sample points in the well-known 36-element sample space for the process, Herman determined the probability of an even sum showing to be 0.50. He expected that an even sum would show in the neighborhood of 500 times for 1000 tosses of his dice and made betting plans accordingly. Herman participated in a friendly game one evening, and after 1000 tosses of his dice an even sum had showed 200 times and Herman was $600 poorer. Disappointed, confused, and angry, Herman raised the following questions.

(a) If mathematics is such a precise subject, how could this happen?
(b) Isn't my conclusion correct?
(c) What went wrong?

6.6. DERIVED EVENTS AND THEIR PROBABILITIES

If A and B are events connected with a random process, then there are other events, derived events so to speak, that can be described in terms of A and B, such as the event that both *A* and *B* occur, the event that either *A* or *B* occurs (that is, at least one of the events *A*, *B* occurs), and the event that *A* does not occur.

To illustrate, consider the die-tossing process for which we take $S = \{1, 2, 3, 4, 5, 6\}$ as our sample space. Let *A* denote the event that an odd number shows, and *B* denote the event that a number greater than 3 shows; then the event that an odd number greater than 3 shows is a translation in this setting of the event that both *A* and *B* occur. The event that an odd number or a number greater than 3 shows is a translation of the event *A* or *B* occurs. The event that an odd number does not show, that is, an even number shows, is a translation of the event that *A* does not occur. In terms of the language of sets, we have

Odd number shows	$\leftrightarrow$	$A = \{1, 3, 5\}$
Number greater than 3 shows	$\leftrightarrow$	$B = \{4, 5, 6\}$
Odd number or number greater than 3 shows	$\leftrightarrow$	$A \cup B = \{1, 3, 4, 5, 6\}$
Odd number greater than 3 shows	$\leftrightarrow$	$A \cap B = \{5\}$
Odd number does not show	$\leftrightarrow$	$A^c = \{2, 4, 6\}$
Number greater than 3 does not show	$\leftrightarrow$	$B^c = \{1, 2, 3\}$

More generally, for any events *A* and *B* connected with a random process, we have the translation from the colloquial language of events to the formal language of sets shown in Table 6.4.

Table 6.4

Colloquial Language	Set Language
Event *A*	Set *A*
Event *B*	Set *B*
Event *A* and *B*	Set $A \cap B$
Event *A* or *B*	Set $A \cup B$
Event not-*A*	Set A^c

We next turn our attention to establishing relationships among derived events $A \cap B$, $A \cup B$, and A^c and the components *A* and *B*.

Theorem 3. If *A* and *B* are events in a finite sample space *S*, then

$$P(A \cup B) = P(A) + P(B) - P(A \cap B)$$

Proof: To help illustrate the analysis, we shall refer to Figure 6.6. By definition, the probability of $A \cup B$ is the sum of the probabilities of the sample points in $A \cup B$. $P(A) + P(B)$ is the sum of the probabilities of the sample points in A plus the sum of the probabilities of the sample points in B. $P(A)$ includes the probabilities of the sample points in $A \cap B$, as does $P(B)$, since $A \cap B$ is part of A and is also part of B. Thus within $P(A) + P(B)$ the probability of $A \cap B$ is counted twice. [In terms of Figure 6.6, the probability of the cross-shaded region is being counted twice in $P(A) + P(B)$.] By subtracting $P(A \cap B)$ from $P(A) + P(B)$, we are left with the sum of the probabilities of the sample points in $A \cup B$, with each probability value appearing only once. Thus

$$P(A \cup B) = P(A) + P(B) - P(A \cap B)$$

In colloquial language, this result is stated as follows:

$$P(A \text{ or } B) = P(A) + P(B) - P(A \text{ and } B)$$

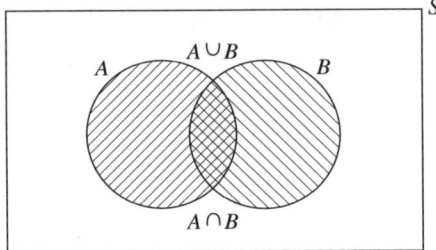

Figure 6.6

Returning to the die-tossing process for a moment, identifying A with the event an odd number shows and B with the event a number greater than 3 shows, and assuming the die is well-balanced, we have

$$P(A \cup B) = P(A) + P(B) - P(A \cap B)$$
$$= \tfrac{3}{6} + \tfrac{3}{6} - \tfrac{1}{6} = \tfrac{5}{6}$$

Two events A and B are said to be **mutually exclusive** or **disjoint** if they have no sample points in common; that is, $A \cap B = \emptyset$. More generally, n events $A_1, A_2, \ldots, A_n$ are said to be **pairwise mutually exclusive** or **pairwise disjoint** if no two of them have any sample points in common. Figure 6.7 shows four pairwise disjoint events.

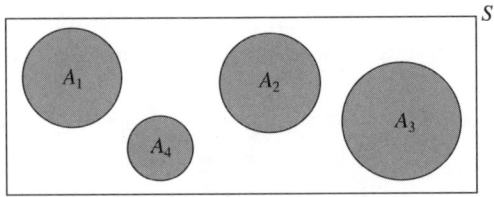

Figure 6.7

Theorem 4. If A and B are disjoint events in a finite sample space, then

$$P(A \cup B) = P(A) + P(B)$$

More generally, if $A_1, A_2, \ldots, A_n$ are pairwise disjoint events in a finite sample space, then

$$P(A_1 \cup A_2 \cup \cdots \cup A_n) = P(A_1) + P(A_2) + \cdots P(A_n)$$

Proof: By definition, the probability of $A \cup B$ is the sum of the probabilities of the sample points in $A \cup B$. $P(A)$ is the sum of the probabilities of the sample points in the A part of $A \cup B$ and $P(B)$ is the sum of the probabilities of the sample points in the B part of $A \cup B$. Thus $P(A) + P(B)$ is the sum of the probabilities of the sample points in $A \cup B$. Since A and B have no points in common, there is no duplication of probabilities when $P(A)$ and $P(B)$ are added. The same sort of argument can be used to establish the more general case involving $A_1, A_2, \ldots, A_n$.

Theorem 5. If A is an event and A^c is its complement, then the probability of one of these events is 1 minus the probability of the other. That is,

$$P(A) = 1 - P(A^c)$$

Proof: A and A^c are complementary events means that $A \cup A^c = S$ and $A \cap A^c = \emptyset$ (see Figure 6.8). Thus we have

$$P(A \cup A^c) = P(A) + P(A^c)$$
$$P(S) = P(A) + P(A^c)$$
$$1 = P(A) + P(A^c)$$
$$P(A) = 1 - P(A^c)$$

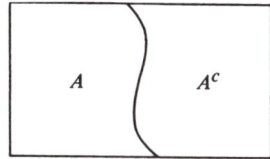

Figure 6.8

EXAMPLE 1

In connection with the process of tossing Susan Reti's pair of dice (one of which is red and the other green), assumed to be well balanced (Rachael's model, p. 210), find the probability that (a) the green die shows 2 or the sum of the numbers showing is less than 5; (b) the green die does not show 2.

Let A denote the event that the green die shows 2, and B denote the event that the sum of the numbers showing is less than 5. Then $A \cup B$ is the event the green die shows 2 or the sum of the numbers showing is less than 5. In terms of the sample space for the process, events A and B are shown in Table 6.5. Therefore, $A \cap B = \{(1, 2), (2, 2)\}$ and $P(A \cap B) = \frac{2}{36}$. We have

$$P(A \cup B) = P(A) + P(B) - P(A \cap B)$$
$$= \frac{6}{36} + \frac{6}{36} - \frac{2}{36} = \frac{10}{36} = 0.278$$

Table 6.5

	(1, 1)	(1, 2)	(1, 3)	(1, 4)	(1, 5)	(1, 6)
	(2, 1)	(2, 2)	(2, 3)	(2, 4)	(2, 5)	(2, 6)
	(3, 1)	(3, 2)	(3, 3)	(3, 4)	(3, 5)	(3, 6)
	(4, 1)	(4, 2)	(4, 3)	(4, 4)	(4, 5)	(4, 6)
	(5, 1)	(5, 2)	(5, 3)	(5, 4)	(5, 5)	(5, 6)
	(6, 1)	(6, 2)	(6, 3)	(6, 4)	(6, 5)	(6, 6)

Since the events the green die does not show 2 and the green die shows 2 are complementary, we have

$$P(A^c) = 1 - P(A)$$
$$= 1 - \frac{6}{36} = \frac{30}{36} = 0.833$$

Situations often arise for which the best attack on the problem of finding the probability of an event A is through the backdoor approach of finding the probability of its complement A^c, and using Theorem 5 to obtain the probability of A.

EXERCISES

1. For the process of tossing a die, assumed to be well balanced, find the probability that (a) an even number or a number greater than 2 shows; (b) an odd number or a number less than 4 shows.

2. In a certain probability model, $P(A) = \frac{2}{5}$, $P(B) = \frac{1}{10}$, and $P(A \cup B) = \frac{2}{5}$. Find $P(A \cap B)$.

3. In connection with the process of tossing a pair of dice discussed in Example 1, find the probability that (a) the red die shows 3 or the sum of the numbers showing is greater than 7; (b) the sum of the numbers showing is less than or equal to 7; (c) the red die shows a number other than 3; (d) the red die shows 3 or the green die shows 2; (e) the sum of the numbers showing does not equal 3.

6.7. INTERPRETATIONS OF PROBABILITY

Historically the study of the stability exhibited by long-run relative frequencies of events connected with games of chance played a fundamental role in the birth and early development of the theory of probability. But having been born, probability theory began a separate mathematical life of its own. This mathematical life begins with the concept of probability model, and consists of the theorems and definitions that are built up on the basis of this concept. Theorems 1 through 5 and the definitions of complementary and mutually exclusive events, considered in the preceding two sections, are humble examples of theorems and concepts that are part of this mathematical life; other theorems and concepts are considered in the chapters that follow. Although the relative-frequency point of view suggested the formulation of many concepts of the mathematical theory of probability, this interpretation, and any other interpretation for that matter, are not part of the internal structure of this mathematical theory. The relationship between a mathematical theory and an envisioned interpretation of the theory that suggested many of the theory's concepts is somewhat similar to the relationship between parents and child. The parents give birth to the child and influence the child's development. Although we may hear comments about physical and temperament similarities between parents and child, we recognize the child as a separate and distinct individual. In a similar sense, a mathematical theory is a structure separate from an envisioned application or interpretation that may have played the role of parents in giving rise to the theory.

Since a given interpretation of a mathematical theory is a structure separate from the theory itself, it might be possible and useful to interpret the theory's concepts in other ways. Such is the case with probability theory. Another interpretation that has received much attention in recent years is one in which the probability value assigned to an event is interpreted as a quantitative measure of an individual's degree of belief in the occurrence of the event. An ardent New York Mets fan might say that the Mets have a 90 percent chance of taking the pennant next year; translation: the probability that the Mets take the pennant next year is 0.90. A less ardent Mets admirer might say that the Mets have a 10 percent chance of taking the pennant next year. A business man might say that there's an 80 percent chance that the sales volume of the firm will top $5 million this year. Individuals often have beliefs or opinions about possible outcomes connected with situations in which the outcome is not certain. Such an individual sometimes finds it useful to assign a value between 0 and 1, inclusive, to a possible outcome as a quantitative measure of his feelings about the likelihood of occurrence of the outcome. A strong opinion about the occurrence of an event is reflected by the assignment of a value close to 1 to the event; a strong opinion about the nonoccurrence of an event is reflected by the assignment of a value close to 0 to the event. Probability values that are assigned to an event from this point of view, or that are interpreted in this way, are called **personal** or **subjective probabilities** and are said to express a person's degree of belief in the occurrence of the event. The point of view of subjective probability admits as meaningful probabilistic assertions about events connected with situations that occur once and cannot be repeated; the relative-frequency interpretation, on the other hand, is only meaningful for random processes that can be repeated a large number of times. The subjective point of view admits as meaningful such statements as the probability that the Smith Company will spend more than $2 million on advertising this year is 0.8, and the probability that Jack will not study for his math exam is 0.9. Such statements are meaningless from the relative-frequency point of view. An important feature, as well as difficulty, of subjective probability is that the assignment of subjective probabilities to outcomes depends very much on the person doing the assigning. Two individuals might assign markedly different subjective probabilities to the same event.

Subjective probability has been criticized along the following lines: if mathematical probability is regarded as a quantitative measure of a person's degree of belief, then the theory of probability is somewhat like a branch of psychology. The final result of a purely subjective interpretation of probability is subjective idealism. To assume that the evaluation of probability only concerns the feelings of the observer implies that conclusions based on probabilistic judgments are deprived of the objective meaning that they have independent of the observer. In defense of subjective probability, its proponents argue that there are many important once-and-only situations, especially in business, where a decision must be reached. By using subjective probabilities to express his judgments (based on data, analysis, experience, etc.) in quantitative terms, the decision maker can employ the machinery of the mathematical theory of probability to arrive at conclusions that are valid with respect to his judgments translated into quantitative terms. It is understood that the decision maker is not arriving at truth. But he does obtain conclusions that are consistent with his judgments, and such conclusions are often helpful for making the decision he is responsible for.

The interpretation of probabilities as measures of degrees of belief remains controversial at this point. Rather than take sides for or against the subjective point of view, it is more important for a student to obtain as thorough an understanding of this point of view as possible and to be able to recognize the subjective use of probability when it appears.

EXAMPLE 1

A. M. Bradley, who is suffering from a rare disease, was told by his friend that, since recent medical statistics show that 75 percent of those who have had the disease recovered, the probability that he will recover is 0.75. How is this probability assignment to Bradley's recovery to be interpreted?

Although a relative-frequency background is involved (recent medical statistics), the focus is on a once-and-only situation, A. M. Bradley's state of health. This by itself is sufficient to exclude a relative-frequency interpretation; to repeat the process a large number of times would entail giving Bradley the disease a large number of times and observing how often he recovers, a procedure that does have its difficulties. Bradley's friend is using background relative-frequency data as a basis for expressing in quantitative terms his degree of belief in Bradley's recovery.

The following example is due to my colleague Irwin Kabus.

EXAMPLE 2.   SUBJECTIVE PROBABILITY IN BANKING

For the last twenty years, the top management of Morgan Guaranty Trust Company has been using a technique called histogramming to quantify and picture the uncertainty that surrounds future interest rates on which the bank's asset/liability decisions are based.

One utilizes the histogram technique by first listing all the possible outcomes he believes may result from some future situation and then by assigning subjective probabilities to each of these outcomes. From observing the width of the range of outcomes and the chance associated with each a reader of the histogram is able to assess how confident the histogram maker feels about his judgments. For example, an individual's feeling about the probable level of the interest rate on a 90-day CD at some future date—say, three months from now—might look like the one shown in Figure 6.9.

Each of the interest rates shown represents an interval extending 1/8% below and 1/8% above it. The percentages of subjective chances (show in the vertical bars) are those that the individual has chosen to spread over the possible rates.

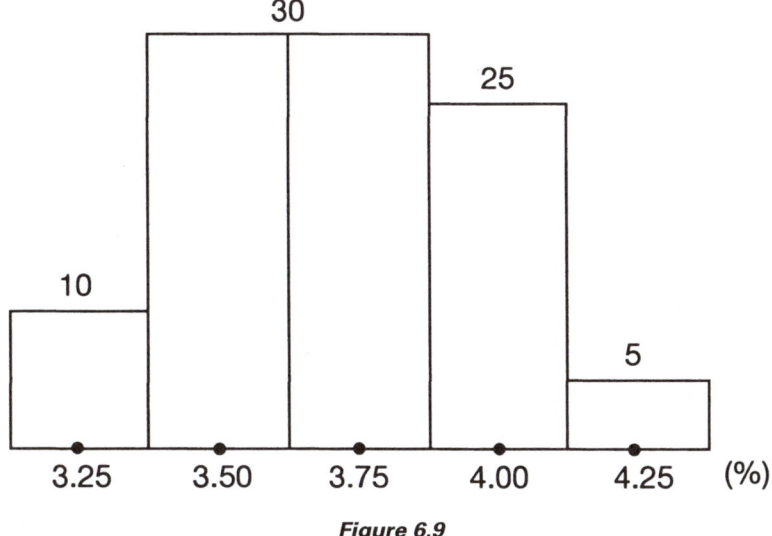

Figure 6.9

Mathematically, this histogram states that the individual feels there is a 10% chance that in three months the 90-day CD rate will be between 3 & 1/8% and 3 & 3/8%, a 30% chance that it will be between 3 & 3/8% and 3 & 5/8%, a 30% chance it will be between 3 & 5/8% and 3 & 7/8%, a 25% chance it will be between 3 & 7/8%

and 4 & 1/8%, and a 5% chance it will be between 4 & 1/8% and 4 & 3/8%. However, and this is of the utmost importance, the histogram is not meant to be read with mathematical precision; rather, it is meant to convey a message in terms that express judgments in numerical form.

Qualitatively, this histogram indicates that the individual feels very confident that in three months the 90-day CD rate will be between 3 & 3/8% and 4 & 1/8%. However, he is very uncertain as to what value, within the range, the rate will take on. There are small chances that the rate could drop to 3 & 1/8% to 3 & 3/8% or rise as high as 4 & 1/8% to 4 & 3/8%, although it is more likely that the lower rate is attained than the higher one.

As is most often the case, top management does not base its decisions solely on the opinion of one individual but on the opinions of several. Thus, it becomes necessary to produce an histogram whose qualitative message is some combination of the thinking of all those individuals involved in the process. Individual histograms can be combined to form a single weighted average histogram, with the weights being assigned by the individual responsible for endorsing the final histogram that top management sees.

In cases where all individual histograms receive equal weights (as is true at Morgan Guaranty) the weighted average becomes a simple arithmetic average. In other cases where individual histograms are weighted differently, the weights are generally based on the qualifications and track records of the individuals producing the histograms.

At Morgan Guaranty the histogram committee consisted of seven key executives from various areas of the bank. After collecting each individual's histogram, a summary sheet was generated which collectively showed the subjective probabilities assigned by each committee member. This summary served as the basis for a discussion (moderated by the analyst coordinating the histogram process) in which those members with differing views, immediately identifiable from the summary sheet, had a chance to defend them. At the end of the discussion the members of the histogram committee had the opportunity to change their histograms if they felt inclined to do so after hearing other opinions. The final histograms were then averaged into one consensus histogram which represented the views of the committee as a whole and was then presented to top management. When the moderator of the committee meeting presented the results to top management it was the qualitative story that was given to them.

References

[1] I. Kabus, "You can Bank on Uncertainty," *Harvard Business Review,* 54,3 (May–June 1976), 95–105.
[2] L. J. Savage, *The Foundations of Statistics* (New York: Wiley and Sons, Inc., June 1976), 95–105.
[3] R. Schlaifer, *Probability and Statistics for Business Decisions* (New York: McGraw Hill, 1959).
[4] R. Schlaifer, *Analysis of Decisions Under Uncertainty* (New York: McGraw Hill, 1969).

EXERCISES

1. It is asserted that the probability of an event E is 0.80. State the relative-frequency and subjective-probability interpretations of this assertion, and describe the main features of these interpretations.

2. After watching a pair of dice being tossed 100 times, an observer commented that the probability of an even sum showing on the 101st toss is 0.85. Is this probability assignment one that is to be interpreted in relative-frequency or subjective-probability terms? Explain.

3. The following comment appeared in an article on natural gas supplies (*New York Times*, Feb. 22, 1977, p. 14): "How much gas is left to be discovered? . . . The last Geological Survey estimate, made two years ago, was this: Given available technology and current economics, there is a 95 percent probability that 322,000 billion cubic feet can be located and produced; there is a 5 percent probability that 655,000 billion cubic feet can be located and produced." How should these probabilistic statements be interpreted? Explain.

4. In a letter to the editor of the *New York Times* (Feb. 28, 1971) on the background of the atomic bomb, Hans Bethe wrote, "By February 1945 it appeared to me and

to other fully informed scientists that there was a better than 90 per cent probability that the atomic bomb would in fact explode. . . . ." How should this probabilistic statement be interpreted? Explain.

5. Eric Roberts, a student at Huxley College, commented that the probability that he will get an A in Sociology this semester is 0.95. Is this probability value to be interpreted in relative frequency or subjective probability terms. How so?

6. On Dec. 29, 1978 two acoustics experts said tests showed a probability of 95 percent or better that a shot was fired from a grassy knoll in Dallas where President John F. Kennedy was assassinated. This testimony was presented before the House Select Committee on Assassinations. (*New York Times*, Dec. 30, 1978). How should this probability statement be interpreted? Explain.

6.8. PROBABILITIES AND ODDS

In many situations that exhibit a random quality, the terms "odds in favor of an event" and "odds against its occurrence" are used. If the probability of an event E is p, where p is unequal to 0 or 1, then the **odds in favor of E** are p to $1 - p$; the **odds against E** are $1 - p$ to p. For example, if the probability that a sum of 7 shows on the toss of a pair of well-balanced dice is $\frac{1}{6}$, the odds in favor of a sum of 7 showing are $\frac{1}{6}$ to $\frac{5}{6}$ or 1 to 5; the odds against a sum of 7 showing are 5 to 1. A translation of this observation into monetary terms is that it would be considered fair for an individual to bet 1¢ against 5¢ (or $1 against $5) that a sum of 7 will show on a toss of a pair of well-balanced dice.

We also have the following relationship: if it is considered equitable to bet x dollars against y dollars that a given event E will occur, then E is being assigned the probability $x/(x + y)$. Thus, for example, if odds of 7 to 3 are given that our favorite team will win the pennant this year, this is equivalent to saying that the probability the team will win the pennant this year is $7/(7 + 3) = 0.70$.

EXERCISES

1. If the odds are 7 to 2 that the home team will win the pennant this year, what is the envisioned probability that this will be the case?

2. An investor feels that the odds are 13 to 5 against the price of his stock going up this week. What is the envisioned probability that this will be the case?

3. If it is asserted that the probability of rain today is 0.90, what are the odds in favor of this event?

4. Find the odds that correspond to the following probability values: **(a)** 0.82, **(b)** 0.95, **(c)** 0.98, **(d)** 0.99, **(e)** 0.10.

5. Archie Goodwin, confidential assistant to the grand master of detection, Nero Wolfe, likes to state opinions in terms of odds. In *Plot it Yourself*, by Rex Stout, Archie observes that it was 20 to 1, or maybe 30 to 1, that Kenneth Rennert did not write the stories being considered (p. 38). What probabilities correspond to these odds?

CHAPTER 7

Equally Likely Outcome Models

As noted in Section 6.3, in the special case of an equally likely outcome probability model, wherein all sample points are assigned the same probability value, probability questions reduce to counting questions. If k is the number of sample points describing an event A and n is the total number of sample points, then the probability of A is the ratio k/n (see Theorem 2, Section 6.3). The counting process may range from trivial to extraordinarily complex, and to facilitate the task of counting we look at two basic principles.

7.1. TOOLS FOR COUNTING

Multiplication Principle. Let us suppose that a task is to be performed and that it can be viewed as a sequence of two procedures where the first can be performed in h ways and, after it has been performed, the second can be performed in k ways; then the two procedures can be performed in the stated order in $h \cdot k$ ways.

Since any one of the h ways in which the first procedure can be performed can be coupled with any of the k ways in which the second procedure can be performed, there are h groups of k possibilities, which gives us $h \cdot k$ outcomes.

More generally, the multiplication principle extends to a sequence of any number of procedures which are to be performed in order.

The challenge to applying the multiplication principle is in seeing a situation from the point of view of a sequence of procedures to be performed in order. Sometimes it is obvious that this is the case, but often this view is more deeply hidden.

EXAMPLE 1

Two books are to be chosen from three, denoted by A, B and C, and arranged next to each other on a bookshelf. **(a)** How many arrangements are possible? **(b)** How many arrangements are possible if second place must be filled by book B?

(a) Two positions are to be filled, which we think of as first place and second place. For first place we can choose any of the three available books and, after this has been done, for second place we can choose any of the two remaining books. This yields $3 \cdot 2 = 6$ possible arrangements, namely, AB, AC, BA, BC, CA and CB.

$$\underbrace{3}_{\text{1st place}} \cdot \underbrace{2}_{\text{2nd place}} = 6$$

(b) If a certain task is to be handled is a special way, it is best to turn to it first. Second place can be filled in one way since it must be filled with B. Turning to first place, we have two options, fill it with A or C. This yields $2 \cdot 1 = 2$ arrangements, AB and CB.

$$\underbrace{2}_{\text{1st place}} \cdot \underbrace{1}_{\text{2nd place}} = 2$$

In both cases the result is obvious, but the approach underlying the analysis is instructive.

EXAMPLE 2

A license plate is to consist of two capital letters followed by three digits. In how many ways can we construct such a license plate?

We can look at this problem from the point of view of five spaces to be filled in order. In first place we may put any of the 26 capital letters, in second place we may put any of the 26 capital letters, in third place we may put any of the 10 digits, as is the case for fourth and fifth places. Thus, there are $26 \cdot 26 \cdot 10 \cdot 10 \cdot 10 = 676{,}000$ ways in which we can construct such a license plate.

$$\underbrace{26}_{\text{1st place}} \cdot \underbrace{26}_{\text{2nd place}} \cdot \underbrace{10}_{\text{3rd place}} \cdot \underbrace{10}_{\text{4th place}} \cdot \underbrace{10}_{\text{5th place}} = 676{,}000$$

EXAMPLE 3

Alice woke up late and rushed to school without having breakfast. Before going to her second class she decided to make a stop in the school cafeteria for a snack consisting of a sandwich and beverage. She found 5 different kinds of sandwiches and 6 different kinds of beverages to choose from. In how many ways can she put together a snack?

We can think of this in terms of two procedures to be performed, choose a sandwich and then choose a beverage, which can be done in $5 \cdot 6 = 30$ ways. Equivalently, choose a beverage and then chose a sandwich, which yields the same result.

> **Addition Principle.** Suppose that a task can be performed in h ways and that a second task can be performed in k ways, where the performance of one of these tasks excludes the performance of the other; then one or the other of these tasks (but not both) can be performed in $h + k$ ways.
>
> More generally, this addition principle extends to a setting involving any number of tasks, where the performance of any one of them excludes the performance of any of the others.

EXAMPLE 4

As the semester moved along Alice found herself more and more frequently getting up late. One day she went for a snack and found, to her consternation, that she had forgotten her wallet and did not have enough loose change to purchase both a sandwich and a beverage. In how many ways can she select a snack if, as previously, there are 5 different kinds of sandwiches and 6 different kinds of beverages?

Because of financial difficulties, Alice has $5 + 6 = 11$ options. She may choose one of the 5 sandwiches or one of the 6 beverages.

EXERCISES

1. In how many ways can 2 of 5 books be chosen and arranged next to each other on a shelf?

2. An encyclopedia of science consists of 7 volumes. **(a)** In how many ways can these volumes be arranged next to each other on a shelf? **(b)** How many of these arrangements are out of order? **(c)** In how many of these arrangements will volume 1 occupy first place and volume 2 occupy second place?

3. A traveler is planning to go from New York to Chicago by plane and make the return trip by bus. There are 5 airlines that have flights at the desired time and 3 bus lines that provide Chicago to New York service. In how many ways can the trip be made?

4. Motors are to pass through two inspection stations. At the first station 2 ratings are possible; at the second station 4 ratings are possible. In how many ways can a motor be marked?

5. Air-conditioners are to be assembled in four stages. At the first stage there are 3 assembly lines, at the second stage 4 assembly lines, at the third stage 4 assembly lines, and at the fourth stage 3 assembly lines. In how many ways can an air-conditioner be routed through the assembly process?

6. In how many ways can 5 people line up at a ticket office?

7. How many 4-digit numbers are there? Note, 0 cannot occupy first place.

8. How many numbers are there between 1000 and 5000, including 1000 and 5000?

9. An examination consists of 8 true-false questions. How many possible different answer sheets can be turned in?

10. Andrius has 7 books, 4 in English and 3 in Lithuanian, which he wants to arrange on a bookshelf. How many arrangements are possible if (a) there are no restrictions; (b) a book in English is to occupy first place and a book in Lithuanian is to occupy second place; (c) the books in English are to occupy the first four places; (d) books in the same language are to be kept together?

11. In how many ways can n people line up at a ticket office?

12. Robert Adams went shopping one day to buy a shirt and pair of shoes. He found 8 shirts and 6 shoe styles to choose from. If he did not have enough funds to pay for both a shirt and shoes, in how many ways could he make a purchase?

13. A signal is a 3-digit sequence of 0's and 1's or a 4-digit sequence of 0's and 1's. How many signals are there?

14. In how many ways can four girls and four boys be seated alternately in a row of eight chairs numbered from 1 to 8 if (a) a boy is to occupy the first chair; (b) either a boy or a girl can occupy the first chair?

15. A telephone dial has 10 holes. How many different signals, each consisting of seven impulses in succession, can be formed (a) if no impulse is to be repeated in any given signal; (b) if repetitions are permitted?

16. In how many ways can a baseball team of 9 players be arranged in batting orders (a) if tradition is followed and the pitcher must bat last; (b) if no restriction is imposed; (c) if a certain 4 players must occupy the first 4 positions in some order?

Permutations

A **permutation** of a set of objects is an arrangement of these objects in some order in a line. If there are n distinct objects, then any arrangement of r of them in some order in a line is called a **permutation of the n distinct objects taken r at a time**. The number of permutations of n distinct objects taken r at a time is denoted by $P(n,r)$ or $_nP_r$.

EXAMPLE 5

Returning to Example 1, which involved choosing 2 books from 3, denoted by A, B, and C, and arranging them next to each other on a shelf, there are $P(3, 2) = 3 \cdot 2 = 6$ permutations of the 3 books taken 2 at a time. The permutations are AB, AC, BA, BC, CA and CB.

> As a computation tool for $P(n, r)$ we have
> $$P(n, r) = n(n - 1) \cdots (n - r + 1),$$

which is the product of the first r integers in descending order starting with n. To establish this result, we note that there are r places to be filled and n objects from which to choose. First place can be filled with any of these n objects. After it has been filled with one of these n objects, there remain $n - 1$ objects available for filling the second place. Once it has been filled, 2 objects will have been used and $n - 2$ choices remain for the third place; once first, second, and third places have been filled, 3 objects will have been used, and $n - 3$ choices remain for the fourth place. More generally, when we come to the rth place, $r - 1$ objects will have been used to fill the previous $r - 1$ places, and $n - (r - 1)$ or $n - r + 1$ choices remain for rth place. From the multiplication principle, the number of ways of filling the r places is $n(n - 1) \cdots (n - r + 1)$.

For example, $P(7, 3)$ is the product of the first 3 integers in decending order beginning with 7, $P(7, 3) = 7 \cdot 6 \cdot 5 = 210$. Also, $P(10, 4) = 10 \cdot 9 \cdot 8 \cdot 7 = 5040$, and $P(48, 2) = 48 \cdot 47 = 2256$.

Since products of consecutive integers arise frequently in counting problems, it is useful to have notation to denote such products. The symbol $n!$ (read "n factorial") is used to stand for the product of all integers from 1 to n inclusive. Thus

$$1! = 1 \qquad\qquad 4! = 4 \cdot 3 \cdot 2 \cdot 1 = 24$$
$$2! = 2 \cdot 1 = 2 \qquad\qquad 5! = 5 \cdot 4 \cdot 3 \cdot 2 \cdot 1 = 120$$
$$3! = 3 \cdot 2 \cdot 1 = 6 \qquad\qquad 6! = 6 \cdot 5 \cdot 4 \cdot 3 \cdot 2 \cdot 1 = 720$$
$$n! = n(n - 1)(n - 2) \cdots 1$$

It is convenient to define $0!$ by

$$0! = 1$$

This definition, although perhaps strange at first sight, is useful in that certain counting formulas can be more easily stated without a need for considering separate special cases.

When $r = n$, we have the following special case: The number of permutations of n distinct objects (taken n at a time) in a line is:

$$P(n, n) = n! = n(n - 1) \cdots 1$$

The computation formulas derived for $P(n, r)$ and $P(n, n)$ follow from the multiplication principle. Many, but not all, counting problems exhibit a structure which permit us to apply these results directly. In other cases which exhibit this structure it is preferable to go back to basic principles to work the problems.

Combinations

Many situations arise in which r distinct objects are to be selected from n without regard to order. A subset or selection of r objects chosen from n distinct objects, without regard to the order in which they were chosen or appear, is called a **combination of n objects taken r at a time**. The number of combinations of n distinct objects taken r at a time is denoted by $C(n, r)$ or $_nC_r$.

EXAMPLE 6

Returning to Examples 1 and 5 which involved choosing 2 books from 3, denoted by A, B, and C, we saw that there are 6 permutations of the 3 books taken 2 at a time, namely AB, AC, BA, BC, CA and CB.

There are 3 combinations of these 3 books taken 2 at a time, namely, $\{A, B\}$, $\{A, C\}$ and $\{B, C\}$. The combination $\{A, B\}$ is the set consisting of A and B, no order implied. This combination gives rise to two permutations, AB, which means books A and B in the order A followed by B, and BA, which means books A and B in the order B followed by A.

> The number of combinations of r objects chosen from n distinct objects is expressed by the following formula:
> $$C(n, r) = \frac{P(n, r)}{r!} = \frac{n(n-1)\cdots(n-r+1)}{r!}$$

To establish this result, consider the related problem of determining $P(n, r)$, and think of the process of forming a permutation of r objects selected from n as being carried out in two stages. The first stage consists of selecting r of n objects without regard to order, which can be done in $C(n, r)$ ways. The second stage consists of arranging the r objects chosen in some order, which can be done in $r!$ ways. From the multiplication principle, the number of ways of selecting r of n objects with regard to order, $P(n, r)$, equals the number of ways of selecting r of n objects without regard to order, $C(n, r)$, times the number of ways of ordering the r objects chosen, $r!$. That is:

$$P(n, r) = C(n, r) \cdot r!,$$

Solving for $C(n, r)$ by dividing both sides by $r!$ yields the desired result.

Thus, for example, the number of combinations of 10 objects taken 3 at a time, is

$$C(10, 3) = \frac{P(10, 3)}{3!} = \frac{10 \cdot 9 \cdot 8}{3 \cdot 2 \cdot 1} = 120.$$

There are 120 ways of choosing 3 of 10 distinct objects without regard to order.

In how many ways can no objects be chosen from n objects? The answer, of course, is 1; just don't choose any, that's the one way. This leads us to define $P(n, 0)$ and $C(n,0)$ as follows:

$$P(n, 0) = C(n, 0) = 1$$

A frequently asked question is, when is order important and when is it not important? When is a situation a permutation situation and when is it a combination situation? We have formulas for computing the number of permutations and combinations inherent in a situation, but there is no formula for deciding which is to used. This is a matter of judgment which requires a careful reading and analysis of the situation. It is where we are most on our own.

EXAMPLE 7

How many lines are determined by 12 points, no three of which lie on the same line?

The problem reduces to determining the number of ways of choosing 2 of 12 points without regard to order, which is $C(12, 2) = 66$. A line is determined by two points, irrespective of order. (The line determined by points P and Q is the same as the one determined by Q and P.) The condition that no three points lie on the same line is essential to avoiding line duplication. If P, Q and R lie on the same line, P and Q, P and R, and Q and R would determine the same line and 66 would overcount the number of lines determined by the 12 points.

EXAMPLE 8

The student Math Society at Ecap University has 25 members. An election is to be held to elect a president, secretary and treasurer from its membership. In how many ways can an election slate be formed if no person may hold more than one office?

The problem reduces to determining the number of ways of choosing 3 distinct club members (no repetitions since no one may hold more than one office) with regard to order, which is $P(25, 3) = 25 \cdot 24 \cdot 23 = 13{,}800$. The order feature is determined by the offices to be filled, which must be distinguished.

EXAMPLE 9

A poker hand of 5 cards is to be dealt from a standard deck of 52 cards. How many such hands contain **(a)** 1 king, **(b)** 2 kings, **(c)** 2 kings and 2 queens?

If our interest in a hand is in terms of its composition and not in the order in which the cards are dealt or arranged (which would be the case in closed poker where no cards in a hand are showing on the table), then it is appropriate to view two problems from a combination rather than permutation point of view.

(a) We can view this problem from the point of view of two procedures that are to be performed: choose 1 of the 4 kings, which can be done in $C(4, 1) = 4$ ways, and then choose 4 other (non-kings) from the remaining 48 non-kings to make up the hand of 5, which can be done in $C(48, 4) = 194{,}580$ ways.

The number of ways of putting together a hand of 5 cards with 1 king is, by the multiplication principle, the product of the number of ways of carrying out both of these procedures, $C(4, 1) \cdot C(48, 4) = 4(194{,}580) = 778{,}320$.

The analysis is summarized in Figure 7.1.

Hand of 5 cards

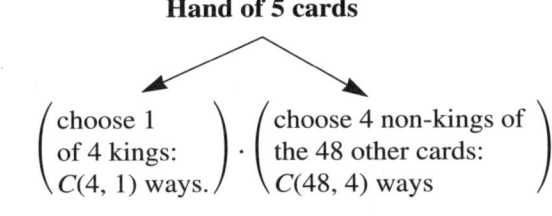

$C(4, 1) \cdot C(48, 4) = 778{,}320$ hands of 5 with 1 king

Figure 7.1

Why do we multiply the components and not add them? The multiplication principle is the appropriate tool because we are carrying out both procedures; we are choosing 1 of 4 kings and we are also choosing 4 of 48 non-kings. It's not a situation where performing one of these procedures excludes us from performing the other. If it were this kind of situation, then the addition principle would be the appropriate tool.

(b) The basic framework developed in (a) applies to (b) as well. The analysis is summarized in Figure 7.2.

Hand of 5 cards

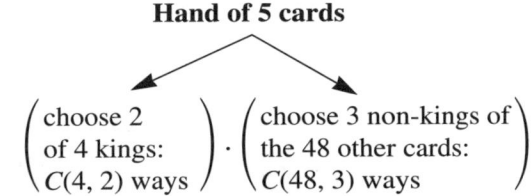

$C(4, 2) \cdot C(48, 3) = 103{,}776$ hands of 5 with 2 kings

Figure 7.2

(c) The analysis, which involves a three way splitting of the hand, is summarized in Figure 7.3.

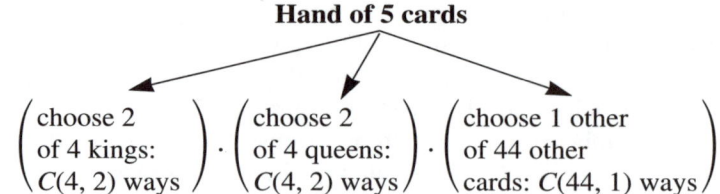

$C(4, 2) \cdot C(4, 2) \cdot C(44, 1) = 1584$ hands of 5 with 2 kings and 2 queens

Figure 7.3

EXAMPLE 10

Two percent of a lot of 100 items are known to be defective. In how many ways can (a) a sample of 3 items be drawn from the lot; (b) a sample of 3 items, all of which are good, be drawn from the lot; (c) a sample of 3, one of which is defective, be drawn from the lot?

(a) Assuming that our interest in the sample is its composition from the point of view of good versus defective items, and not in any order in which the items may appear, the problem reduces to determining the number of ways of choosing 3 of 100 items without regard to order. This is given by $C(100, 3) = 161,700$.

(b) If all items drawn are to be good, they must be drawn from the 98 good ones. There are $C(98, 3) = 152,096$ ways to do this.

(c) To form this sample consider two procedures. The first is to choose 1 of the 2 defectives, which can be done in $C(2, 1) = 2$ ways. The second is to choose the 2 other items needed to make up the sample of 3 from the 98 good ones, which can be done in $C(98, 2) = 4,753$ ways. By the multiplication principle, the number of samples of 3 items containing 1 defective that can be drawn from the lot is $C(2, 1) \cdot C(98, 2) = 9,506$.

EXERCISES

17. Evaluate $P(7, 3)$, $P(12, 4)$, $P(6, 1)$, $C(16, 3)$, $C(18, 2)$ and $C(52, 4)$.

18. A club consisting of 60 members meets to elect 4 officers, president, vice president, secretary and treasurer, from its membership. In how many ways can this slate be formed if no person may occupy more than one position?

19. A guest house with 12 single rooms receives 6 single reservations. In how many ways can these reservations be filled?

20. The committee on sabbaticals at Ecap University has authority to grant 5 sabbaticals for any given year. In how many ways can the sabbaticals be granted if 15 requests are received?

21. How many choices of 3 suits and 4 ties for a trip can be made from a wardrobe of 5 suits and 6 ties?

22. There are 4 vacancies on the state Court of Appeals. In how many ways can these vacancies be filled if 20 names have been placed in nomination?

23. How many permutations of the letters of the English alphabet are there?

24. The Alumni Association of Ecap University has organized a one mile race to be run by 2 faculty, W. J. Adams and H. Lurier, and 3 alumni of Ecap University, J. Ross, M. Tilson and E. Kapp. **(a)** How many possible finishes are there? **(b)** In how many finishes does Adams finish first? **(c)** In how many finishes do alumni finish in the first three places?

25. From a lot of 50 color television sets, a sample of 3 is selected for inspection. There are 4 defective sets in the lot. **(a)** How many samples of 3 of 50 sets are there? How many of these samples contain **(b)** no defective sets; **(c)** 1 defective set; **(d)** 2 defective sets; **(e)** 3 defective sets?

26. There are 12 faculty in the mathematics department and 10 faculty in the economics department of Ecap University. A joint committee of 5 faculty is to be set up to study curriculum questions of interest to both departments. In how many ways can such a committee be formed if the committee **(a)** is to contain 2 members of the mathematics department; **(b)** is to contain the chairperson of both departments; **(c)** is to contain the chairperson of both departments, but is not to contain Professor Adelson of the mathematics department?

7.2. RETURN TO EQUALLY LIKELY OUTCOMES

Five Probability Problems

EXAMPLE 1. THE ONE MILE RACE

The setting of this problem is provided by Exercise 24 of the preceding section. The Alumni Association of Ecap University has organized a one mile race to be run by 2 faculty, W.J. Adams and H. Lurier, and 3 alumni, J. Ross, M. Tilson and E. Kapp.
What is the probability that Adams finishes first?

One approach to this problem is to note that there $5! = 120$ possible finishes, that Adams is first in $4! = 24$ of them, and conclude that the probability that Adams finishes first is $4!$ divided by $5!$, or $1/5$.

This approach, which is based solely on counting, leaves much to be desired. It is based on an underlying assumption of equally likely outcomes, but which outcomes are assumed to be equally likely is not made clear and is left to the imagination of the reader. The absence of an explicitly stated probability model and assumption on which the probability function of the model is based obscures the necessity for a critical examination of the realism of the assumption made and suggests the mistaken view that there is only one probabilistic conclusion possible which is an unassailable truth. This kind of approach, which is far too commonly seen in applications of probability, should be accompanied by a skull and cross-bones to warn the reader that his perspective and understanding are in danger of being poisoned.

To analyze the question posed, we will have to back up and provide a probability model for the process along with a justification for the model's probability function, which is open to scrutiny.

As to notation, let (ALRTK), to take an example, denote the outcome indicated by the order, Adams (1st), Lurier (2nd), Ross (3rd), Tilson (4th), Kapp (5th). We take as our sample space S the outcomes expressed by all permutations of A, L, R, T and K. There are $5! = 120$ sample points in S.

If all five runners are in comparable physical condition, age and running experience, then this would make reasonable an assignment of equal probabilities of 1/120 to the 120 sample points in S, from which it would follow as a valid conclusion that the probability Adams finishes first is 1/5.

Some observers have argued, however, that the five runners are not in comparable physical condition, that Adams tires quickly when the temperature is over 75°F, that the weather forecast is for an 80°F day when the race is to be run, and that it is therefore unrealistic to assign the same probability value to all finishes.

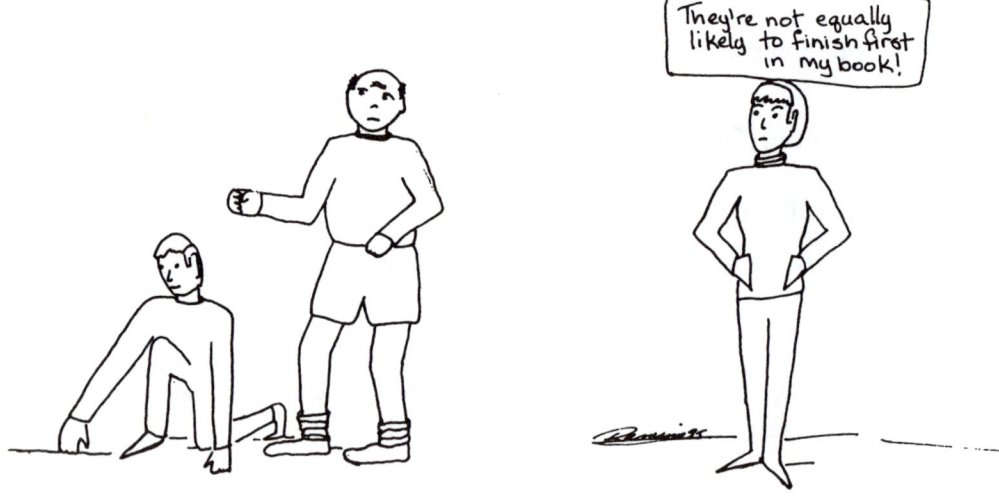

EXAMPLE 2. THE POKER HAND

We return to the setting provided by Example 9 of the preceding section. A poker hand of 5 cards is to be dealt from a standard deck of 52 cards. Find the probability that such a hand contains **(a)** 1 king, **(b)** 2 kings, **(c)** 2 kings and 2 queens, **(d)** 2 kings or 2 queens.

These questions are more complex than those posed in Example 9 in that there we were counting; no probability questions and therefore structure is involved. To address these probability questions we must introduce a probability model, which means that we need a sample space and probability function.

The question of a sample space is addressed in Exercise 7 of Section 6.2 (p. 203). While many sample spaces can be formulated for this process, not every one is suitable for our undertaking. The sample space must be rich enough to describe the events we seek to study. We must also look ahead to the probability function P that we intend to formulate and the assumptions it is to reflect.

Looking ahead to reflecting by P an unbiased dealing of the hand and keeping in mind that our interest in a hand is in terms of its composition as opposed to the order in which the cards are dealt or arranged, we take as our sample space S the collection of all unordered hands of 5 cards that can be dealt from 52 cards.

$$S = \{(C_1,C_2,C_3,C_4,C_5), (C_1,C_2,C_3,C_4,C_6), \cdots, (C_{48},C_{49},C_{50},C_{51},C_{52})\}$$

The sample point (C_1,C_2,C_3,C_4,C_5), for example, is the event that the hand dealt consists of cards C_1, C_2, C_3, C_4, and C_5, dealt in some order and arranged as one desires. The number of sample points in S is equal to the number of ways of choosing 5 of 52 cards without regard to order, $C(52, 5) = 2{,}598{,}960$.

The assumption that the hand will be dealt in an unbiased manner (from a well shuffled deck) is best reflected by the probability function P which assigns the same value, $1/C(52, 5)$, to all $C(52, 5)$ sample points in S.

We have:

$$P(C_1,C_2,C_3,C_4,C_5) = \cdots = P(C_{48},C_{49},C_{50},C_{51},C_{52}) = \frac{1}{C(52, 5)}$$

With our probability model in place, we are ready to address the questions posed. Since this is an equally likely outcome model, probability questions reduce to counting questions.

(a) From Example 9, part (a), of Section 7.1 (p. 236) we have:

$$P(1 \text{ king}) = \frac{\text{Nu. of hands of 5 with 1 king}}{\text{Nu. of sample points}}$$

$$= \frac{C(4, 1) \cdot C(48, 4)}{C(52, 5)}$$

$$= 0.30$$

(b) From Example 9, part (b), of Section 7.1 we have:

$$P(2 \text{ kings}) = \frac{C(4, 2) \cdot C(48, 3)}{C(52, 5)})$$

$$= 0.040$$

(c) From Example 9, part (c), of Section 7.1 we have:

$$P(2 \text{ kings and 2 queens}) = \frac{C(4, 2) \cdot C(4, 2) \cdot C(44, 1)}{C(52, 5)}$$

$$= 0.00061$$

(d) The "or" in the description of the event 2 kings or 2 queens are drawn gives this problem another dimension. To handle this dimension we must employ Theorem 3 discussed in Section 6.6 (p. 217); for two events A and B,

$$P(A \cup B) = P(A) + P(B) - P(A \cap B),$$

or in colloquial language:

$$P(A \text{ or } B) = P(A) + P(B) - P(A \text{ and } B)$$

Thus:

$$P(2 \text{ kings or 2 queens}) = P(2 \text{ kings}) + P(2 \text{ queens})$$
$$- P(2 \text{ kings and 2 queens})$$

$$= \frac{C(4, 2) \cdot C(48, 3)}{C(52, 5)} + \frac{C(4, 2) \cdot C(48, 3)}{C(52, 5)} - \frac{C(4, 2) \cdot C(4, 2) \cdot C(44, 1)}{C(52, 5)}$$

$$= 0.040 + 0.040 + 0.00061$$

$$= 0.0794$$

EXAMPLE 3. DECISION MAKING BASED ON RANDOM SAMPLING

A consulting firm is engaged in studying various decision-making criteria in connection with accepting a shipment of items based on the results of a sample that is randomly chosen from the shipment. One part of this study envisions a shipment of 120 items with 5 percent defectives along with an inspection procedure that calls for drawing 3 items at random from the shipment and determining whether they are defective or good. The decision criteria calls for accepting the shipment if the sample drawn contains no defectives, and rejecting the shipment if the sample drawn contains one or more defectives. The problem is to find the probability that such a shipment is accepted, find the probability that the shipment is rejected, and interpret the results obtained.

Our first task is to set up a sample space for the sampling procedure and define a probability function on the subsets of that sample space that reflects the nature of random sampling. For notational convenience, let us think of the 120 items as tagged. Let S denote the set of events expressed by all combinations of 3 items that can be chosen from 120 items. That is,

$$S = \{(I_1, I_2, I_3), (I_1, I_2, I_4), \ldots, (I_{118}, I_{119}, I_{120})\}$$

where the sample point (I_1, I_2, I_3) is the event that the sample chosen contains items I_1, I_2, and I_3, and so on. The number of sample points in S is equal to the number of ways of choosing 3 of 120 items without regard to order, which is $C(120, 3) = 280{,}840$. The envisioned randomness of the selection leads us to take as our probability function P the one that assigns the same value, $1/C(120, 3)$ or $1/280{,}840$, to all sample points in S. That is,

$$P(I_1, I_2, I_3) = \cdots = P(I_{118}, I_{119}, I_{120}) = \frac{1}{C(120, 3)} = \frac{1}{280{,}840}$$

Since the shipment is accepted only if the sample drawn contains no defectives, we have

$$P(\text{shipment accepted}) = \frac{\text{number of samples with no defectives}}{C(120,3)}$$

Since 5 percent of the 120 items are estimated to be defective, 6 items are defective and 114 are good. The number of samples with no defectives equals the number of ways of choosing 3 of the 114 good items, which is $C(114, 3) = 240{,}464$. Thus

$$P(\text{shipment accepted}) = \frac{C(114, 3)}{C(120, 3)} = \frac{240{,}464}{280{,}840} = 0.856$$

Since "shipment accepted" and "shipment rejected" are complementary events, we have

$$P(\text{shipment rejected}) = 1 - P(\text{shipment accepted})$$
$$= 1 - 0.856$$
$$= 0.144$$

If the sampling procedure is repeated a large number of times under the envisioned conditions, we can expect the shipment to be accepted approximately 85.6 percent of the time.

EXAMPLE 4. THE SAMPLING DISTRIBUTION OF THE MEAN

Consider a population Q of numerical values arising from some concern of interest (the lifetimes of steel-belted tires produced by a manufacturer, for example, or the grades on the last Certified Public Accounting exam, to take another example). Q has a mean value, usually denoted in statistics by the Greek letter μ (read moo). If Q is small or of modest size, $Q = \{2, 4, 6, 8\}$, for example, the problem of finding μ can be handled in a simple, direct manner. For $Q = \{2, 4, 6, 8\}$,

$$\mu = \frac{2+4+6+8}{4} = 5.$$

If Q is large or indeterminate in the sense that we can talk about its values but we are not quite sure what they all are, then we seek to get a grip on μ by taking an unbiased sample, computing its mean—denoted by $\bar{x}$, and using $\bar{x}$ as an estimate for μ.

Different samples giving rise to a variety of sample means may arise. If the sample mean $\bar{x}$ is to be used as an estimate for μ and $\bar{x}$ differs according to the particular sample drawn, then clearly we need a grip on how $\bar{x}$ may vary. This grip is provided by a function, called the **sampling distribution of the mean**, which describes in probability terms how $\bar{x}$ may vary.

As an illustration we return to the population $Q = \{2, 4, 6, 8\}$ (with $\mu = 5$) and consider the process of drawing a sample of size 2 from Q. Our first problem is to formulate a probability model for this process.

Since our interest is in the sample mean, which only depends on the identity of the numbers to be averaged (and not their order) we take as our sample space S the collection of all **unordered samples** of 2 that can be drawn from Q.

$$S = \{(2, 4), (2, 6), (2, 8), (4, 6), (4, 8), (6, 8)\}$$

How is the sample to be drawn from S? Our basic assumption, here and in all such statistical sampling situations, is that the sample is drawn at random from the underlying population. In the case at hand this leads to P defined by:

$$P(2, 4) = P(2, 6) = \cdots = P(6, 8) = \frac{1}{C(4, 2)} = \frac{1}{6}$$

The following is a list of the sample means arising from the samples in S.

$\bar{x}(2, 4) = 3$ $\qquad\qquad$ $\bar{x}(4, 6) = 5$
$\bar{x}(2, 6) = 4$ $\qquad\qquad$ $\bar{x}(4, 8) = 6$
$\bar{x}(2, 8) = 5$ $\qquad\qquad$ $\bar{x}(6, 8) = 7$

The function $p(x)$ defined by

$$p(x) = P(\bar{x} = x), \text{ where } x = 3, 4, \cdots, 7$$

is called the **sampling distribution of the sample mean** $\bar{x}$. $p(x)$ describes the probability behavior of $\bar{x}$. In this case we have:

$$p(3) = P(\bar{x} = 3) = P(2, 4) = 1/6$$
$$p(4) = 1/6 \qquad\qquad p(6) = 1/6$$
$$p(5) = 2/6 \qquad\qquad p(7) = 1/6$$

Thus, the probability that $\bar{x}$ gives us μ directly on target (5) is 1/3; the probability that $\bar{x}$ is within 1 unit of μ ($4 \leq \bar{x} \leq 6$) is $p(4) + p(5) + p(6) = 2/3$.

For further discussion of the sampling distribution of the mean see, for example, W. J. Adams, I. Kabus, M. P. Preiss, *Statistics: Basic Principles with Applications* (Dubuque: Kendall/Hunt Pub. Co., 1994), Ch. 9.

EXAMPLE 5. POPULATION ESTIMATION PROBLEMS

How many fish are in your favorite lake? How many raccoons are in your neighborhood? How many animals of your favorite kind are in the game reserve or national park? More generally, how many "whatever" are in your region of interest?

One approach to problems of this sort is based on what is called the **capture-release-recapture** method. We illustrate it by considering a fish population estimation problem, but the approach is applicable to the other situations noted as well.

We begin by catching a certain number of fish from the lake—100, say. These fish are tagged so as to be identifiable if caught again and are thrown back into the lake. We wait for a reasonable time to elapse to allow the fish to disperse (maybe a few days) and then catch another batch of fish, 200, say, and make note of how many in this batch were caught before. Let us suppose that one fish was twice caught.

Let N denote the number of fish in the lake. Our problem is to estimate N. To do this we set up a probability model for the experiment of catching the second batch of 200 fish. As our sample space we take the events represented by the collection of all batches (combinations) of 200 fish that can be selected from N. There are $C(N, 200)$ such batches. Based on the assumption that all fish in the lake have the same likelihood of being caught, we take as our probability function P the one that assigns the same value, $1/C(N, 200)$, to each sample point. We next determine the probability of catching 1 marked fish. This probability is equal to

$$\frac{\text{number of ways of catching 1 marked fish}}{C(N, 200)}$$

The number of ways of catching 1 marked fish in a batch of 200 fish is equal to the number of ways of catching 1 of 100 marked fish, $C(100, 1)$, times the number of ways of catching 199 of $N - 100$ unmarked fish, $C(N - 100, 199)$, which yields the product $C(100, 1) \cdot C(N - 100, 199)$. Thus

$$P(\text{1 marked fish is caught}) = \frac{C(100, 1) \cdot C(N - 100, 199)}{C(N, 200)} \qquad (7.1)$$

246 Finite Mathematics, Models, and Structure

The right side of (7.1) depends on N. It varies as different numbers are substituted for N. Of special interest is that number which when substituted for N makes (7.1) assume its maximum value. This value is called the maximum likelihood estimate of N; that is, the **maximum likelihood estimate of N** is that number which maximizes the probability of catching the number of marked fish that were actually caught in the second batch.

We shall now show that the maximum likelihood estimate of N is 20,000. The right side of (7.1) is a function of N, which we shall denote by $P(N)$.

$$P(N) = \frac{C(100, 1) \cdot C(N - 100, 199)}{C(N, 200)} \tag{7.2}$$

We seek a positive integer value of N such that

$$P(N - 1) \leq P(N) \quad \text{and} \quad P(N) \geq P(N + 1) \tag{7.3}$$

or, equivalently,

$$\frac{P(N - 1)}{P(N)} \leq 1 \quad \text{and} \quad \frac{P(N)}{P(N + 1)} \geq 1 \tag{7.3}$$

Our first task is to determine and simplify $P(N - 1)$, $P(N)$, and $P(N + 1)$. From (7.2) we obtain

$$P(N) = \frac{\dfrac{100(N - 100)\ldots(N - 298)}{199 \cdot 198 \ldots 1}}{\dfrac{N(N - 1)\ldots(N - 199)}{200 \cdot 199 \ldots 1}}$$

$$= \frac{100(N - 100)\ldots(N - 298)}{199 \cdot 198 \ldots 1} \cdot \frac{200 \cdot 199 \ldots 1}{N(N - 1)\ldots(N - 199)}$$

$$= \frac{100(200)(N - 100)\ldots(N - 298)}{N(N - 1)\ldots(N - 199)} \tag{7.4}$$

For $P(N - 1)$ we have

$$P(N - 1) = \frac{C(100, 1) \cdot C(N - 1 - 100, 199)}{C(N - 1, 200)}$$

$$= \frac{\dfrac{100(N - 101)\ldots(N - 299)}{199 \cdot 198 \ldots 1}}{\dfrac{(N - 1)\ldots(N - 200)}{200 \cdot 199 \ldots 1}}$$

Inverting and simplifying yields

$$P(N - 1) = \frac{100(200)(N - 101)\ldots(N - 299)}{(N - 1)\ldots(N - 200)} \tag{7.5}$$

For $P(N + 1)$ we have

$$P(N + 1) = \frac{C(100, 1) \cdot C(N + 1 - 100, 199)}{C(N + 1, 200)}$$

$$= \frac{\dfrac{100(N - 99)\ldots(N - 297)}{199 \cdot 198 \ldots 1}}{\dfrac{(N + 1)\ldots(N - 198)}{200 \cdot 199 \ldots 1}}$$

Inverting and simplifying yields

$$P(N + 1) = \frac{100(200)(N - 99)\ldots(N - 297)}{(N + 1)\ldots N - 198)} \tag{7.6}$$

From (7.4) and (7.5) we have

$$\frac{P(N - 1)}{P(N)} = \frac{\dfrac{100(200)(N - 101)\ldots(N - 299)}{(N - 1)\ldots(N - 200)}}{\dfrac{100(200)(N - 100)\ldots(N - 298)}{(N - 1)\ldots(N - 199)}}$$

Inverting and canceling like terms yields

$$\frac{P(N - 1)}{P(N)} = \frac{100(200)(N - 101)\ldots(N - 299)}{(N - 1)\ldots(N - 200)} \cdot \frac{N(N - 1)\ldots(N - 199)}{100(200)(N - 100)\ldots(N - 298)}$$

$$\frac{P(N - 1)}{P(N)} = \frac{(N - 299)N}{(N - 200)(N - 100)} \tag{7.7}$$

From (7.4) and (7.6) we have

$$\frac{P(N)}{P(N + 1)} = \frac{\dfrac{100(200)(N - 100)\ldots(N - 298)}{N(N - 1)\ldots(N - 199)}}{\dfrac{100(200)(N - 99)\ldots(N - 297)}{(N + 1)\ldots(N - 198)}}$$

Inverting and canceling like terms yields

$$\frac{P(N)}{P(N + 1)} = \frac{100(200)(N - 100)\ldots(N - 298)}{N(N - 1)\ldots(N - 199)} \cdot \frac{(N + 1)\ldots(N - 198)}{100(200)(N - 99)\ldots(N - 297)}$$

$$\frac{P(N)}{P(N + 1)} = \frac{(N - 298)(N + 1)}{(N - 199)(N - 99)} \tag{7.8}$$

From (7.3), (7.7), and (7.8), our problem reduces to finding N such that

$$\frac{(N-299)N}{(N-200)(N-100)} \leq 1 \quad \text{and} \quad \frac{(N-298)(N+1)}{(N-199)(N-99)} \geq 1$$

From the first of these conditions we obtain

$$(N-299)N \leq (N-200)(N-100)$$
$$N^2 - 299N \leq N^2 - 300N + 20{,}000$$
$$N \leq 20{,}000$$

From the second of these conditions we obtain

$$(N-298)(N+1) \geq (N-199)(N-99)$$
$$N^2 - 297N - 298 \geq N^2 - 298N + 19{,}701$$
$$N \geq 19{,}999$$

Thus

$$19{,}999 \leq N \leq 20{,}000$$

We may take 19,999 or 20,000 as our maximum-likelihood estimate of the size of the fish population.

More generally, the maximum-likelihood function that corresponds to catching k fish from the lake, tagging them and throwing them back, and catching a second batch of n fish that is observed to contain r tagged fish is defined by

$$P(N) = \frac{C(k, r) \cdot C(N-k, n-r)}{C(N, n)}$$

In our example $k = 100$, $n = 200$, and $r = 1$. By an analysis similar to the preceding one, it can be shown that the maximum-likelihood estimate of N is characterized by

$$\frac{nk}{r} - 1 \leq N \leq \frac{nk}{r}$$

There have been more black bear sightings around the town of Charlotte and its residents have come to raise questions about the size of this population in the surrounding forest.

At the request of the town's council an applied statistics team at the local college undertook the project of obtaining a maximum likelihood estimate of the bear population. Six bears were caught, tagged, and released. Shortly thereafter 5 bears were caught and it was found that 1 had been previously caught.

In this situation $k = 6$, $n = 5$, and $r = 1$. Thus, the maximum-likelihood estimate of the bear population is:

$$N = \frac{nk}{r} = \frac{5(6)}{1} = 30$$

It is important to keep in mind that this maximum-likelihood estimate is based on the assumption that the second group of five bears caught is a random sample of the bear population. For the estimate of 30 to be viable it is essential that the manner in which the second sample of five bears is caught be a "close approximation" of the randomness required by the theory.

Random Selections in Theory and Practice

Consider a population of size N, $Q = \{x_1, x_2, ..., x_N\}$, and let us suppose that an unordered sample of size n is to be chosen at random from this population. As we have noted, when we say that a sample is to be chosen at random we have in mind the idea that there is to be no bias, deliberate or inadvertent, which favors certain samples of size n being chosen over others. The sampling procedure is to be an equal opportunity procedure; no favoritism.

The probability assignment P which best reflects random sampling is the one which assigns the same value, $1 / C(N, n)$, to each of the $C(N, n)$ unordered samples of size n that arise from Q. While easily envisioned, random selections are not easily achieved in practice, especially when the population being sampled from is large. Yet, it is essential to closely approximate in reality the random selections envisioned in theory if the results deduced from theory are to be applicable to reality. This is particularly urgent in statistics, a subject whose theoretical framework is to a large extent based on the assumption of random sampling.

250 Finite Mathematics, Models, and Structure

EXERCISES

1. In connection with the process of dealing a poker hand of 5 cards discussed in Example 2, find the probability that the hand dealt contains (**a**) 3 kings, (**b**) 3 kings and 1 ace, (**c**) 3 kings or 1 ace, (**d**) 3 kings or 2 aces, (**e**) 2 clubs, (**f**) 2 clubs and 3 spades, (**g**) 2 clubs or 3 spades.

2. In connection with the process of dealing a poker hand of 5 cards discussed in Example 2, consider the following approach to finding the number of ways of dealing a hand that contains at least 1 king. To obtain a hand with at least 1 king, choose 1 of 4 kings, which can be done in $C(4, 1) = 4$ ways, and then choose 4 of the remaining 51 cards, which can be done in $C(51, 4) = 249{,}900$ ways. By the multiplication principle, the number of hands with at least 1 king $C(4,1) \cdot C(51, 4) = 4(249{,}900) = 999{,}600$. Is this approach correct? Explain. How else can be problem be approached?

3. With respect to the probability model developed for the decision-making problem (Example 3) based on random sampling, find the probability that the sample drawn contains (**a**) 1 defective, (**b**) 2 defectives, (**c**) 3 defectives, (**d**) 1 or 2 defectives.

4. On the basis of the probability model developed for the decision-making situation, we can expect samples free of defectives to be drawn approximately 85.6 percent of the time. Let us suppose that samples free of defectives are obtained in 70 percent of the samples drawn on 400 repetitions of the sampling procedure. How are we to account for the discrepancy between the obtained 70 percent and the predicted 85.6 percent?

In Exercises 5–10 set up a probability model for the process described, state the assumption which underlies the probability function of your model, and determine the probability values required.

5. From a lot of 50 color television sets, a sample of 3 is selected for inspection. There are 4 defective sets in the lot, (see Sec. 7.1, Exercise 25, p. 000). Determine the probability that the sample drawn contains (**a**) no defective sets; (**b**) 1 defective set; (**c**) 2 defective sets; (**d**) 1 or 2 defective sets.

6. A store's file of 90 accounts contains 12 delinquent accounts and 78 nondelinquent accounts. An auditor plans to choose a sample of 4 accounts for examination. Determine the probability that the sample drawn has (**a**) no delinquent accounts; (**b**) 1 delinquent account; (**c**) 1 or 2 delinquent accounts.

7. A poker hand of 7 cards is dealt from a standard deck of 52 cards.
   Find the probability that the hand dealt contains (**a**) 3 aces, (**b**) 3 aces and 2 queens, (**c**) 3 aces or 2 queens, (**d**) 4 spades, (**e**) 4 spades and 2 clubs, (**f**) 4 spades or 2 clubs.

8. A university student body of 2000 consists of 800 male students and 1200 female students. Four percent of the male students and 5 percent of the female students are studying psychology. Two students are chosen from the student body to be inter-

viewed. Find the probability that (**a**) both students are studying psychology, (**b**) one of the two students is studying psychology, (**c**) neither student is studying psychology.

9. Two books are chosen at random from 8 books and placed on a shelf. Find the probability that (**a**) a certain book, call it A, occupies first place, (**b**) two particular books, A and B, are next to each other, (**c**) A and B are not next to each other.

10. The letters s, p, e, v, and k are written separately on five slips of paper and the slips are put into a bag and stirred. A person then draws the slips from the bag, one after the other, and arranges them in a line according to the order in which they were drawn. Each such arrangement is called a word. Find the probability that in the word formed (**a**) k occupies first place and e occupies fifth pace, (**b**) v occupies third place, (**c**) p and v are together, (**d**) p and v are not together.

11. For the purpose of obtaining a maximum-likelihood estimate of the fish population of Lake Mark, 300 fish were caught, tagged, and released into the lake. Shortly thereafter, 500 fish were caught from the lake, and it was found that 2 had been previously caught. Let N denote the size of the fish population.

    (**a**) Set up a probability model for the experiment of catching the 500 fish.

    (**b**) With respect to the given conditions, what is meant by the maximum-likelihood estimate of the fish population?

    (**c**) Determine the maximum-likelihood function for this situation.

    (**d**) Find the maximum-likelihood estimate of the size of the fish population.

12. Judging from complaints about damage due to raccoons in the community of East Beach, the raccoon population had increased significantly during the last five years along with the human population. But how large was it? This is what the Community Council wanted to find out. They commissioned a team headed by Irving Fine to obtain a maximum likelihood estimate of the raccoon population in East Beach.

    Twenty raccoons were caught, tagged, and released. Shortly thereafter 15 were caught and it was found that 3 had been previously caught.

    (**a**) Based on these results, what is the maximum likelihood estimate of the raccoon population in Eash Beach?

    (**b**) What assumption(s) underlie this estimate?

    (**c**) From the damage that her property had sustained, Janet Reed felt that the estimate was too low. What aspects of the Fine team's analysis should she look over most carefully in order to satisfy herself that the estimate is realistic or present a credible case that it is not? Explain.

    (**d**) Is it possible that the estimate is much too high? Explain.

13. Fred Bass caught 9 fish, 3 of which were smaller than the law permits to be caught. A game warden inspects the catch by selecting 2 fish at random from Fred's bag and examining them. Some questions of interest to the fisherman are: What is the probability that no undersized fish are selected? What is the probability that at least one undersized fish is selected? To answer these questions, the fisherman set up a probability model by taking as a sample space $S = \{f_1, f_2, f_3, f_4,$

$f_5, f_6, f_7, f_8, f_9\}$, where f_1 is the event that fish 1 is selected, f_2 is the event that fish 2 is selected, and so on, and taking as his probability function P the one that assigns the same value, $\frac{1}{9}$, to each of the sample points. From this probability model the fisherman concluded that the probability that no undersized fish are selected is $C(6, 2)/C(9,2) = \frac{5}{12}$. and that the probability that at least one under sized fish is selected is $1 - \frac{5}{12} = \frac{7}{12}$. Is Fred's analysis correct? Explain. How would you analyze the problem.

14. Population $Q = \{2, 4, 6, 8, 10\}$. Consider the process of drawing a sample of size 2 at random from Q.
    (a) Define a probability model for this process.
    (b) What assumption underlies your model?
    (c) Define the sampling distribution of $\bar{x}$.
    (d) Find: (i) $P(5 \leq \bar{x} \leq 7)$, (ii) $P(\bar{x} > 6)$

15. Population $Q = \{5, 6, 7, 8, 9, 10\}$. Consider the process of drawing a sample of size 2 at random from Q. Same questions as stated for Exercise 14.

16. A lot of 20 items is known to contain 2 defectives. Consider an inspection procedure that consists of selecting 2 items at random from the shipment, one after the other, where the first item selected is not replaced before the second one is drawn. Al Williams was interested in finding the probability of the event B that the sample drawn contains 2 good items, and set up a probability model with sample space $S = \{G_1G_2, G_1D_2, D_1G_2, D_1D_2\}$, where G_1G_2 is the event that the first and second items drawn are good, G_1D_2 is the event that the first item drawn is good and the second item drawn is defective, and so on. Mr. Williams assigned equal probabilities of $\frac{1}{4}$ to these sample points and found the probability to be $\frac{1}{4}$ that the sample drawn contains 2 good items.
    (a) Is Mr. Williams's conclusion valid? Explain.
    (b) State the relative-frequency interpretation of Mr. Williams's conclusion.
    (c) Take 20 pennies, 2 that are new and shiny (to represent the 2 defective items in the lot) and 18 that have lost their luster (to represent the 18 good items in the lot), put them in a bag, shake the bag, and, without peeking, draw 2 pennies from the bag, one after the other. Repeat the process 300 times, record the occurrence of event B, and find the relative-frequency of B for the 300 repetitions of the process. Compare the result obtained with the relative-frequency interpretation of Mr. Williams's conclusion.
    (d) Does the result obtained affect the validity of Mr. Williams's conclusion? Explain.
    (e) Is Mr. Williams's probability model realistic? Explain.
    (f) How would you set up a probability model for the selection process?
    (g) Find the probability that the sample drawn contains 2 good items in your probability model, and interpret your result in relative-frequency terms.
    (h) Do the findings obtained in (c) support the results obtained in (g)? Explain.

CHAPTER 8

Conditional Probability Models

8.1. RETURN TO A PRODUCTION PROCESS SITUATION

Let us reconsider the Bokson Company's production process situation first encountered in Example 3 of Section 6.5 (p. 213). The Bokson Company makes television tubes in two plants, B1 and B2. The weekly output is 3000 tubes, with 1800 tubes being produced in plant B1, of which 1 percent are defective, and 1200 tubes being produced in plant B2, of which 2 percent are defective. A tube is selected at random from the week's output. One question of interest is, what is the probability that a defective tube is chosen?

To address this question we set up a probability model by taking as our sample space $S = \{t_1, t_2, \cdots, t_{3000}\}$, where t_1 is the event that tube 1 is chosen, t_2 is the event that tube 2 is chosen, and so on. Based on the envisioned randomness of the selection, we took as our probability function P the one defined by:

$$P(t_1) = P(t_2) = \cdots = P(t_{3000}) = \frac{1}{3000} \tag{8.1}$$

On the basis of this probability model we have that if B_1 is the event that a defective tube made by plant B1 is chosen, B_2 is the event that a defective tube made by plant B2 is chosen, and B is the event that a defective tube is chosen, then:

$$P(B_1) = \frac{18}{3000} \quad P(B_2) = \frac{24}{3000} \quad P(B) = \frac{42}{3000}$$

Suppose we are informed that the tube chosen is defective and are then asked for the probability that it was made by plant B1. A probability model is defined based on the existing information picture. We have been given additional information and the question facing us is this:

> How should we modify our probability model so as to take into account this additional information?

One way is to keep the same sample space $S = \{t_1, t_2, \cdots, t_{3000}\}$ but "suitably" modify the probability function P defined in (8.1). Let P_B denote the new probability function to be defined. To simplify our notation let us suppose that tubes $t_1, t_2, \cdots, t_{42}$ are the defective ones (with $t_1, \cdots, t_{18}$ made by plant B1 and $t_{19}, \cdots, t_{42}$ made by plant B2) so that the event B that a defective tube is chosen is described by:

$$B = \{t_1, t_2, \cdots, t_{42}\}$$

How should P_B be defined? To be informed that event B has occurred is to exclude the occurrence of the sample points in $B^c = \{t_{43}, t_{44}, \cdots, t_{3000}\}$. Thus, P_B should assign 0 to these sample points. As to $t_1, t_2, \cdots, t_{42}$, the drawing is still being made at random so that these sample points should be equally-likely in the new model. This leads us to assign 1/42 to each one. Note, this can be achieved in terms of probability function P by defining P_B on B as follows:

$$P_B(t_1) = \frac{P(t_1)}{P(B)}, \quad P_B(t_2) = \frac{P(t_2)}{P(B)}, \cdots, P_B(t_{42}) = \frac{P(t_{42})}{P(B)}$$

In each case the numerator is 1/3000, and the denominator is 42/3000, so that their ratio is the required 1/42 value.

In summary then, to take into account the additional information in the form that an event B has occurred (in this case B being the event that the tube chosen is defective) we modify P on S by defining P_B as follows on $S = \{t_1, t_2, \cdots, t_{42}, t_{43}, \cdots, t_{3000}\}$:

$$P_B(t_i) = \frac{P(t_i)}{P(B)} = \frac{1}{42}, \text{ for } t_i \text{ in } B = \{t_1, \cdots, t_{42}\}$$

$$P_B(t_i) = 0, \text{ for } t_i \text{ in } B^c = \{t_{43}, \cdots, t_{3000}\}$$

(8.2)

Consider the event $B_1 = \{t_1, t_2, \cdots, t_{18}\}$ that a defective tube made by plant B1 is chosen.

$$P_B(B_1) = \frac{P_B(t_1)}{P(B)} + \cdots + \frac{P_B(t_{18})}{P(B)} = \frac{1}{42} + \cdots + \frac{1}{42} = \frac{18}{42} \tag{8.3}$$

Let us also note that B_1 may be expressed by $B_1 = A_1 \cap B$, where A_1 is the event that a tube made by plant B1 is chosen and B, of course, is the event that a defective tube is chosen. Taking the intersection of an event of interest, in this case A_1, with B cuts out the sample points in the B^c part of sample space S, and gives us the same result obtained in (8.3):

$$\frac{P(A_1 \cap B)}{P(B)} = \frac{18/3000}{42/3000} = \frac{18}{42}.$$

More generally, if A is any event, then the modified probability value of A, $P_B(A)$, also denoted by $P(A/B)$, and called the **conditional probability of A given B**, is given in terms of P by:

$$P_B(A) = P(A/B) = \frac{P(A \cap B)}{P(B)} \tag{8.4}$$

8.2. CONDITIONAL PROBABILITY

Guided by our observations concerning the production process situation considered in Section 8.1, let us suppose that we have under study a random process for which a probability model with sample space S and probability function P has been formulated. The information picture changes in that an event B has occurred. Prompted by (8.2) we take into account this additional information by modifying P to P_B on $S = \{s_1, s_2, \ldots, s_n\}$ as follows:

$$P_B(s_i) = \frac{P(s_i)}{P(B)}, \text{ for } s_i \text{ in } B$$

$$P_B(s_i) = \frac{P(s_i)}{P(B)}, \text{ for } s_i \text{ in } B^c \tag{8.5}$$

As suggested by (8.4), if A is any event, then $P_B(A)$, also denoted by $P(A/B)$, and called the **conditional probability of A given B**, is given in terms of P by:

$$P_B(A) = P(A/B) = \frac{P(A \cap B)}{P(B)} \qquad (8.6)$$

EXAMPLE 1

For the toss of a pair of well-balanced dice, what is the probability that a sum of 7 shows given that an odd sum shows?

Our probability model is the usual one with 36 outcomes, each of which is assigned probability $\frac{1}{36}$. Let B denote the event that an odd sum shows, and A denote the event that a sum of 7 shows. Then $A \cap B$ is the event that an odd sum equal to 7 shows; that is, $A \cap B$ is the event that a sum of 7 shows. Our problem is to find $P(A/B)$. We have

$$P(A/B) = \frac{P(A \cap B)}{P(B)} = \frac{6/36}{18/36} = \frac{1}{3}$$

EXERCISES

1. If $P(A) = 0.50$, $P(B) = 0.40$, and $P(A \cap B) = 0.10$, find **(a)** $P(A \cup B)$, **(b)** $P(A/B)$, **(c)** $P(B/A)$.

2. If $P(A/B) = P(B/A)$ and $P(A) = 0.30$, find $P(B)$.

3. For the toss of a well-balanced die, **(a)** what is the probability that 4 shows given that an even number shows? **(b)** What is the probability that a number greater than 3 shows given that an odd number shows? **(c)** What is the probability that an even number shows given that a number greater than 3 shows?

4. A factory producing light bulbs employs two machines, I and II, in the production process. The daily output is 1000 bulbs, with machine I producing 400 bulbs of which 1 percent are defective, and machine II producing 600 bulbs of which 0.5 percent are defective. A bulb is selected at random from the day's output.
   (a) Set up a probability model for the selection process.
   (b) If a bulb chosen at random is found to be defective, what is the probability that it was made by machine I?

5. Suppose that the die considered in Example 1 were governed by the model $S = \{1, 2, 3, 4, 5, 6\}$, $P(1) = 1/9$, $P(2) = 2/9$, $P(3) = 1/9$, $P(4) = 2/9$, $P(5) = 1/9$, $P(6) = 2/9$. Find the probability that an odd number shows given that a number greater than 3 shows in this model.

6. The Parks company manufactures cloth, which is subject to three classifications: perfect, acceptable, and defective. Three machines, I, II, and III, are used in the production process, and the total weekly output is 6000 pieces, with machine I producing 2000 pieces of which 2 percent are defective and 7 percent are acceptable, machine II producing 2500 pieces of which 1 percent are defective and 10 percent are acceptable, and machine III producing 1500 pieces of which 3 percent are defective and 5 percent are acceptable. A piece of cloth is selected at random from the weekly output.

   (a) Set up a probability model for the selection process and find the probability that (1) the cloth selected is defective; (2) the cloth selected is acceptable.

   (b) If the cloth selected is found to be defective, what is the probability that (1) it was made by machine I; (2) it was made by machine II?

   (c) If the cloth selected is found to be acceptable, what is the probability that (1) it was made by machine II; (2) it was made by machine III?

8.3. BAYES'S THEOREM

To introduce Bayes's Theorem we return to the Bokson Company's production process problem considered in Section 8.1.

With respect to the given background, the process is to select a tube at random from the week's output. We have the following probability model for this process:

$$S = \{t_1, \cdots, t_{3000}\}, P(t_1) = \cdots = P(t_{3000}) = \frac{1}{3000}$$

Let us assume that the tube selected is found to be defective, and consider the problems of finding the probabilities that it was (a) made by plant B1, (b) made by plant B2. To analyze these problems, we introduce the following events:

B: defective tube is chosen.

A_1: tube made by plant B1 is chosen.

A_2: tube made by plant B2 is chosen.

In connection with problems like this the events A_1 and A_2 are sometimes viewed as possible causes of event B, and the probabilities $P(A_1)$ and $P(A_2)$ are often called the **prior probabilities** of events A_1 and A_2 (prior to selecting a tube and observing its nature—defective or good). The conditional probabilities $P(A_1/B)$, $P(A_2/B)$ are often called the **posterior probabilities** of the possible causes A_1 and A_2, given that event B has occurred, occurs, or will occur. They are post of after-the-experiment weights assigned to the possible causes A_1 and A_2.

The problem is to determine the following conditional probabilities:

$P(A_1/B)$, the probability that the tube was made by plant B1 given that the tube chosen was defective.

$P(A_2/B)$, the probability that the tube was made by plant B2 given that the tube chosen was defective.

By definition of conditional probability, we have:

$$P(A_1/B) = \frac{P(A_1 \cap B)}{P(B)} \qquad (8.7)$$

$$P(A_2/B) = \frac{P(A_2 \cap B)}{P(B)} \qquad (8.8)$$

Rather than use (8.7) and (8.8) to obtain the desired conditional probabilities directly, we shall derive other useful relations that culminate in the statement of a special case of Bayes's theorem. Let us first observe that A_1 and A_2 are mutually exclusive (a tube cannot be made by both plant B1 and plant B2), and that their union is sample space S (a tube is made by either plant B1 or B2). These relationships among A_1, A_2, and the sample space S are shown in Figure 8.1; A_1 and A_2 are said to form a partition of S. Let us observe that event B can be expressed in terms of A_1 and A_2 by

$$B = (A_1 \cap B) \cup (A_2 \cap B)$$

where $A_1 \cap B$ and $A_2 \cap B$ are mutually exclusive (see Figure 8.2). $A_1 \cap B$ is the event that a defective tube made by plant B1 is drawn, and $A_2 \cap B$ is the event that a defective tube made by plant B2 is drawn. Thus we have:

$$P(B) = P(A_1 \cap B) + P(A_2 \cap B) \qquad (8.9)$$

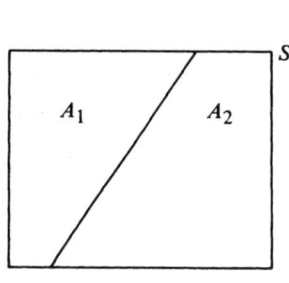

Figure 8.1

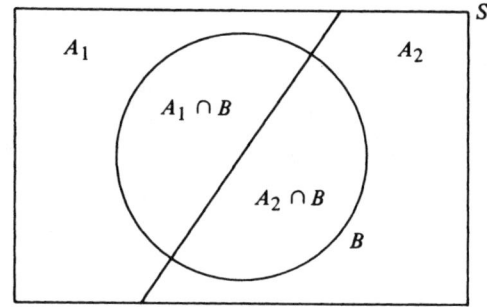

Figure 8.2

By replacing $P(B)$ in (8.7) and (8.8) by the right side of Equation (8.9), we obtain:

$$P(A_1/B) = \frac{P(A_1 \cap B)}{P(A_1 \cap B) + P(A_2 \cap B)} \qquad (8.10)$$

$$P(A_2/B) = \frac{P(A_2 \cap B)}{P(A_1 \cap B) + P(A_2 \cap B)} \qquad (8.11)$$

Also, by definition of conditional probability,

$$P(B/A_1) = \frac{P(A_1 \cap B)}{P(A_1)}, \qquad P(B/A_2) = \frac{P(A_2 \cap B)}{P(A_2)}$$

from which we obtain;

$$P(A_1 \cap B) = P(A_1) \cdot P(B/A_1),$$
$$P(A_2 \cap B) = P(A_2) \cdot P(B/A_2).$$

Substituting these results for $P(A_1 \cap B)$ and $P(A_2 \cap B)$ in (8.10) and (8.11) yields:

$$P(A_1/B) = \frac{P(A_1) \cdot P(B/A_1)}{P(A_1) \cdot P(B/A_1) + P(A_2) \cdot P(B/A_2)}$$

$$P(A_2/B) = \frac{P(A_2) \cdot P(B/A_2)}{P(A_1) \cdot P(B/A_1) + P(A_2) \cdot P(B/A_2)}$$

These results, which are special cases of Bayes's theorem, can be expressed by means of a probability tree such as the one shown in Figure 8.3. In terms of this probability tree, $P(A_1/B)$ is the ratio of the probability product associated with the first branch to the sum of the probability products associated with all branches of the tree; $P(A_2/B)$ is the ratio of the probability product associated with the second branch to the sum of the probability products associated with all branches of the tree.

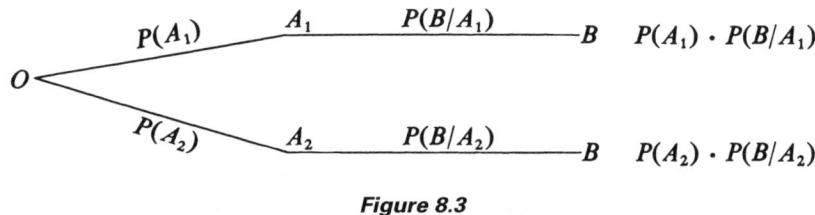

Figure 8.3

In terms of the Bokson Company's production process, the probabilities needed to employ Bayes's theorem have the following meanings:

1. $P(A_1)$, the probability that a tube made by plant B1 is chosen.
2. $P(A_2)$, the probability that a tube made by plant B2 is chosen.
3. $P(B/A_1)$, the probability that the tube is defective given that it was made by plant B1.
4. $P(B/A_2)$, the probability that the tube is defective given that it was made by plant B2.

On the basis of the randomness of the tube selection and the data provided, we define these probabilities as follows:

$$P(A_1) = \frac{1800}{3000} = 0.6 \text{ (1800 of the 3000 tubes were made by plant B1)}$$

$$P(A_2) = \frac{1200}{3000} = 0.4 \text{ (1200 of the 3000 tubes were made by plant B2)}$$

$P(B/A_1) = 0.01$ (1% of plant B1's output is defective)

$P(B/A_2) = 0.02$ (2% of plant B2's output is defective)

Figure 8.4 expresses the data in probability-tree terms. Thus:

$$P(A_1/B) = \frac{0.006}{0.014} = 0.43$$

$$P(A_2/B) = \frac{0.008}{0.014} = 0.57$$

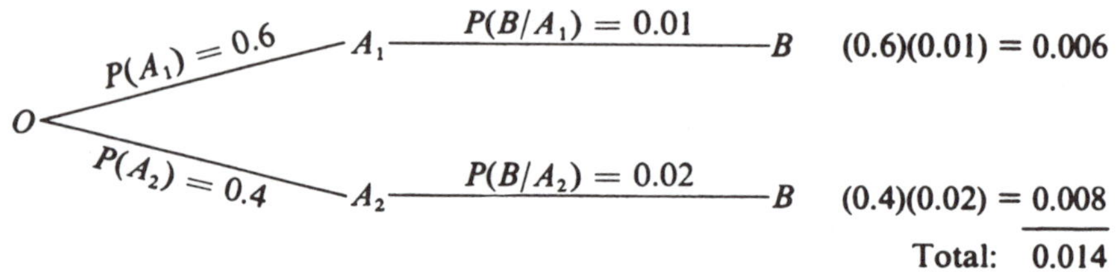

```
                P(B/A₁) = 0.01
    P(A₁) = 0.6  A₁ ─────────────── B   (0.6)(0.01) = 0.006
  O<
    P(A₂) = 0.4  A₂ ─────────────── B   (0.4)(0.02) = 0.008
                P(B/A₂) = 0.02
                                         Total:  0.014
```

Figure 8.4

If the tube chosen was defective, it was made by plant B1 with probability 0.43; it was made by plant B2 with probability 0.57.

Bayes's Theorem. More generally, consider a probability model with sample space S and probability function P. Let $A_1, A_2, \ldots, A_n$ denote a partition of S (that is, the A's are pairwise disjoint and their union is S), and B an event with nonzero probability. Then:

$$P(A_1/B) = \frac{P(A_1) \cdot P(B/A_1)}{P(A_1) \cdot P(B/A_1) + P(A_2) \cdot P(B/A_2) + \cdots + P(A_n) \cdot P(B/A_n)}$$

Similar results hold for $P(A_2/B), \ldots, P(A_n/B)$. For example, $P(A_2/B)$ is obtained from the above by replacing its numerator by $P(A_2) \cdot P(B/A_2)$ and retaining the denominator.

As has been noted in connection with a special case of Bayes's theorem, the events $A_1, A_2, \ldots, A_n$ are sometimes viewed as possible causes of event B, and the probabilities $P(A_1)$, $P(A_2)$, ..., $P(A_n)$ are often called the **prior probabilities** of the possible causes $A_1, A_2, ..., A_n$ (probabilities assigned prior to performing the process and observing outcome B). The conditional probabilities $P(A_1/B)$, $P(A_2/B)$, ..., $P(A_n/B)$ are often called the **posterior probabilities** of the possible causes $A_1, A_2, ..., A_n$ given that event B has occurred. These are post experiment probabilities assigned to the possible causes $A_1, A_2, ..., A_n$. Bayes's theorem is a foundtion stone for what has come to be called Bayesian analysis or Bayesian statistics, an aspect of modern decision theory under uncertainty that has received much attention in recent years. It should be noted that in the past much confusion in thought has surrounded the use of Bayes's theorem. This was primarily due to the fact that Bayes's theorem involves an inverse sort of reasoning from observed outcome to a statement about a possible cause, a situation that unfortunately offers an open invitation to extreme and misguided interpretations. There is no question about the validity of Bayes's theorem as a mathematical statement, but caution must be exercised in its application and interpretation. Bayes's theorem is so named in honor of the Reverend Thomas Bayes (1702-1761), an English clergyman and mathematician, whose pioneering paper on this subject was published posthumously in 1763.

A Reliability Question

The following data are available on the reliability of a blood test for detecting the presence of a certain disease. Of people with the disease, 90 percent of the blood test examinations detected the disease and 10 percent went undetected. Of the people who did not have the disease, 98 percent of those given the blood test were correctly diagnosed as not having the disease, and 2 percent were incorrectly diagnosed as having the disease. In a certain population it is estimated that 2 percent have the disease. A person is selected at random from the population. If, on the basis of the blood test, it is concluded that he has the disease, what is the probability that he actually has the disease?

To analyze this problem, we must be very careful to properly identify the event whose occurrence is given (designated by B in the notation that we have been using), and the event whose probability we are seeking, given the occurrence of B. For this situation, we introduce B, A_1, and A_2 as follows:

B: the blood test indicates the presence of the disease.

A_1: a person with the disease is selected.

A_2: a person not having the disease is selected.

The problem is to determine $P(A_1/B)$, the probability that the person selected has the disease given that the blood test indicates the presence of the disease. For this situation we have the following background:

$P(A_1)$, the probability that the person selected has the disease.

$P(A_2)$, the probability that the person selected does not have the disease.

$P(B/A_1)$, the probability that the blood test indicates the presence of the disease given that the person selected has the disease.

$P(B/A_2)$, the probability that the blood test indicates the presence of the disease given that the person selected does not have the disease.

The randomness of the selection and the data given leads us to the following assignment of probability values.

$P(A_1) = 0.02$ (2% of the population are estimated to have the disease.)

$P(A_2) = 0.98$ (98% of the population do not have the disease.)

$P(B/A_1) = 0.9$ (Of those with the disease, 90% were correctly diagnosed as having the disease.)

$P(B/A_2) = 0.02$ (Of those not having the disease, 2% were incorrectly diagnosed as having the disease.)

Figure 8.5 gives us a probability-tree view of this situation.

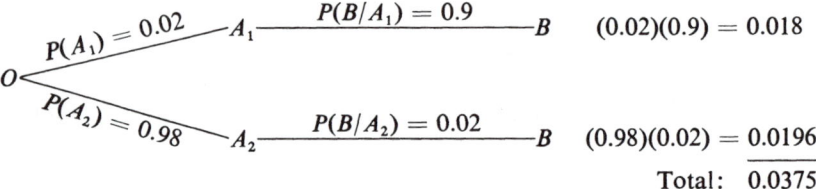

Figure 8.5

From Bayes's theorem we obtain:

$$P(A_1/B) = \frac{0.018}{0.0376} = 0.479$$

The relative-frequency interpretation of this result is that, over the long run, of the people selected at random and diagnosed as having the disease, less than one half (approximately 47.9 percent) will actually have the disease.

A Marketing Problem

In response to an antipollution drive the Starr Company is planning to market a new cleaning product that does not exhibit pollution side effects. The marketing department of the Starr Company initially assigned a probability of 0.5 to the product being a big seller, a probability of 0.3 to the product being a fair seller, and a probability of 0.2 to the product being a poor seller. A marketing test was planned and carried out. If the product is a big seller, it is estimated that the probability of selling between 7000 and 10,000 units in the test is 0.6. If the product is a fair seller, it is estimated that the probability of selling between 7000 and 10,000 units in the test is 0.9. If the product is a poor seller, the probability of selling between 7000 and 10,000 units is estimated to be 0.2. If

9000 units were sold in the marketing test, what is the probability that the cleaning product will be **(a)** a fair seller; **(b)** a big seller; **(c)** a fair seller or big seller?

To analyze this problem, we introduce the following events:

B: between 7000 and 10,000 units were sold in the test.

A_1: the product is a big seller.

A_2: the product is a fair seller.

A_3: the product is a poor seller.

The problem is to determine $P(A_2/B)$, the probability that the product is a fair seller given that between 7000 and 10,000 units were sold in the test; $P(A_1/B)$, the probability that the product is a big seller given that between 7000 and 10,000 units were sold in the test. For this situation we have the following background:

$P(A_1)$, the probability that the product is a big seller.

$P(A_2)$, the probability that the product is a fair seller.

$P(A_3)$, the probability that the product is a poor seller.

$P(B/A_1)$, the probability that between 7000 and 10,000 units are sold in the test given that the product is a big seller.

$P(B/A_2)$, the probability that between 7000 and 10,000 units are sold in the test given that the product is a fair seller.

$P(B/A_3)$, the probability that between 7000 and 10,000 units are sold in the test given that the product is a poor seller.

The given data lead us to the following assignment of probability values:

$$P(A_1) = 0.5, \quad P(A_2) = 0.3 \quad P(A_3) = 0.2$$
$$P(B/A_1) = 0.6 \quad P(B/A_2) = 0.9 \quad P(B/A_3) = 0.2$$

A probability-tree view of this situation is provided by Figure 8.6.

From Bayes's theorem we have:

(a) $P(A_2/B) = 0.27/0.61 = 0.44$.
(b) $P(A_1/B) = 0.3/0.61 = 0.49$

```
                            P(B|A₁) = 0.6
        P(A₁) = 0.5    A₁ ─────────────── B₁   (0.5)(0.6) = 0.3
       ╱
      ╱ P(A₂) = 0.3        P(B|A₂) = 0.9
   O ─────────────── A₂ ─────────────── B₂   (0.3)(0.9) = 0.27
      ╲
       ╲ P(A₃) = 0.2       P(B|A₃) = 0.2
                      A₃ ─────────────── B₃   (0.2)(0.2) = 0.04
                                                   Total:  0.61
```

Figure 8.6

(c) The probability of a fair or big seller, given that between 7000 and 10,000 units were sold in the test is 0.44 + 0.49 = 0.93

Considering the once-and-only nature of this situation, the probability assignments made and the results obtained are to be interpreted in subjective terms. The value 0.93 for the probability of a fair or big seller is a valid conclusion of initial judgments translated into quantitative terms [$P(A_1) = 0.5$, $P(B/A_1) = 0.6$, and so on], and may be helpful to the management of the Starr Company for making a decision on whether or not to market the new cleaning product on a major scale. It is important to keep in mind that the value 0.93, as a quantitative expression of the product's potential as a fair or big seller, does not go beyond the initial inputs. Never forget the GIGO principle; garbage in, garbage out.

EXERCISES

1. In connection with the discussion of a reliability question, suppose that a person is selected at random from the population and, on the basis of the blood test, it is concluded that he does not have the disease. What is the probability that he has the disease? What is the relative-frequency interpretation of your answer?

2. The following data are available on the reliability of a certain X-ray detection procedure for lung cancer. Of those people with lung cancer who were tested, 98 percent of the cases were detected, whereas 2 percent were undetected; 99 percent of those who did not have lung cancer who were tested were correctly diagnosed as not having lung cancer, whereas 1 percent were misdiagnosed. In a certain highly populated industrial center it is estimated that 3 percent have lung cancer. A person is selected at random from the population and tested.

    (a) If he is diagnosed as having lung cancer, what is the probability that he actually has the disease?

    (b) If the person is diagnosed as not having lung cancer, what is the probability that he has the disease?

    (c) What is the relative-frequency interpretation of your conclusions?

3. The following data are available on the reliability of a test to determine when the level of mercury contamination in fish exceeds what is considered to be a permissible level. Of the fish with excess mercury contamination that were tested, 99 percent of the cases were detected, whereas 1 percent went undetected; 96 percent of the fish that were tested whose mercury content did not exceed the permissible level were correctly diagnosed as being within the permissible level, whereas 4 percent were misdiagnosed. In a certain lake it is estimated that 6 percent of the fish population have a mercury content that exceeds the permissible level. A fish is caught from the lake and tested.

    (a) If the mercury content of the fish is diagnosed as excessive, what is the probability that it is not excessive?

(b) If the mercury content of the fish is diagnosed as within the permissible level, what is the probability that it is excessive?

(c) What is the relative-frequency interpretation of your conclusions?

4. In connection with the discussion of the Starr Company's marketing problem, given that 9000 units were sold in the marketing test, what is the probability that the cleaning product is a poor seller?

5. In a factory producing ball bearings, three machines, M1, M2, and M3, are used in the production process. The total daily output is 5000 ball bearings. Machine M1 produces 35 percent of the output, of which 2 percent are defective, machine M2 produces 25 percent, of which 2 percent are defective, and machine M3 produces 40 percent of the output, of which 1 percent are defective. If a ball bearing drawn at random from the day's output is found to be defective, what is the probability that **(a)** it was made by machine M1; **(b)** it was made by machine M2; **(c)** it was made by machine M3?

6. Messages sent through a communications channel consist of a sequence of 0's and 1's. Because of distortion effects, a transmitted 0 is sometimes received as a 1 and a transmitted 1 is sometimes received as a 0. The probability of distortion when 1 is sent is 0.02, and the probability of no distortion when 1 is sent is 0.98. The probability of distortion when 0 is sent is 0.05, with the probability of no distortion being 0.95. At a certain time it is estimated at the receiving end that the probability of 1 being sent is 1/3 and the probability of 0 being sent is 2/3. **(a)** If at this time 1 is received, what is the probability that 1 was sent? **(b)** If 0 is received, what is the probability that 1 was sent?

7. The Borg Company is planning to market an inexpensive computer especially designed for classroom use. Their sales department initially estimated that the probability of a big seller is 0.5, the probability of a good seller is 0.3, and the probability of a poor seller is 0.2. A market test was planned and carried out. It was estimated by the sales department that, if the computer is a big seller, the probability of selling more than 200 units in the test is 0.9; if the computer is a good seller, it was estimated that the probability of selling more than 200 units in the test is 0.5; and if the computer is a poor seller, it was estimated that the probability of selling more than 200 units in the test is 0.1. If more than 200 units were sold in the marketing test, what is the probability that **(a)** the computer is a big seller; **(b)** the computer is a good seller; **(c)** the computer is a poor seller; **(d)** the computer is a big or good seller?

8. Experience indicates that 30 percent of the entering students at Ecap University have insufficient mathematics preparation and need a noncredit review course in the fundamentals of algebra. To help make mathematics placement decisions, a proficiency exam has been developed, and it is estimated that a student with sufficient mathematics preparation will pass the exam with probability 0.95, whereas a student with insufficient preparation will pass the exam with probability 0.1.

(a) If a student passes the exam, what is the probability that he has sufficient mathematics preparation?

(b) If a student fails the exam, what is the probability that he has sufficient mathematics preparation?

8.4. MARKOV CHAINS

The Concept of Markov Chain

To introduce the concept of Markov chain, consider an atomic particle in random motion along a circle (Figure 8.7). Suppose the particle is initially located at one of the four positions, 1 let us say, shown in the circle. We will call this initial position of the particle the **initial state**.

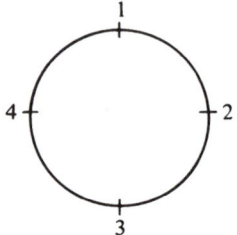

Figure 8.7

Let E_i denote the event that the particle is in position i, more generally called state i. The initial state E_1 of the particle is position 1.

From state E_1 (position 1) the particle can jump clockwise to position 2, or counterclockwise to position 4. This movement is called trial 1; if the particle jumps to position 2, the outcome is state E_2; if it jumps to position 4, the outcome is state E_4. Assuming the particle moved to state E_2 in trial 1, in trial 2 it may move to state E_3 or to state E_1. If it moves to state E_3, then the outcome of trial 2 is E_3.

For any trial, the probability that the particle jumps clockwise to a neighboring state is, let us assume, 3/5; the probability that it jumps counterclockwise to a neighboring state is 2/5.

The particle's behavior has the following characteristics.

1. Each movement (trial) can result in one of $n = 4$ states: E_1, E_2, E_3, E_4.
2. The state the particle occupies next depends only on the state it is in now, and not on any previous states.
3. The probability with which the particle moves to any state from its current state is known and is independent of the trial number.

The probability that a transition is made from state E_i to state E_j on a trial, denoted by p_{ij}, is called a **transition probability**. The transition probabilities define the behavior of the particle. They are conveniently displayed in a square array called a **transition matrix**. The transition matrix for the particle is:

$$T = \begin{array}{c} \\ E_1 \\ E_2 \\ E_3 \\ E_4 \end{array} \begin{array}{cccc} E_1 & E_2 & E_3 & E_4 \end{array} \\ \left[\begin{array}{cccc} 0 & 3/5 & 0 & 2/5 \\ 2/5 & 0 & 3/5 & 0 \\ 0 & 2/5 & 0 & 3/5 \\ 3/5 & 0 & 2/5 & 0 \end{array} \right]$$

The entry in row 1, column 1, $p_{11} = 0$, is the probability that the particle goes from state E_1 to E_1 on any one trial; more generally, the entry in row i, column j, p_{ij}, is the probability that the particle goes from state i to state j on any one trial.

This system illustrates a system known as an homogenous Markov chain.

> More generally, an **homogeneous Markov chain** is a structure consisting of a sequence of trials with the following properties:
>
> 1. Each trial leads to one of n outcomes $E_1, E_2, \ldots, E_n$, called **states**.
> 2. The probability that state E_i occurs on the kth trial depends only on the state in the preceding $(k - 1)$th trial, and not on the states of the system in earlier trials.

The matrix T of transition probabilities, describes the basic characteristics of an homogeneous Markov chain.

$$T = \begin{array}{c} \\ E_1 \\ E_2 \\ \cdot \\ \cdot \\ \cdot \\ E_n \end{array} \begin{array}{cccc} E_1 & E_2 & \cdots & E_n \end{array} \\ \left[\begin{array}{cccc} p_{11} & p_{12} & \cdots & p_{1n} \\ p_{21} & p_{22} & \cdots & p_{2n} \\ \cdot & \cdot & & \cdot \\ \cdot & \cdot & & \cdot \\ \cdot & \cdot & & \cdot \\ p_{n1} & p_{n2} & \cdots & p_{nn} \end{array} \right]$$

The Markov chain structure, named for Andrei Andreyevich Markov (1856-1922) who initiated a systematic study of such processes in the early part of this century, has found important applications in physics, biology and business. The application of the Markov chain model to a problem in marketing also provides us with an interesting case study of the danger of forcing a model onto a situation where it is not a good fit.

The Credit State of Beta City*

To illustrate, we turn to the credit state of Beta City. Borrowers owing $500,000 or more are assigned one of five credit states by the Last National Bank: 1 unsatisfactory, 2 poor, 3 satisfactory, 4 good, 5 excellent. Credit-state information is used to determine whether or not a borrower will be extended additional credit and the interest rate. Time is measured in monthly units, $t = 1,2,3,4,5,6$. The situation is reviewed every 6 months. A homogeneous Markov chain based on the following matrix of transition probabilities was formulated by the bank's credit department to describe the credit situation of Beta City, one of its largest borrowers.

$$\text{From state} \begin{array}{c} \\ 1 \\ 2 \\ 3 \\ 4 \\ 5 \end{array} \overset{\text{To state}}{\begin{bmatrix} 1 & 2 & 3 & 4 & 5 \\ 0.5 & 0.3 & 0.2 & 0 & 0 \\ 0.3 & 0.3 & 0.3 & 0.1 & 0 \\ 0.2 & 0.2 & 0.3 & 0.2 & 0.1 \\ 0.1 & 0.2 & 0.4 & 0.2 & 0.1 \\ 0 & 0.1 & 0.2 & 0.2 & 0.5 \end{bmatrix}} = T_1$$

For this Markov chain, we have the following transition probabilities:

$p_{11} = 0.5$ the probability of remaining in an unsatisfactory credit state in the next time period is 0.5

$p_{12} = 0.3$ the probability of passing from an unsatisfactory to a poor credit state in one time period is 0.3

$p_{13} = 0.2$ the probability of passing from an unsatisfactory to a satisfactory credit state in one time period is 0.2

$p_{22} = 0.3$ the probability of remaining in a poor credit state in the next time period is 0.3

$p_{43} = 0.4$ the probability of passing from a good to a satisfactory credit state in one time period is 0.4

We now address ourselves to questions of the following sort: what is the probability that a system described by a homogeneous Markov chain passes from state 2 to state 4 in two steps; that is, what is $P(E_4^{n+2}/E_2^n)$? More generally, what is $P(E_j^{n+2}/E_i^n)$, the probability that the system passes from state i to state j in two steps? What is $P(E_j^{n+3}/E_i^n)$, the probability that the system passes from state i to state j in three steps? We denote such probabilities by means of the following notation:

$p_{24}(2) = P(E_4^{n+2}/E_n^2)$, the probability that the system passes from state 2 to state 4 in two steps

$p_{ij}(2) = P(E_j^{n+2}/E_i^n)$, the probability that the system passes from state i to state j in two steps

---

*This example requires knowledge of matrix multiplication, which is taken up in Section 5.1.

$p_{24}(3) = P(E_4^{n+3}/E_2^n)$, the probability that the system passes from state 2 to state 4 in three steps

$p_{ij}(3) = P(E_j^{n+3}/E_i^n)$, the probability that the system passes from state i to state j in three steps

$p_{ij}(4) = P(E_j^{n+4}/E_i^n)$, the probability that the system passes from state i to state j in four steps, and so on

The **matrix of transition probabilities after two steps, T_2**, is defined as follows:

$$\text{From state} \begin{array}{c} \\ 1 \\ 2 \\ \cdot \\ \cdot \\ \cdot \\ N \end{array} \overset{\text{To state}}{\begin{bmatrix} 1 & 2 & \cdots & N \\ p_{11}(2) & p_{12}(2) & \cdots & p_{1N}(2) \\ p_{21}(2) & p_{22}(2) & \cdots & p_{2N}(2) \\ \cdot & \cdot & & \cdot \\ \cdot & \cdot & & \cdot \\ \cdot & \cdot & & \cdot \\ p_{N1}(2) & p_{N2}(2) & \cdots & p_{NN}(2) \end{bmatrix}} = T_2$$

T_2 is related to T_1 in a remarkably simple way:

$$T_2 = T_1 \cdot T_1 = T_1^2$$

The **matrix of transition probabilities after three steps, T_3**, is defined by

$$\text{From state} \begin{array}{c} \\ 1 \\ 2 \\ \cdot \\ \cdot \\ \cdot \\ N \end{array} \overset{\text{To state}}{\begin{bmatrix} 1 & 2 & \cdots & N \\ p_{11}(3) & p_{12}(3) & \cdots & p_{1N}(3) \\ p_{21}(3) & p_{22}(3) & \cdots & p_{2N}(3) \\ \cdot & \cdot & & \cdot \\ \cdot & \cdot & & \cdot \\ \cdot & \cdot & & \cdot \\ p_{N1}(3) & p_{N2}(3) & \cdots & p_{NN}(3) \end{bmatrix}} = T_3$$

T_3 can be determined from T_1 by means of the following relation:

$$T_3 = T_1 \cdot T_2 = T_1^3$$

The matrices of transition probabilities after four steps, five steps, and so on, are defined in an analogous way. These matrices can be determined from T_1 by means of the following relations:

$$T_4 = T_1 \cdot T_3 = T_1^4$$
$$T_5 = T_1 \cdot T_4 = T_1^5$$
$$T_n = T_1 \cdot T_{n-1} = T_1^n$$

The matrix of transition probabilities after two steps for Beta City is

$$\text{From state} \begin{array}{c} 1 \\ 2 \\ 3 \\ 4 \\ 5 \end{array} \begin{bmatrix} 0.38 & 0.28 & 0.25 & 0.07 & 0.02 \\ 0.31 & 0.26 & 0.28 & 0.11 & 0.04 \\ 0.24 & 0.23 & 0.29 & 0.14 & 0.10 \\ 0.21 & 0.22 & 0.30 & 0.16 & 0.11 \\ 0.09 & 0.16 & 0.27 & 0.19 & 0.29 \end{bmatrix} = T_2$$

with To state columns 1, 2, 3, 4, 5.

Thus, for example, we have

$p_{11}(2) = 0.38$ the probability of going from an unsatisfactory credit state back to an unsatisfactory credit state in two time periods is 0.38

$p_{12}(2) = 0.28$ the probability of going from an unsatisfactory to a poor credit state in two time periods is 0.28

$p_{24}(2) = 0.11$ the probability of going from a poor to a good credit state in two time periods is 0.11

Consider the problem of finding $p_{24}(3)$, the probability of going from a poor to a good credit state in three time periods. Since $p_{24}(3)$ is the entry in the second row and fourth column of matrix T_3, which equals the inner product of row 2 of matrix T_1 and column 4 of matrix T_2, we obtain

$$p_{24}(3) = (0.3)(0.07) + (0.3)(0.11) + (0.3)(0.14)$$
$$+ (0.1)(0.16) + (0)(0.19)$$
$$= 0.112$$

Considering the once-and-only nature of this situation, the probability assignments made and the results obtained are to be interpreted in subjective terms. The value 0.112 for the probability of going from a poor to a good credit state in three time periods is a valid conclusion of initial judgments translated into quantitative terms as expressed by matrix T_1.

Markov for Marketing?

The late 1950s and 1960s saw the development of Markov chain models to describe consumer brand choice behavior. A starting point for many of these investigations is the view that a brand loyalty and brand switching matrix of probabilities can be constructed from data on sequences of consumer purchases.

$$T_1 = \begin{bmatrix} p_{11} & p_{12} & \cdots & p_{1n} \\ p_{21} & p_{22} & & p_{2n} \\ \vdots & \vdots & & \vdots \\ p_{n1} & p_{n2} & \cdots & p_{nn} \end{bmatrix}$$

The value p_{11}, for example, expresses the probability that the consumer, having bought brand 1 in the last period, will also purchase brand 1 in the next period. More generally, p_{ij} expresses the probability that the consumer, having bought brand i in the last period, will purchase brand j in the next period.

An early application of this sort is one undertaken by Benjamin Lipstein [8] concerning the test marketing of a new margarine, fictitiously called Electra, in the Chicago area from November 1958 to May 1959. In Lipstein's study the possible states a margarine buyer could be in were the following:

E_1: Electra Brand E_4: Aunt Mary's brand

E_2: Gloria E_5: Meadowlark brand

E_3: B-R Stores brand E_6: All other brands

E_7: Did not buy margarine during time period

Lipstein's paper contains the brand loyalty and brand switching matrix of transition probabilities (Table 8.1) which represents the situation in the margarine market in Chicago shortly after the introduction of the new brand Electra.

But are Markov chain models realistic for the study of consumer brand choice behavior? A critical appraisal was given by A.S.C. Ehrenberg [1], who expressed the view that "frequent public reference to Markov-brand switching models had not been matched by an obvious array of published demonstration of their practical effectiveness." On the basis of a detailed discussion and analysis, Ehrenberg concluded that

> the failure of the Markov brand-switching model to live up to its earlier public reputation need not be surprising if seen as an example of misguided but perhaps

Table 8.1

Next period \ Original Period	Electra	Gloria	B-R	Aunt Mary's	Meadow-lark	Other	Did not buy
Electra	.12	.05	.03	.02	.04	.03	.05
Gloria	.05	.25	.02	.05	.01	.05	.03
B-R	.07	.03	.21	.01	.03	.03	.04
Aunt Mary's	.04	.02	.05	.23	.02	.04	.01
Meadowlark	.03	.02	.03	.04	.22	.05	.02
Other	.28	.26	.26	.25	.30	.23	.28
Did not buy	.41	.37	.40	.40	.38	.57	.57

understandable enthusiasm for forcing an attractively simple piece of college mathematics (stationary Markov theory) onto repeat-buying and brand-switching data while:

1. Omitting to ensure that the data are of a technically suitable form to be modeled by the model.
2. Omitting to examine the crucial assumption involved.
3. Omitting any self-critical appraisal of the various concepts and analytical steps in the approach.
4. Omitting to gather any generalized empirical knowledge of repeat-buying and brand-switching behavior as such.

William F. Massey and Donald G. Morrison [12] expressed agreement with many of Ehrenberg's arguments that the simple Markov chain does not fit all, or even many, real brand-switching situations, but felt that Ehrenberg had been too harsh in his judgment. They expressed the view that the basic Markovian approach is fruitful and

should not be abandoned. In a reply [2], Ehrenberg took issue with Massey and Morrison and again raised the question, "Can we not bury Markov for Marketing?"

The Ehrenberg-Massey-Morrison exchange should serve to remind us that mathematical models can only provide us with valid conclusions with respect to our assumptions. If the assumptions are unrealistic, then the mathematical model does not properly fit the situation, and to try to force a fit can be as counterproductive and painful as forcing a pair of size 8 shoes on feet that require size 10.

As to Lipstein's study, it follows from his transition matrix that over the long term Electra brand would end up with about 4% of the margarine market. However, six months or so after its introduction Electra succeeded in capturing about 12% of the market. Electra had been effective in building up the percentage of buyers who having purchased Electra in one period, remained loyal to it in the next period, so that the 0.12 value in the first row, first column of Table 8.1 would have to be revised to 0.23. The need to change transition probabilities had not been taken into account in Lipstein's model.

References

[1] A.S.C. Ehrenberg. "An Appraisal of Markov Brand-Switching Models," *Journal of Marketing Research,* vol. 2, no. 4 (Nov. 1964), pp. 347-362.

[2] _____. "On Clarifying M and M," *Journal of Marketing Research,* vol. 5, no. 2 (May 1968), pp. 228-29.

[3] Jean E. Draper and Larry H. Nolan. "A Markov Chain Analysis of Brand Preference." *Journal of Advertising Research,* vol. 4, no. 3 (September 1964), pp. 33-39.

[4] Frank Harary and Benjamin Lipstein. "The Dynamics of Brand Loyalty: A Markovian Approach." *Operations Research,* vol. 10., no. 1 (January-February 1962), pp. 19-40.

[5] Jerome D. Herniter and John F. Mag. "Customer Behavior as a Markov Process." *Operations Research,* vol. 9, no. 1 (January-February 1961), pp. 105-122.

[6] Ronald A. Howard. "Stochastic Process Models of Consumer Behavior." *Journal of Advertising Research,* vol. 3, no. 3 (September 1963), pp. 35-42. Reprinted in *Marketing Models: Quantitative Applications,* edited by R. L. Day and L. J. Parsons, pp. 104-117. Scranton, Pa.: Intext Educational Publishers, 1971.

[7] Benjamin Lipstein. "The Dynamics of Brand Loyalty and Brand Switching." *Proceedings of the Fifth Annual Conference of the Advertising Research Foundation* (1959), pp. 101-108.

[8] _____. "Tests for Test Marketing," *Harvard Business Review*, vol. 76 (March-April, 1961), pp. 365-369.

[9] _____. "A Mathematical Model of Consumer Behavior." *Journal of Marketing Research*, vol. 2, no. 3 (August 1965), pp. 259-265. Reprinted in *Marketing Models: Quantitative Applications*, edited by R. L. Day and L. J. Parsons, pp. 65-79.

[10] P. A. Longton and B. T. Warner. "A Mathematical Model for Marketing." *Metra*, vol. 1 (September 1962), pp. 297-310.
[11] Richard B. Maffei. "Brand Preferences and Simple Markov Processes." *Operations Research*, vol. 8, no. 2 (March-April 1960), pp. 210-218.
[12] William Massey and Donald Morrison. "Comments on Ehrenberg's Appraisal of Brand-Switching Models." *Journal of Marketing Research*, vol. 5, no. 2 (May 1968), pp. 225-227.
[13] George P. H. Styan and Harry Smith Jr. "Markov Chains Applied to Marketing." *Journal of Marketing Research*, vol. 1, no. 1 (February 1964), pp. 50-55.

An Application to Genetics

Let us consider a situation in Mendelian genetics where the appearance of a certain trait is governed by a pair of genes of two types, say B and b. Trait B is said to be dominant if an individual carrying the pair of genes Bb exhibits trait B. Trait b is said to be recessive under these conditions. Trait B will be exhibited if either of the genetic combinations BB or Bb occur in the individual. An individual of the BB type is said to be dominant, and an individual of type Bb is said to be hybrid. Trait b will be exhibited only if the individual is carrying the combination bb. Such an individual is said to be recessive. Johann Mendel (1822-1884) carried out experiments with varieties of peas in which one of the traits observed was seed color of two kinds, yellow and green. Yellow was dominant and green was recessive.

In the crossing of two animals or plants the offspring inherits one gene from each parent. Thus, for example, if a dominant (gene type BB) is crossed with a hybrid (gene type Bb), there are four possible genetic combinations for the offspring: BB, BB, Bb, and Bb. If a hybrid (gene type Bb) is crossed with a hybrid (gene type Bb), the possible combinations for the offspring are BB, Bb, Bb, and bb; and if a recessive (gene type bb) is crossed with a hybrid (gene type Bb), the possible combinations for the offspring are Bb, Bb, bb, and bb. A basic assumption in genetics is that in crossing two individuals the offspring is as likely to have one of the four possible genetic combinations as it is to have any of the others. In probabilistic terms this assumption is translated into the following:

1. If a dominant is crossed with a hybrid, the probability of a dominant offspring is $\frac{1}{2}$ and the probability of a hybrid offspring is $\frac{1}{2}$.
2. If a hybrid is crossed with a hybrid, the probability of a dominant offspring is $\frac{1}{4}$, the probability of a hybrid offspring is $\frac{1}{2}$, and the probability of a recessive offspring is $\frac{1}{4}$.
3. If a recessive is crossed with a hybrid, the probability of a hybrid offspring is $\frac{1}{2}$ and the probability of a recessive offspring is $\frac{1}{2}$.

Consider a process of four continued crossings carried out in the following way. An individual of unknown genetic structure (as far as traits B and b are concerned) is crossed with a hybrid. The offspring is crossed with a hybrid, the offspring of this union is crossed with a hybrid, and so on. A total of four such crossings is made. If the offspring of the first crossing is hybrid, what is the probability that **(a)** the outcome of the third crossing is hybrid, and **(b)** the outcome of the fourth crossing is hybrid? What is the relative-frequency interpretation of these conclusions?

This process can be described by a homogeneous Markov-chain model in which the states are E_d^1, the event that the offspring from the first crossing is dominant; E_d^2, the event that the offspring from the second crossing is dominant; E_d^3, E_d^4, E_h^1, the event that the offspring from the first crossing is hybrid; E_h^2, E_h^3, E_h^4, E_r^1, E_r^2, E_r^3, E_r^4. The matrix of transition probabilities is

$$\begin{array}{c} \\ \text{From} \\ \text{state} \end{array} \begin{array}{c} \\ d \\ h \\ r \end{array} \overset{\displaystyle \text{To state}}{\begin{array}{ccc} d & h & r \end{array}} \begin{bmatrix} \frac{1}{2} & \frac{1}{2} & 0 \\ \frac{1}{4} & \frac{1}{2} & \frac{1}{4} \\ 0 & \frac{1}{2} & \frac{1}{2} \end{bmatrix} = T_1$$

In row 1 we have p_{11}, the probability that dominant crossed with hybrid yields a dominant offspring, is $\frac{1}{2}$; p_{12}, the probability that dominant crossed with hybrid is hybrid, is $\frac{1}{2}$; p_{13}, the probability that dominant crossed with hybrid is recessive, is 0. In row 2 we have p_{21}, the probability that hybrid crossed with hybrid is dominant, is $\frac{1}{4}$, and so on.

(a) The probability that the outcome of the third crossing is hybrid, given that the outcome of the first crossing is hybrid, is $p_{22}(2)$, the probability of passing from a hybrid state to a hybrid state in two stages. $p_{22}(2)$ is the entry in row 2, column 2 of matrix T_2, which equals the inner product of row 2 of matrix T_1 and column 2 of matrix T_1. Thus

$$p_{22}(2) = \tfrac{1}{4} \cdot \tfrac{1}{2} + \tfrac{1}{2} \cdot \tfrac{1}{2} + \tfrac{1}{4} \cdot \tfrac{1}{2} = \tfrac{1}{2}$$

The relative-frequency interpretation of this result is that, if the crossings are carried out as described a large number of times and the offspring of the first crossing is hybrid, then approximately 50 percent of the offspring from the third crossing will be hybrid.

(b) The probability that the outcome of the fourth crossing is hybrid, given that the outcome of the first crossing is hybrid, is $p_{22}(3)$, the probability of passing from a hybrid state to a hybrid state in three stages. $p_{22}(3)$ is the entry in row 2, column 2 of matrix T_3, which is the inner product of row 2 of matrix T_1 and column 2 of matrix T_2. Now

$$T_2 = \begin{bmatrix} \frac{3}{8} & \frac{1}{2} & \frac{1}{8} \\ \frac{1}{4} & \frac{1}{2} & \frac{1}{4} \\ \frac{1}{8} & \frac{1}{2} & \frac{3}{8} \end{bmatrix}$$

Thus

$$p_{22}(3) = \frac{1}{4} \cdot \frac{1}{2} + \frac{1}{2} \cdot \frac{1}{2} + \frac{1}{4} \cdot \frac{1}{2} = \frac{1}{2}$$

The relative-frequency implication is that, with respect to the stated conditions, over the long run we can expect approximatley 50 percent of the offspring from the fourth crossing to be hybrid.

EXERCISES

1. A certain system under study is described by an homogeneous Markov chain with transition matrix

$$T_1 = \begin{bmatrix} \frac{1}{2} & 0 & \frac{1}{2} \\ \frac{1}{3} & \frac{2}{3} & 0 \\ 0 & \frac{1}{2} & \frac{1}{2} \end{bmatrix}$$

What is the probability that the system (a) passes from state 1 to state 3 in two steps; (b) passes from state 2 to state 1 in two steps; (c) passes from state 2 to state 1 in three steps; (d) passes from state 3 to state 1 in three steps; (e) passes from state 2 to state 2 in three steps?

2. In connection with the Beta City credit situation, determine the probability of (a) passing from a poor to satisfactory credit state in three time periods, (b) passing from a satisfactory to good credit state in three time periods, (c) passing from a poor to good credit state in four time periods, (d) passing from a good to excellent credit state in four time periods.

3. An homogeneous Markov-chain model based on the following matrix of transition probabilities was formulated to describe the credit situation of Alpha City, the twin of Beta City across the river.

$$T_1 = \begin{bmatrix} 0.4 & 0.4 & 0.2 & 0 & 0 \\ 0.2 & 0.4 & 0.2 & 0.1 & 0.1 \\ 0.1 & 0.3 & 0.3 & 0.2 & 0.1 \\ 0 & 0.1 & 0.5 & 0.3 & 0.1 \\ 0 & 0 & 0.2 & 0.3 & 0.5 \end{bmatrix}$$

Determine the probability of (a) passing from a poor to a satisfactory credit state in two time periods, (b) passing from a satisfactory to a good credit state in two time periods, (c) passing from a satisfactory to a good credit state in three time periods, (d) passing from a good to an excellent credit state in three time periods.

4. Consider a particle in random motion along a circle under the following conditions.

   1. The particle can occupy any one of four positions, 1, 2, 3, 4, called states.
   2. Time is measured in discrete units, $t = 1, 2, 3, \ldots, 10$.

3. At time $t = 1$ the particle is in one of the aforementioned states, and at each subsequent time value it is found in a neighboring state. The probability that the particle jumps clockwise is $\frac{1}{2}$, and the probability that it jumps counterclockwise is $\frac{1}{2}$.

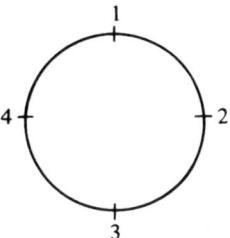

(a) Write the transition matrix for the Markov chain describing this random motion.

(b) What is the probability that the particle goes from state 2 to state 4 in three time steps?

(c) If the particle is in state 2, what is the probability that it is found in state 2 four time steps later?

5. Consider a particle in random motion along a circle under conditions 1 and 2 as stated in Exercise 4 and the following condition 3: at time $t = 1$ the particle is in one of its states, and at each subsequent time value it is found either in the same state or a neighboring state. The probability that it remains in the same state is $\frac{1}{6}$; it jumps clockwise to a neighborning state with probability $\frac{2}{6}$, and jumps counterclockwise to a neighboring state with probability $\frac{3}{6}$.

(a) Write the transition matrix for the Markov chain describing this motion.

(b) Find the probability that the particle (1) goes from state 1 to state 4 in two time steps, (2) goes from state 1 to state 3 in two time steps, (3) is found in state 1 two steps later if it started there.

(c) Find the probability that the particle (1) goes from state 1 to state 4 in three time steps, (2) goes from state 1 to state 2 in three time steps, (3) is found in state 3 three time steps later if it starts there.

6. The following questions refer to the genetics situation discussed in the preceding section.

(a) If the offspring of the first crossing is dominant, find the probability that (1) the outcome of the third crossing is hybrid, (2) the outcome of the third crossing is dominant, (3) the outcome of the fourth crossing is hybrid, (4) the outcome of the fourth crossing is dominant.

(b) If the offspring of the first crossing is recessive, find the probability that (1) the outcome of the third crossing is recessive, (2) the outcome of the third crossing is hybrid, (3) the outcome of the third crossing is dominant, (4) the outcome of the fourth crossing is recessive, (5) the outcome of the fourth crossing is hybrid.

7. A process of continued crossings is carried out in the following way. An individual of unknown genetic structure (as far as traits B and b are concerned) is crossed with a dominant. The offspring is crossed with a dominant. The offspring of this union is crossed with a dominant. A total of five such crossings are made.

   (a) Write the matrix of transition probabilities of the homogeneous Markov chain describing this process.

   (b) If the offspring of the first crossing is hybrid, find the probability that (1) the outcome of the third crossing is hybrid, (2) the outcome of the fourth crossing is hybrid, (3) the outcome of the fifth crossing is hybrid.

CHAPTER 9

Bernoulli Trial Models

9.1. INDEPENDENT EVENTS

The term independent events suggests events with the property that the occurrence or nonoccurrence of one has no influence of the occurrence or nonoccurrence of the other. Such an intuitive notion of independent events is not precise enough to serve as a mathematical definition of this concept. We use it as a guide to develop a precise condition which captures its spirit.

If A and B are independent in the intuitive sense described, then it makes sense to reflect this by the condition

$$P(A/B) = P(A), \qquad (9.1)$$

which says that the probability assigned to A is not to be modified if we are given the additional information that B has occurred. Now:

$$P(A/B) = \frac{P(A \cap B)}{P(B)} \qquad (9.2)$$

Replacing $P(A/B)$ in (9.1) by the right side of (9.2) yields.

$$\frac{P(A \cap B)}{P(B)} = P(A),$$

from which we obtain:

$$P(A \cap B) = P(A) \cdot P(B) \tag{9.3}$$

This is the precise condition that will do the job.

Two events A and B connected with a probability model with probability function P are said to be **independent** if condition (9.3) is satisfied. Our intuitive understanding of this condition is that the occurrence or nonoccurrence of one of the events A and B does not influence the occurrence or nonoccurrence of the other.

EXAMPLE 1

For the process of tossing a pair of dice (one red and the other green), assumed to be fair, are the events A, an even sum shows, and B, the sum showing is less than 5, independent?

We take as our probability model the "standard" one with 36 sample points and probability function P which assigns 1/36 to each one.

For a view of the sample points that describe A, B and $A \cap B$ (an even sum less than 5 shows), refer to Table 9.1. The probabilities of these events are:

$$P(A) = \frac{1}{2}, \quad P(B) = \frac{1}{6}, \quad P(A \cap B) = \frac{1}{9}.$$

Table 9.1

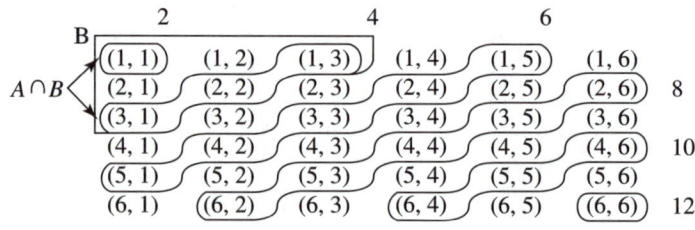

Since $P(A \cap B) = 1/9$ is not equal to $P(A) \cdot P(B) = 1/12$, events A and B are not independent.

Our interpretation of this is that the occurrence or nonoccurrence of one of these events does influence the occurrence or nonoccurrence of the other.

Three events, A, B, and C in a probability model with sample space S and probability function P are said to be **independent** if

$$P(A \cap B \cap C) = P(A) \cdot P(B) \cdot P(C)$$

and all pairs of events selected from A, B, and C are independent. That is:

$$P(A \cap B) = P(A) \cdot P(B), \quad P(A \cap C) = P(A) \cdot P(C)$$
$$P(B \cap C) = P(B) \cdot P(C)$$

More generally, n events $A_1, A_2, \ldots, A_n$ are said to be **independent** if

$$P(A_1 \cap \ldots \cap A_n) = P(A_1) \cdots P(A_n),$$

and this product condition holds for all sub-collections of events chosen from $A_1, \ldots, A_n$ as well.

Our intuitive interpretation of this more general condition is much the same as previously described; namely, that the occurrence or nonoccurrence of any of $A_1, A_2, \ldots, A_n$ does not influence the occurrence or nonoccurrence of the others.

A Property of Independence

Our intuitive interpretation of the independence of $A_1, A_2, \ldots, A_n$ includes the idea that the nonoccurrence of any of these events (which is described by their complements) has no influence on the occurrence or nonoccurrence of the others, but this property is not directly reflected in the definition of independence. While this is the case, the condition involving the nonoccurrence of any of $A_1, A_2, \ldots, A_n$ not influencing the occurrence or nonoccurrence of the others is present in the sense of the following theorem.

If $A_1, A_2, \ldots, A_n$ is a collection of independent events, then the collection of events obtained by replacing any number of the A's by their complements is also independent.

Thus, for example, if A and B are independent, with $P(A) = 1/3$ and $P(B) = 1/4$, then A^c and B^c are independent and

$$P(A^c \cap B^c) = P(A^c) \cdot P(B^c) = \frac{2}{3} \cdot \frac{3}{4} = \frac{1}{2}.$$

EXERCISES

1. In connection with tossing a well balanced die, are the events an even number shows and a number greater than 3 shows independent?

2. In connection with tossing a pair of well balanced dice (one red and one green),
   (a) are the events red shows less than 3 and green shows less than 4 independent?
   (b) Are the events red shows less than 3 and the sum showing is 7 independent?

3. Events A, B and C are independent, where $P(A) = 1/3$, $P(B) = 1/4$ and $P(C) = 1/5$. Find (a) $P(A \cap B^c)$; (b) $P(A \cap B \cap C)$; (c) $P(A \cap B^c \cap C^c)$; (d) $P(A^c \cap B \cap C^c)$.

9.2. BERNOULLI TRIALS

It is sometimes natural and productive to view a random process as a sequence of repetitions of some basic process. For example, the tossing of a coin five times in succession may be viewed as a sequence of five repetitions of the basic process of tossing a coin once, firing at a target six times may be viewed as a sequence of six repetitions of the basic process of firing at the target, administering an antibiotic to ten people may be viewed as a ten fold repetition of the basic process of administering the antibiotic. A repetition of a basic process is called a **trial**.

The problem we address here is that of formulating a probability model for a sequence of trials under the following conditions.

1. On each trial attention is focused on the occurrence or nonoccurrence of a certain event E. The occurrence of E on a given trial is called **success**; the occurrence of E^c is called **failure**.
2. The probability of success is the same for each trial; call this probability value p. The probability of failure for each trial is $1 - p$, which we shall denote by q.
3. The outcome of a trial (success or failure) does not influence the outcome of any other trial.

A sequence of repetitions of an experiment carried out under these conditions is sometimes called a sequence of Bernoulli trials in honor of Jacob Bernoulli (1654–1705), who initiated a systematic study of such processes in the last part of the seventeenth century.

EXAMPLE 1

Consider the process of tossing a coin, assumed to be well balanced, five times in succession. Take for E the event that head shows on a trial, with assigned probability $p = 1/2$. The third Bernoulli trial condition is satisfied since the outcome of any toss of the coin does not influence the outcome of any other toss. We might be influenced about how the coin will fall on a subsequent toss based on its behavior on previous tosses, but the coin has no memory and cannot be influenced by such. The probability assignment of 1/2 for head showing on any toss is realistic for a well balanced coin.

Sequences of coin tossings are realisticly described by Bernoulli trial conditions.

The constancy of p and the independence condition are the ones most likely to give trouble in practice and must be looked at most carefully.

The problem we now address is that of constructing a probability model for a sequence of n Bernoulli trials. If a sequence of n repetitions of the underlying experiment results in event E occurring on trial 1, not occurring on trial 2, occurring on trial 3, not occurring on trial 4, and so on, then we have a run of successes and failures that can be represented by the sequence

$$\underbrace{SFSF\ldots SF}_{n \text{ terms}}$$

The outcome of any sequence of n repetitions of the experiment can be described by a sequence of n S's and F's, and every sequence of n S's and F's describes a way in which a sequence of n repetitions of the experiment could occur. Accordingly, we take as our sample space T the events described by all sequences of n S's and F's.

To define p on the sample points of T, let S_1 denote the event success on trial 1, S_2 the event success on trial 2, ..., S_n the event success on trial n, F_1 the event failure on trial 1, F_2 the event failure on trial 2, ..., F_n the event failure on trial n. To reflect the condition that the probabilities of success and failure are the same for each trial, p and $q = 1 - p$, respectively, we impose on p the condition:

$$P(S_1) = P(S_2) = \cdots = P(S_n) = p \qquad (1)$$

$$P(F_1) = P(F_2) = \cdots = P(F_n) = q \qquad (2)$$

To reflect the condition that success or failure on any one trial does not influence the outcome on any other trial, we require that any n events chosen from $S_1, S_2, \ldots, S_n, F_1, F_2, \ldots, F_n$ with different subscripts (that is, when S_k is chosen, F_k is to be left out, and vice versa) be independent. These conditions uniquely define P on the sample points of T. Consider, for example, the sequence

$$\underbrace{SS\ldots S}_{k \text{ terms}} \underbrace{F\ldots F}_{n-k \text{ terms}}$$

whose first k terms are S's and whose last $n - k$ terms are F's. Then:

$$\underbrace{SS\ldots S}_{k \text{ terms}} \underbrace{F\ldots F}_{n-k \text{ terms}} = S_1 \cap S_2 \cap \ldots \cap S_k \cap F_{k+1} \cap \ldots \cap F_n$$

This condition, the independence of the n events $S_1, S_2, \ldots, S_n, F_1, F_2, \ldots, F_n$, and conditions (1) and (2) yield:

$$P(SS\ldots SF\ldots F) = P(S_1 \cap S_2 \cap \ldots \cap S_k \cap F_{k+1} \cap \ldots \cap F_n)$$
$$= P(S_1) \cdot P(S_2) \cdots P(S_k) \cdot P(F_{k+1}) \cdots P(F_n)$$
$$= \underbrace{p \cdot p \cdots p}_{k \text{ terms}} \cdot \underbrace{q \cdots q}_{n-k \text{ terms}}$$
$$= p^k q^{n-k}$$

Each S in the sequence yields a p in the product $p^k q^{n-k}$ and each F in the sequence yields a q in this product. More generally, if Z is any sequence with k S's and $n - k$ F's, we have:

$$P(Z) = p^k q^{n-k}$$

How many sequences have k S's and $n - k$ F's? To obtain such a sequence, we choose k places in the sequence and fill them with S's; the remaining places will be filled with F's. The number of ways of doing this equals the number of ways of choosing k of n places without regard to order, which is $C(n, k)$.

Let X denote the number of successes in a sequence of n trials. Since $C(n, k)$ sequences have k successes and $n - k$ failures, the probability of k successes in n trials is:

Thus:
$$P(X = k) = C(n, k) p^k q^{n-k}, \text{ for } k = 0, 1, 2, \ldots, n$$

$$P(X = 0) = C(n, 0) p^0 q^n = q^n$$
$$P(X = 1) = C(n, 1) p^1 q^{n-1}$$
$$P(X = 2) = C(n, 2) p^2 q^{n-2}$$
$$\vdots$$
$$P(X = n) = C(n, n) p^n q^0 = p^n$$

In summary, the Bernoulli trial conditions 1, 2 and 3 lead to a class of Bernoulli trial models. To specify a model in this class we must specify the number of trials n, the event E which is the subject of our focus, and the probability p of the occurrence of E on a trial. Since the occurrence of E on a trial is called success, p expresses the probability of success on each trial.

Of particular interest is the event that k successes occur in n Bernoulli trials. If X denotes the number of successes in n Bernoulli trials, then the probability of k successes is written as $P(X = k)$, which is given by:

$$P(X = k) = C(n, k) p^k q^{n-k}, \text{ for } k = 0, 1, 2, \ldots, n, \text{ where } q = 1 - p.$$

In many ways this is the easier part of our analysis. The harder part is to apply Bernoulli trial models to real world problems. What makes this harder, at least in certain problems, is that we must make a judgment call as to whether a Bernoulli trial model is a realistic fit for the random process being studied; judgment calls of this type are often difficult and may be rather controversial.

EXAMPLE 2

Records show that a marksman hit the bullseye two-thirds of the time. To win a prize he must hit the bullseye at least 4 times in 5 firings. What is the probability that he will a prize?

Our first task is to set up a probability model for the process, which is to fire at the target five times. Since we have a situation which can clearly be viewed as a sequence of 5 repetitions of the basic process of firing at the target, the Bernoulli trial model defined by $n = 5$, $E =$ the event that the marksman hits the bullseye, $p = 2/3$ with $q = 1/3$, emerges as a strong possibility, depending on the assumptions involved.

The assignment $p = 2/3$ for the probability of success on each firing is based on long term statistical evidence of how well the marksman performed in the past. This is "reasonable" and the best we can do, but it does assume that the conditions under which the firings take place are uniform and that the marksman is warmed up and performing at his usual level.

As to the independence condition, we are assuming that the outcome of any firing has no affect, or at worst a negligible affect, on any other firing.

If we agree that these conditions are realistic, then we emerge with the Bernoulli trial model defined by $n = 5$, $E =$ the event that the marksman hits the bullseye, $p = 2/3$ with $q = 1/3$, as our model for the firings. Let X denote the number of successes in 5 trials. The probability that the marksman wins a prize is $P(X \geq 4) = P(X = 4) + P(X = 5)$, which is given by:

$$C(5, 4)(2/3)^4(1/3) + C(5, 5)(2/3)^5$$
$$= 0.329 + 0.132$$
$$= 0.461$$

The relative frequency interpretation of this result is that if the sequence of five firings took place a large number of times under the assumed conditions, then the marksman would win a prize approximately 46.1% of the time.

EXAMPLE 3

Records kept by Larry's Apparel Shop indicate that 20% of the shoppers who come into the store to browse actually make a purchase. What is the probability that among 100 shoppers who examine Larry's merchandise at least 30 will make a purchase?

We can identify each of the 100 shoppers with a trial, so that a sequence of 100 trials corresponding to the 100 shoppers emerges. We have $n = 100$. Take for E the event that a shopper makes a purchase. From the statistical data available on the percentage of shoppers who make a purchase, it seems reasonable to take $p = 0.20$.

As to how realistic it is to take $p = 0.20$ for E for all trials, that is, for all shoppers, and to what extent the independence condition is satisfied by the trials, we should keep in mind that shoppers are sometimes acquainted and are influenced by the purchase behavior of friends or relatives. If we are willing to assume that such interaction can be regarded as negligible, then we are led to the Bernoulli trial model defined by $n = 100$, $E =$ the event that a shopper makes a purchase, $p = 0.20$ with $q = 0.80$. Let X denote the number of successes (purchases made) in 100 trials. We require $(P(X \geq 30) = P(X = 30) + \cdots + P(X = 100)$, which is given by:

$$C(100, 30)(0.20)^{30}(0.80)^{70} + \ldots + C(100, 100)(0.20)^{100}$$

It can be shown that this value is approximately 0.0087.

This conclusion is valid with respect to our model, no-more and no-less. If our feeling is that this model does not reflect reality sufficiently closely, then it would be foolish to implement its conclusion. The wiser course would be to try to refine the model or develop another which better reflects reality.

Approximation of Bernoulli Trial Probabilities

As Example 3 makes clear, Bernoulli trial models may give rise to substantial computation problems, even in an era which boasts of powerful computation machinery. Abraham de Moivre (1667–1754) recognized this in the period immediately following the publication of Bernoulli's *Ars Conjectandi* (The Art of Conjecture) in 1713 and set himself the problem of developing a tool to approximate Bernoulli trial probabilities. de Moivre's efforts met with success and his work on this problem first appeared in a privately printed pamphlet dated November 12, 1733. de Moivre's work marks the first appearance of what is today called a normal curve approximation for Bernoulli trial probabilities. We turn to these matters in Sections 9.3 and 9.4.

EXERCISES

Set up Bernoulli trial models for the following situations, state the underlying assumptions and comment on their realism, and determine or state (in combinatorial form) the probabilities asked for. In each case express a judgment on the suitability of a Bernoulli trial model for the process in question. Is a more realistic model available for any of the situations described? If so, state the model, comment on its realism, and determine the probabilities required.

1. A coin is tossed 5 times in succession. Find the probability that **(a)** head shows in 2 of the 5 tosses; **(b)** head shows in at least 3 of the 5 tosses.

2. Records show that a slot machine, known as stingy to its patrons, has hit the jackpot 1% of the time over its life of ten years. If you play the machine 20 times, what is the probability that you will win at least once?

3. A die is tossed 6 times in succession. What is the probability that an even number shows at least 4 times?

4. Consider an examination that consists of 10 multiple-choice questions. Each question allows four choices, with only one answer being correct. If a student guesses the answer for each question, what is the probability that he will pass the exam (6 or more correct answers)?

5. According to company records, 1% of the output of light bulbs in a certain mass production process are defective. What is the probability that of 1,000 light bulbs produced, no more than 15 are defective?

6. The fruit fly *Drosophila* has been widely used for genetic studies and has furnished much of the evidence for modern genetic theory. It has been shown that mutations can be induced in *Drosophila* by subjecting the flies to X-rays, ultraviolet radiation, or any high-energy radiation, as well as chemical compounds such as those in the mustard gas family. Evidence suggests that 0.01% of large batches of *Drosophila* subjected to high-energy radiation develop a mutation. What is the probability that of 20,000 *Drosophila* subjected to high-energy radiation, at least one will develop a mutation?

7. A recently developed antibiotic has been found to cause an allergic reaction in one out of 200 people injected. Find the probability that fewer than 100 reactions occur among 10,000 people injected with the antibiotic.

9.3. NORMAL CURVES

The normal curves are bell-shaped, with a single peak, and symmetric with respect to the line that is drawn through the peak. All normal curves lie above the horizontal axis; as we move along the horizontal axis away from the value at which a curve peaks, the curve gets closer and closer to the horizontal axis, but does not intersect it. The value at which a curve peaks is called its **mean.** Figure 9.1 shows three normal curves with the same mean, denoted by μ. These curves differ in a characteristic, called the **standard deviation** of the curve, which is a measure of the concentration of a curve about

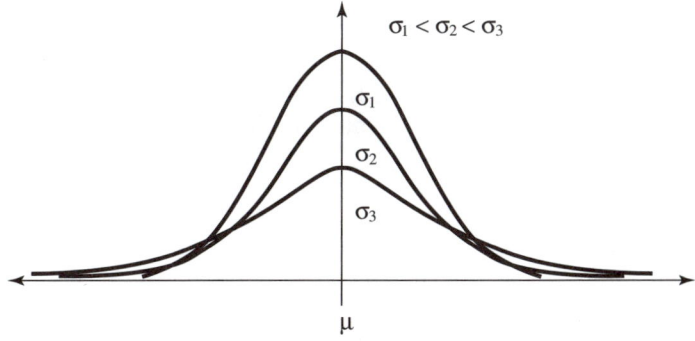

Figure 9.1

its mean; the smaller the standard deviation, the higher the peak of the curve, and the more concentrated is the curve about its mean. The standard deviations σ_1, σ_2, and σ_3 of the three normal curves shown in Figure 9.1 satisfy $\sigma_1 < \sigma_2 < \sigma_3$.

A normal curve is completely determined by its mean and standard deviation. The general equation of the normal curve with mean μ and standard deviation σ is

$$y = \frac{1}{\sigma\sqrt{2\pi}} e^{-(x-\mu)^2/2\sigma^2}$$

where π, the ratio of the circumference of a circle to its diameter, is approximately 3.14159, and e is a constant whose approximate value is 2.71828. When values for μ and σ are given, a specific normal curve is determined; for that normal curve values of y can be determined by substituting values for x. The role played by μ and σ for the normal curves is analogous to the role played by the parameters m and b for the lines defined by $y = mx + b$. When values for m and b are given, a specific line is determined; for that line values of y can be determined by substituting values for x.

Fortunately, our work with normal curves does not require us to plot a curve from its equation; a quickly drawn sketch of a normal curve which shows its general character and mean will suffice. Our concern is with determining areas of regions which have a normal curve as their upper boundary. At first sight it might seem that this would require that we determine normal curve areas for each μ and σ that arise in practice. Once again we can breath a sigh of relief that this is not the case. We can restate an area determination problem that arises for any normal curve in terms of an area determination problem for a comparable region of a normal curve adopted as a standard. This **standard normal curve**, as it is called, has mean $\mu = 0$ and standard deviation $\sigma = 1$.

Given any normal curve with mean μ and standard deviation σ we transfer the scene to the standard normal curve by carrying out a change of scale by means of the formula:

$$z = \frac{x - \mu}{\sigma}$$

In this new z-scale (see Figure 9.2) z tells us in terms of standard deviation units how many standard deviations the x-value lies above or below its mean μ. For $z = 1$, for example, we have:

$$1 = \frac{x - \mu}{\sigma}$$
$$\sigma = x - \mu$$
$$x = \mu + \sigma$$

That is, $z = 1$ corresponds to x being one standard deviation unit above the mean μ in the x-scale, which is what Figure 9.2 shows.

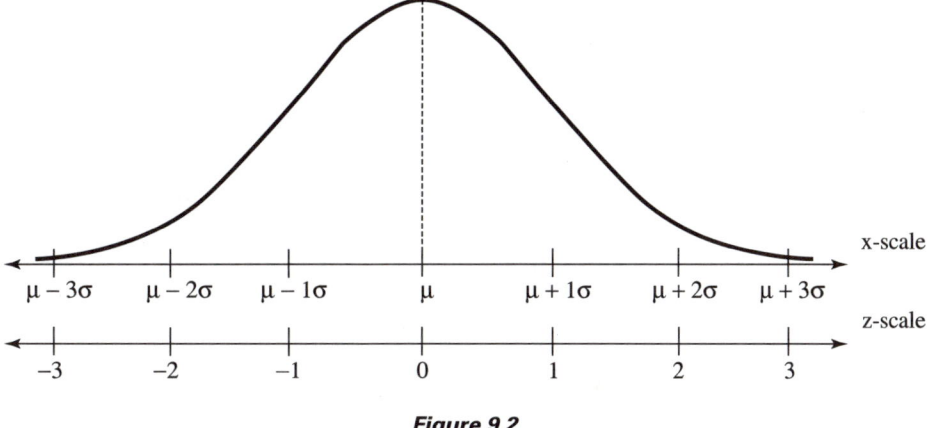

Figure 9.2

Areas of Regions Determined by the Standard Normal Curve

We first show how areas of regions determined by the standard normal curve can be obtained and then show how area determination problems with respect to other normal curves can be shifted to the standard normal curve.

Areas of regions defined by the standard normal curve have been tabulated and Table A (page 378) gives area values to four places. A short version of this table is reproduced here as Table 9.1. Table A specifies the standard normal curve area of the shaded region shown in Figure 9.3 for values of z from 0.00 to 3.09. The z values in our abbreviated Table 9.1 range from 0.00 to 1.19. The values in both tables give the area, to four places, of the region between the mean $\mu = 0$ and the given value of z. To obtain the area of the region between 0 and 0.63, for example (see Figure 9.4), we move across the row labeled 0.6 until we reach the column headed by 3, the last digit of 0.63. This yields 0.2357 as the area of our region, correct to four places.

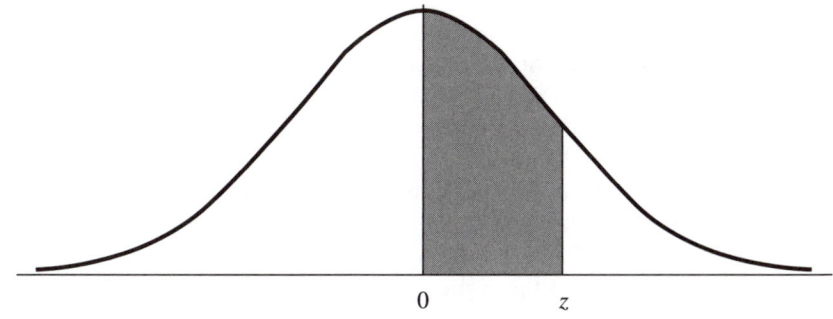

Figure 9.3

Table 9.1

z	.00	.01	.02	.03	.04	.05	.06	.07	.08	.09
0.0	.000	.0040	.0080	.0120	.0160	.0199	.0239	.0279	.0319	.0359
0.1	.0398	.0438	.0478	.0517	.0557	.0596	.0636	.0675	.0714	.0753
0.2	.0793	.0832	.0871	.0910	.0948	.0987	.1026	.1064	.1103	.1141
0.3	.1179	.1217	.1255	.1293	.1331	.1368	.1406	.1443	.1480	.1517
0.4	.1554	.1591	.1628	.1664	.1700	.1736	.1772	.1808	.1844	.1879
0.5	.1915	.1950	.1985	.2019	.2054	.2088	.2123	.2157	.2190	.2224
0.6	.2257	.2291	.2324	.2357	.2389	.2422	.2454	.2486	.2517	.2549
0.7	.2580	.2611	.2642	.2673	.2704	.2734	.2764	.2794	.2823	.2852
0.8	.2881	.2910	.2939	.2967	.2995	.3023	.3051	.3078	.3106	.3133
0.9	.3159	.3186	.3212	.3238	.3264	.3289	.3315	.3340	.3365	.3389
1.0	.3413	.3438	.3461	.3485	.3508	.3531	.3554	.3577	.3599	.3621
1.1	.3643	.3665	.3686	.3708	.3729	.3749	.3770	.3790	.3810	.3830

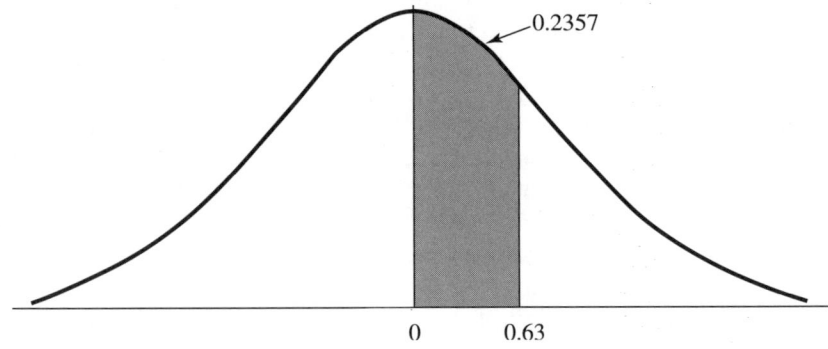

Figure 9.4

To obtain the area value corresponding to a negative value of z, $z = -1.18$, for example, we make use of the fact that the area of the region between 0 and -1.18 is the same as the area of the region between 0 and 1.18 because of the symmetry of the curve, and look up the area corresponding to $z = 1.18$. We thus find that the area of our region, shown in Figure 9.5, is 0.3810.

The area of the region under each of the normal curves is 1. In terms of the standard normal curve, the area of the region to the right of the line of symmetry, determined by $\mu = 0$, is 0.5000 as is the area of the region to the left of $\mu = 0$. To obtain the area of a region to the right of a positive value of z, $z = 0.94$, for example (see Figure 9.6) we subtract the area of the unwanted region between 0 and 0.94, 0.3264, from 0.5000 to obtain 0.1736.

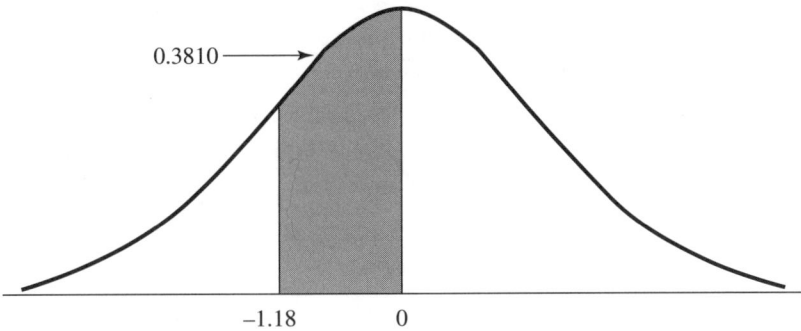

Figure 9.5

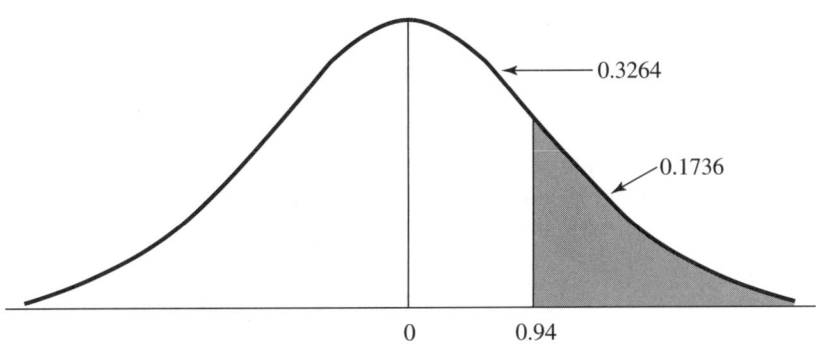

Figure 9.6

To obtain the area of a region to the left of a positive value of $z, z = 0.89$, say (see Figure 9.7), simply add 0.5000 to 0.3133, the tabular value for $z = 0.89$. This yields 0.8133 as the area of the region to the left of $z = 0.89$.

Also of importance are problems which lead to the normal curve area of a region between two given values of z. If both z values are on the same side of the mean, that

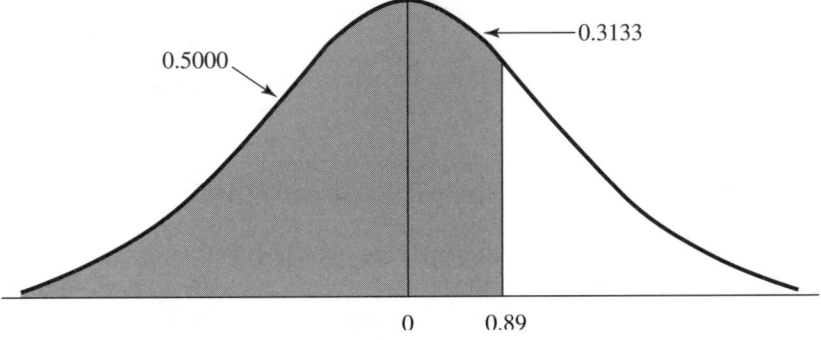

Figure 9.7

is, both are positive or both are negative, then the area of the region between them is the difference in their tabular values. For example, the area of the region shown in Figure 9.8 determined by $z_1 = 0.72$ and $z_2 = 1.13$, is $0.3708 - 0.2642 = 0.1066$.

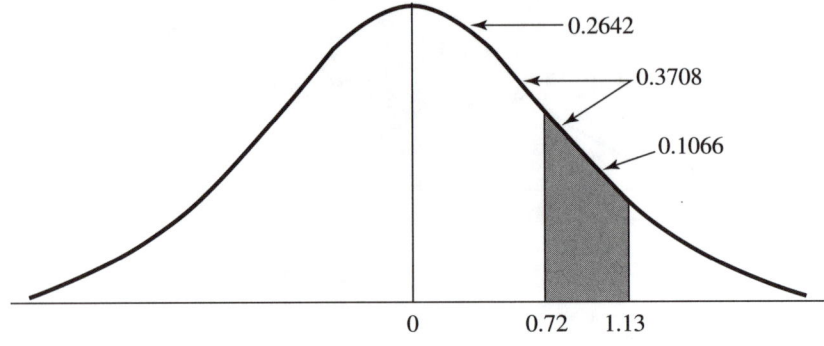

Figure 9.8

If the two z values are on opposite sides of the mean, that is, one is positive and one is negative, then the area of the region between them is the sum of their tabular values. For example, the area of the region shown in Figure 9.9, determined by $z_1 = -0.64$ and $z_2 = 1.18$, is $0.2389 + 0.3810 = 0.6199$.

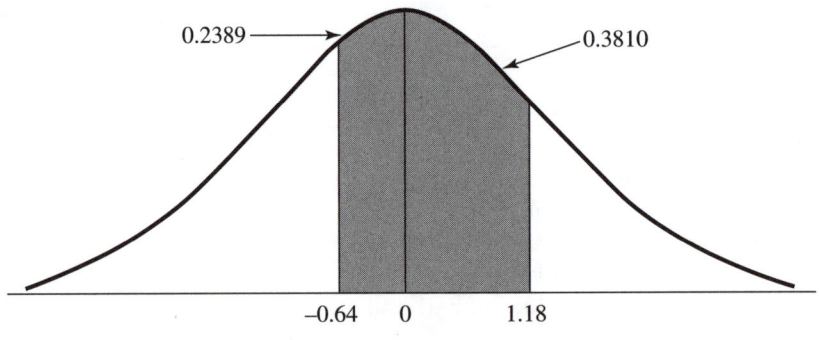

Figure 9.9

With all such area determination problems it is a good idea to make a diagram showing the region of concern. It then becomes much clearer how to proceed.

Areas of Regions Determined by Normal Curves

More generally, consider the general normal curve with mean μ and standard deviation σ and let us suppose that we are required to find the area of the region bounded by x_1 and x_2 as shown in Figure 9.10(a). To do this we shift the scene to the standard normal curve, determine the z_1 and z_2 values corresponding to x_1 and x_2. from the relations

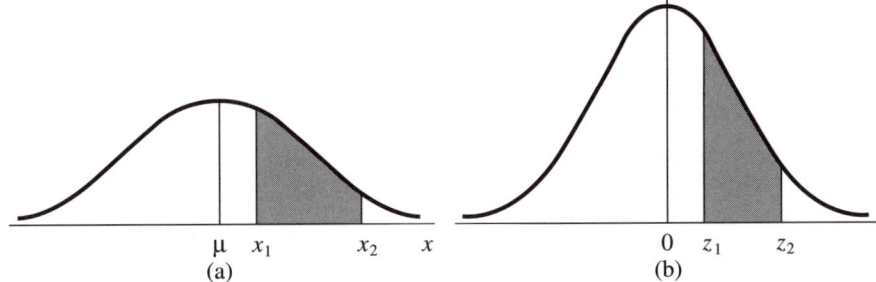

Figure 9.10

$$z_1 = \frac{x_1 - \mu}{\sigma}, \quad z_2 = \frac{x_2 - \mu}{\sigma},$$

and find the area of this comparable region bounded by z_1 and z_2 (see Figure 9.10(b)). The area of this standard normal curve region is equal to the area of the original region (see Figure 9.10(a)).

EXAMPLE 1

For the normal curve with mean $\mu = 4$ and standard deviation $\sigma = \sqrt{2}$, find the area of the region bounded by $x_1 = 4.5$ and $x_2 = 5.5$, shown in Figure 9.11.

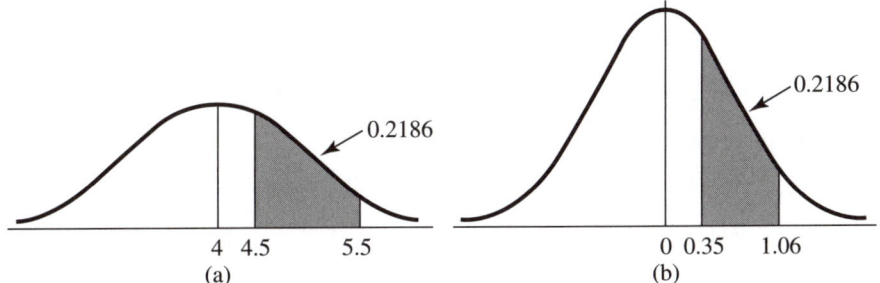

Figure 9.11

Corresponding to $x_1 = 4.5$ and $x_2 = 5.5$ we have:

$$z_1 = \frac{4.5 - 4}{\sqrt{2}} = \frac{0.5}{1.414} = 0.35$$

$$z_2 = \frac{5.5 - 4}{\sqrt{2}} = \frac{1.5}{1.414} = 1.06$$

The region under the standard normal curve between these z values, shown in Figure 9.11(b), has area $0.3554 - 0.1368 = 0.2186$. Thus the region under the given normal curve between $x_1 = 4.5$ and $x_2 = 5.5$ has area 0.2186.

EXERCISES

Find the area of the region under the standard normal curve that is defined by the following bounds.

1. Between 0 and $z = 2.38$.
2. Between 0 and $z = -1.46$.
3. To the right of $z = 1.56$.
4. To the left of $z = 2.04$.
5. To the right of $z = -1.20$.
6. Between $z_1 = -0.84$ and $z_2 = 1.58$.
7. Between $z_1 = 0.82$ and $z_2 = 1.98$.
8. Between $z_1 = -1.96$ and $z_2 = 1.90$.
9. To the left of $z = -1.24$.
10. To the right of $z = -0.50$.
11. Between $z_1 = -2.02$ and $z_2 = -0.44$.
12. To the left of $z = -1.08$.

Find the area of the region under the normal curve defined by the following values of μ and σ with respect to the given bounds.

13. $\mu = 12$, $\sigma = 4$, between $x_1 = 13$ and $x_2 = 16$.
14. $\mu = 20$, $\sigma = 6$, to the left of $x = 23$.
15. $\mu = 52$, $\sigma = 10$, between $x_1 = 50$ and $x_2 = 55$.
16. $\mu = 110$, $\sigma = 5$, to the right of $x = 115$.
17. $\mu = 64$, $\sigma = 2$, between $x_1 = 66$ and $x_2 = 68$.
18. $\mu = 82$, $\sigma = 6$, between $x_1 = 80.2$ and $x_2 = 83.3$.
19. $\mu = 100$, $\sigma = 4.28$, to the right of $x = 97.3$.
20. $\mu = 32.2$, $\sigma = 2.46$, to the left of $x = 30.1$.
21. $\mu = 4$, $\sigma = \sqrt{2}$, to the right of $x = 4.5$.
22. Show that the area of the region under the normal curve with mean μ and standard deviation σ
    (a) between $x_1 = \mu - \sigma$ and $x_2 = \mu + \sigma$ is 0.6826;
    (b) between $x_1 = \mu - 2\sigma$ and $x_2 = \mu + 2\sigma$ is 0.9544;

(c) between $x_1 = \mu - 3\sigma$ and $x_2 = \mu + 3\sigma$ is 0.9972.

Draw diagrams.

In summary, these results say that for any normal distribution roughly 68% of the area is within one standard deviation of the mean, roughly 95% of the area is within two standard deviations of the mean, and roughly 99.7% of the area is within these standard deviations of the mean.

9.4. NORMAL CURVE APPROXIMATION OF BERNOULLI TRIAL PROBABILITIES

Consider a Bernoulli trial process with $n = 8$ trials and probability $p = \frac{1}{2}$ of success on each trial. If we let X denote the number of successes in 8 trials, then

$$P(X = k) = C(8, k)(\tfrac{1}{2})^k(\tfrac{1}{2})^{8-k}$$

For $k = 0, 1, \ldots, 8$, we obtain

$P(X = 0) = 0.0039 \qquad P(X = 5) = 0.2188$

$P(X = 1) = 0.0312 \qquad P(X = 6) = 0.1094$

$P(X = 2) = 0.1094 \qquad P(X = 7) = 0.0312$

$P(X = 3) = 0.2188 \qquad P(X = 8) = 0.0039$

$P(X = 4) = 0.2734$

These probabilities can be described geometrically as areas by means of a **histogram** by locating the possible values of X on the x-axis and then, centered on each of these values, drawing a rectangle whose base has length 1 and whose height is the probability of the k value on which the rectangle is centered. We obtain the histogram shown in Figure 9.12. The rectangle centered at 0 has area 0.0039, which is the probability that $X = 0$; the rectangle centered at 1 has area 0.0312, which is the probability that $X = 1$, and so on.

If we superimpose on this histogram the normal curve with mean $\mu = np = 8(\tfrac{1}{2}) = 4$ and standard deviation

$$\sigma = \sqrt{np(1-p)} = \sqrt{8(\tfrac{1}{2})(\tfrac{1}{2})} = \sqrt{2}$$

we obtain a remarkably good fit, shown in Figure 9.13. The possibility of estimating Bernoulli trial probabilities by areas of regions under a normal curve is certainly suggested. For instance, consider the probability that $X = 5$, for which we have

$$P(X = 5) = 0.2188$$

In terms of the histogram, this probability is the area of the rectangle centered at 5 whose base extends over the interval from 4.5 to 5.5. From Example 1 of Section 9.3 we have that the area of the region under the superimposed normal curve (with $\mu =$

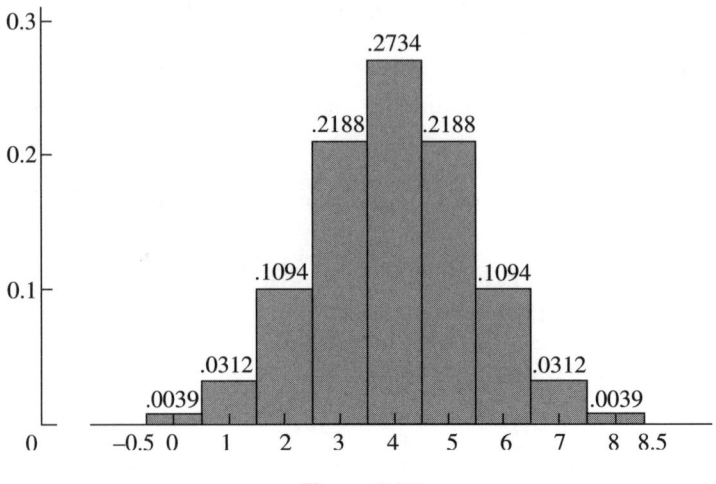

Figure 9.12

4, $\sigma = \sqrt{2}$) between $x_1 = 4.5$ and $x_2 = 5.5$ is 0.2186, which differs from 0.2188 by 0.0002.

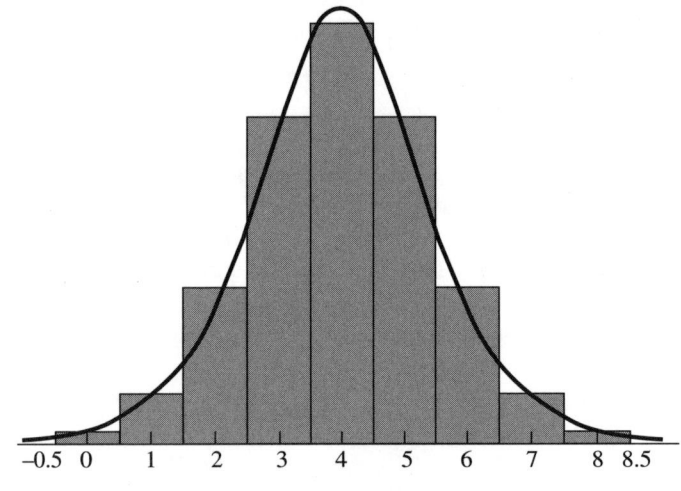

Figure 9.13

The probability that $X \geq 5$ is the sum of the areas of the histogram rectangles centered at 5, 6, 7, and 8. We have

$$P(X \geq 5) = 0.2188 + 0.1094 + 0.0312 + 0.0039$$
$$= 0.3633$$

The area under the superimposed normal curve to the right of $x = 4.5$ is 0.3632 (see Exercise 21, Sec. 9.3), which differs from 0.3633 by 0.0001.

The accuracy of these approximations suggests that the normal curves, under appropriate conditions, can be used to estimate Bernoulli trial probabilities; this is indeed the case. More generally, consider a Bernoulli trial process with n trials, where the probability of success on each trial is p. Let X denote the number of successes in n trials.

$$P(X = k) = C(n, k)p^k(1 - p)^{n-k}, \text{ for } k = 0, 1, \ldots, n$$

We shall more formally state a theorem in a moment, but the basic idea is the following: Suppose that we wish to estimate $P(X = k)$ for some value of k, 70 let us say, and that conditions are such that we have a green light to proceed. In geometric terms $P(X = 70)$ is the area of a rectangle with height $P(X = 70)$ whose base is centered at 70 and extends from 69.5 to 70.5 (see Figure 9.14(a)). Our problem is to approximate the area of this rectangle, which is shown in Figure 9.14(b) along with the "appropriate" normal curve arc which appears over the base from 69.5 to 70.5. The region defined by the normal curve differs from the rectangle in that it includes A, which is not part of the rectangle and does not include B, which is part of the rectangle. If the areas of A and B are approximately equal, and under suitable conditions they are, then the gain is balanced by the loss, and the area of the normal curve region is approximately that of the area of the rectangle, which is $P(X = 70)$, the probability of 70 successes.

Figure 9.14

If instead of 70 successes we talk more generally about c successes, then the probability rectangle whose area is to be approximated is centered at c with base extending from $c - 1/2$ to $c + 1/2$.

To take another example if we wish to approximate $P(c \leq X \leq d)$, the probability that the number of successes is between c and d, inclusive, then we are looking at a sequence of rectangles, the first centered at c and the last centered at d, the sum of whose areas is to be approximated. To include the areas of the first and last of these rectangles in our estimate we must take as our base the interval bounded by $c - 1/2$ and $d + 1/2$ (see Figure 9.15).

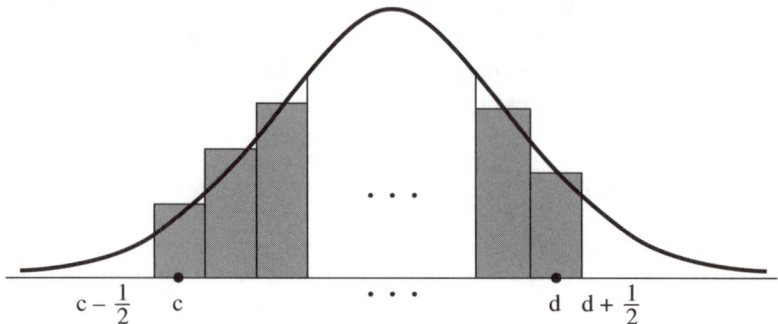

Figure 9.15

There are a number of variations on this basic theme which these examples hopefully indicate. More formally, we have the following theorem which was in part established by Abraham de Moivre and refined by Pierre Simon de Laplace.

de Moivre - Laplace Theorem

If np and nq are both at least 5 (our green light), then the number of successes in n Bernoulli trials, and Bernoulli trial probabilities in general, may be approximated by areas of normal curve regions, the normal curve being determined by mean μ and and standard deviation σ defined by:

$$\mu = np, \quad \sigma = \sqrt{npq}$$

If, for example, we seek to estimate $P(c \leq X \leq d)$, the normal curve region is bounded by $c - 1/2$ and $d + 1/2$. If we seek to estimate $P(x \leq c)$, the normal curve region is the one to the left of $c + 1/2$, and so on.

Normal curve approximations of Bernoulli trial probabilities are best when p and q are close to 1/2. If p or q is very small, then n has be to sufficiently large to overcome the smallness of p or q to allow use of the normal curve approximation. This is what the guideline conditions $np \geq 5$, $nq \geq 5$ give us. If one of np and nq is less than 5, the normal curve approximation should ordinarily not be used. In such cases the Poisson distribution, which we shall not take up here, often provides excellent approximations.

EXAMPLE 1 RETURN TO LARRY'S APPAREL SHOP

Records kept by Larry's Apparel Shop indicate that 20% of the shoppers who come into the store to browse actually make a purchase. What is the probability that among 100 shoppers who examine Larry's merchandise at least 30 will make a purchase?

Our first task is to set up a probability model for this process. In examining this situation in Example 3 of Section 9.2 (page 285) we were led to take for this process a Bernoulli trial model with $n = 100$, $E =$ the event that a shopper makes a purchase, $p = 0.20$. If we let X denote the number of purchases made in 100 trials, then the problem is to determine $P(X \geq 30)$.

To determine the feasibility of a normal curve approximation of $P(X \geq 30)$, we must examine np and nq.

$$np = 100(0.20) = 20, \qquad nq = 100(0.80) = 80$$

Since both values are greater than or equal to 5, we have a green light and may proceed.

The normal curve that we bring to this situation has mean $\mu = np = 20$ and standard deviation $\sigma = \sqrt{npq} = 4$. Our problem is to determine the area of the region to the right of $x_1 = 29.5$ (see Figure 9.16(a)). The z value corresponding to $x_1 = 29.5$ is $z_1 = 2.38$ (see Figure 9.16(b)). The area of the unwanted standard normal curve region between 0 and 2.38 is 0.4913, so that the area of the region is to the right of 2.38 is $0.5000 - 0.4913 = 0.0087$.

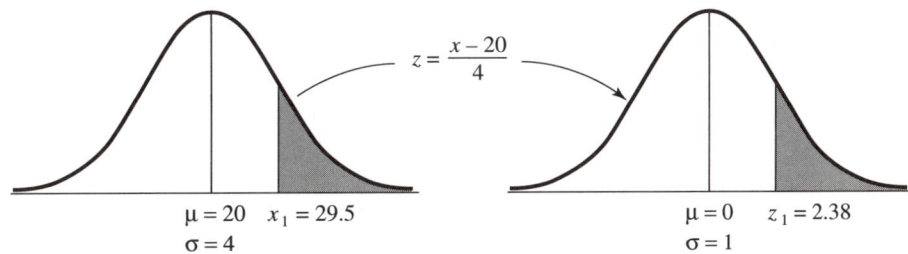

Figure 9.16(a)   **Figure 9.16(b)**

Thus, the probability, in terms of our Bernoulli trial model, that among 100 shoppers who examine Larry's merchandise at least 30 will make a purchase is 0.0087. If we think in terms of groups of 100 shoppers, then at least 30 out of 100 will make a purchase, for many groups of 100 (the long run in this case), approximately 0.87% of the time.

The Importance of the Normal Curves

The importance of the normal curves as an approximation tool in probability and statistics goes far beyond the problems we have considered in estimating Bernoulli trial probabilities. The use of the normal curves to approximate Bernoulli probabilities was pioneered by Abraham de Moivre (1667–1754) as early as 1733. Their use was considerably extended by Pierre Simon Laplace (1749–1827) and they played a central role in the theory of error investigations of Karl Friedrich Gauss (1777–1855). As a result the normal curves are sometimes referred to as the Laplacean or Gaussian distributions.

From the time of Laplace and Gauss until the late 1930's the central problem of probability and statistics was that of determining precise mathematical conditions under which a probability distribution can be approximated by means of a normal curve. Answers to this problem are given in a number of theorems that are known as central limit theorems.* The de Moivre-Laplace theorem belongs to this class of theorems.

EXERCISES

1. Let X denote the number of successes in connection with a Bernoulli trial model with $n = 100$ and $p = 2/5$. Find **(a)** $P(X = 30)$, **(b)** $P(X = 25)$, **(c)** $P(X \leq 50)$, **(d)** $P(X > 30)$, **(e)** $P(20 \leq X \leq 40)$.

2. Let X denote the number of successes in connection with a Bernoulli trial model with $n = 200$ and $p = 0.30$. Find **(a)** $P(X = 60)$, **(b)** $P(X = 70)$, **(c)** $P(X \geq 60)$, **(d)** $P(40 \leq X \leq 50)$, **(e)** $P(X > 40)$.

In Exercises 3–8 state the probability model employed and the assumptions made.

3. A recently developed antibiotic for a spectrum of strep infections has been found to cause an allergic reaction in 5% of the people who take it. Find the probability that of 10,000 people given the antibiotic:

    (a) At least 551 allergic reactions will occur.

    (b) At most 490 allergic reactions will occur.

    (c) Between 480 and 510 allergic reactions will occur.

4. In the mass production of shafts for engine mounts at the MacNeil Company the probability that a defective shaft is produced is 0.02. For a production run of 12,000 shafts find the probability that:

    (a) No more than 200 defectives are produced.

    (b) At least 220 defectives are produced.

    (c) Between 230 and 260 defectives are produced.

5. Find the probability that a student will get at least 70 correct answers on a 100 question true - false test in economics if he answers each question by flipping a balanced coin and recording T if head shows and F if tail shows.

6. Dubin Outlets estimates that 2% of its accounts receivable cannot be collected. Find the probability that of 600 accounts receivable:

    (a) No more than 10 will not be collectable.

    (b) No more than 15 will not be collectable.

    (c) At least 20 will not be collectable.

---

*A historical account of the origin and evolution of the central limit theorems is presented in the author's book *The Life and Times of the Central Limit Theorem* (New York: Kaedmon Pub. Co., 1974).

7. A multicomponent flu vaccine has been found to be effective on 99% of the people inoculated. Of 500,000 who are inoculated, find the probability that:
   (a) Fewer than 2000 come down with the flu.
   (b) More than 4900 come down with the flu.
   (c) Between 4950 and 5100 come down with the flu.

8. An advertising agency claims that 22% of television viewers watch the new hit program *Happy Days and Foolish Nights*. Assuming this to be the case, if 300 television viewers are selected, what is the probability that:
   (a) At least 65 watch the program?
   (b) At most 80 watch the program?
   (c) Between 75 and 100 watch the program?

SELF-TESTS FOR CHAPTERS 6–9

Allow 90 or so minutes for each self-test. Go over each one before proceeding to the next.

Self-Test 1

1. Hasty Harry obtained from a friend of a friend a die whose behavior, he was told, is described by the probability model with sample space $S = \{1, 2, 3, 4, 5, 6\}$ and probability function P defined by:

   $$P(1) = P(3) = P(5) = 1/11;\ P(4) = 2/11,\ P(2) = P(6) = 3/11$$

   On the basis of this probability model Harry determined the probability of an even number showing to be 8/11. He expected that if his die were tossed 1100 times an even number would show $(8/11)(1100) = 800$ times, approximately. Harry participated in a "friendly' game one evening and after 1100 tosses of his die an even number had showed 239 times and Harry owed 672 dollars. He could not understand what had gone wrong and posed the following questions which should be answered in appropriate detail.
   (a) Does this evidence mean that my conclusion isn't valid after all? Explain.
   (b) If mathematical reasoning is as precise as it is reputed to be, how could this happen? Explain.
   (c) Isn't my conclusion true? Explain.

2. The following is a probability model for one of Dapper Dan's dice: $S = \{1, 2, 3, 4, 5, 6\}$,

   $$P(1) = \frac{1}{10},\ P(2) = \frac{2}{10},\ P(3) = \frac{1}{10},\ P(4) = \frac{3}{10},\ P(5) = \frac{1}{10},\ P(6) = \frac{2}{10}$$

Let A denote the event than an even number shows.
   (a) Find $P(A)$.
   (b) State the relative frequency interpretation of the conclusion obtained in (a).
   (c) In tossing Dan's die 2000 times an even number showed 800 times. Does this evidence establish that the conclusion reached in (a) is not valid? Explain.
   (d) Is the conclusion reached in (a), interpreted in relative frequency terms, true? Explain.
   (e) Find the probability that an odd number shows given that a number less than 4 shows.
   (f) Are the events "a number greater than 2 shows" and "an even number shows" independent? Explain.

3. Arnold Williamson believes that the probability that he will receive an A in the statistics course he is taking this semester is 0.95. How is the probability value to be interpreted? Explain.

4. In the production of electronic tubes the Marks Company employs three machines, I, II, and III. Machine I produces 30 percent of the output, of which 1 percent are defective, machine II produces 35 percent of the output, of which 0.5 percent are defective, and the remainder is produced by machine III, with 1.5 percent being defective. A tube is selected at random from the weekly output. If it is found to be defective, what is the probability that **(a)** it was made by machine I; **(b)** it was made by machine II; **(c)** it was made by machine III?

5. The Philosophy Club of Ecap University has 20 members. In how many ways can the positions of president, vice-president, treasurer, and secretary be filled if no person is to hold more than one position?

6. In LOTTO you must pick 6 numbers out of 44. The order in which the numbers are chosen is not significant. In how many ways can this be done?

7. To determine whether to accept a large lot of items offered for sale, a buyer takes a sample (with replacement) of seven items from the lot and inspects them. If more than one defective is found, she rejects the entire lot; otherwise, she accepts the lot. What is the probability that the buyer accepts the lot when it contains 5 percent defectives? What assumption underlies your probability model?

Self-Test 2

1. The following is a probability model for tossing another of Dapper Dan's dice. $S = \{1, 2, 3, 4, 5, 6\}$;

$$P(1) = P(2) = \frac{3}{16}, P(3) = \frac{4}{16}, P(4) = \frac{1}{16}, P(5) = \frac{2}{16}, P(6) = \frac{3}{16}$$

   (a) Find the probability that a number greater than 3 shows.

(b) State the relative frequency interpretation of the result obtained in answer to (a).

(c) If an odd number shows, what is the probability that a number greater than 3 shows?

(d) Are the events "a number greater than 2 shows" and "an even number shows" independent? Explain.

(e) In tossing Dan's die 800 times a number greater then 3 showed 496 times. Dan, who expected to be winning around $200 at this point, found himself $192 in the red. He could not understand what had gone wrong and posed the following two questions which should be answered in appropriate detail.

   (i) Does this evidence mean that the probability value obtained in (a) for a number greater than 3 showing isn't valid?

   (ii) If mathematical reasoning is as precise as it is reputed to be, how could I end up with $192 in the red instead of $200 in the black? Explain.

2. A measles vaccine has been found to be ineffective on 1 out of 1000 people who are given it. Plans call for administering the vaccine to 60,000 people in Bell City.

(a) Describe a probability model for this process.

(b) On what assumption(s) is your model based?

(c) Is your assumption realistic? Explain.

(d) In terms of your model state (but do not compute) the probability that there are at most 70 cases where the measles vaccine is ineffective.

(e) Is it possible to approximate the probability of at most 70 vaccine failures by means of a normal curve estimate? Explain.

(f) Assuming that the answer to (e) is yes, determine a normal curve estimate for the probability of at most 70 vaccine failures.

3. Population $Q = \{2, 6, 10\ 14, 16\}$. Consider the process of choosing a sample of size 2 at random from Q.

(a) Define a probability model for this process.

(b) What assumption underlies your model?

(c) Determine the sampling distribution of $\bar{x}$

(d) Find $P(8 \leq \bar{x} \leq 11)$.

4. The Charles Steel works, a manufacturer of steel plates, devised an inexpensive X-ray technique for detecting flaws in the plates. The following data are available on the reliability of the technique. Of the plates that were known to contain flaws that were tested, 98.5 percent of the cases were detected, and 1.5 percent went undetected; 97 percent of the flawless plates tested were correctly diagnosed as flawless, and 3 percent were misdiagnosed. It is estimated that 3 percent of the weekly output contain flaws. A plate is selected at random from the weekly output and examined by means of the X-ray technique.

(a) If a plate is diagnosed as containing a flaw, what is the probability that it actually contains a flaw?

(b) If a plate is diagnosed as not containing a flaw, what is the probability that it contains a flaw?

5. For the process of tossing a pair of dice, let E denote the event the sum of the numbers showing is even, and let F denote the event that the sum of the numbers showing is less than 6. Is $S = \{E, F\}$ a sample space for this process? Explain.

6. Define five sample spaces for the process of dealing a hand of six cards from a standard deck of 52 cards.

7. The Jansen Company makes light bulbs. The total daily output is 10,000 bulbs with 20 percent of this output being produced by plant J1, of which 0.5 percent are defective; 30 percent being produced by plant J2, of which 2 percent are defective; and 50 percent being produced by plant J3, of which 1 percent are defective. A bulb is selected at random from the day's output.

   (a) Construct a probability model for this selection process.

   (b) What assumption underlies your probability function?

   (c) Find the probability that (1) a defective bulb made by plant J2 is selected, (2) a good bulb made by plant J3 is selected, (3) a defective bulb is selected.

8. J. J. Johnson was interested in finding the probability of the event E that head shows twice on two successive tosses of a well-balanced coin. To do so, Johnson set up the following probability model for tossing the coin twice in succession:
   Sample space: $S = \{0, 1, 2\}$, where 0 is the event that no head shows, 1 is the event that head shows once, and 2 is the event that head shows twice.
   Probability function: $P(0) = P(1) = P(2) = \frac{1}{3}$.
   Johnson found that $P(E) = \frac{1}{3}$.

   (a) Is Johnson's conclusion valid? Explain.

   (b) Is Johnson's conclusion true? Explain.

9. A university population of 5000 students consists of 2000 males and 3000 females; 10 percent of the male students are studying accounting and 6 percent of the female students are studying accounting. A student is selected from this population.

   (a) Set up a probability model for this selection process.

   (b) What assumption underlies your probability model?

   (c) Find the probability that (1) a female student who is studying accounting is selected, (2) a male student who is not studying accounting is selected, (3) a student who is studying accounting is selected.

10. It is asserted that the probability of an event E is 0.95. State the relative-frequency and subjective-probability interpretations of this assertion, and describe the main features of these interpretations.

Self-Test 3

1. On the morning news the radio commentator says that "there is a 75 percent probability of rain today." How should this probabilistic statement be interpreted? Explain. What are the odds in favor of this event?

2. M. N. Vries was told, in a discussion with her brother, that since the most recent mortality statistics show that out of every 100,000 births 60,000 live to age 70 or longer, the probability that she will live to age 70 or longer is 60,000/100,000 = 0.60. How should this probability assignment be interpreted? Explain. What odds correspond to this probability assignment?

3. Bob Sommers feels that the odds are 9 to 2 against his getting an A on his next math test. What is the envisioned probability that this will be the case?

4. Evaluate $P(5, 5)$, $P(5, 2)$, $P(8, 3)$, $C(14, 2)$, $C(20, 3)$, and $C(30, 4)$.

5. Twelve universities belong to a football conference. How many games should be arranged in order to have each team play every other team once?

6. How many permutations are there of 4 cards chosen from a standard deck of 52 cards?

7. Find the number of permutations of the letters a, b, v, g, and n taken three at the time; that is, find the number of "three-letter words" with distinct letters that can be formed from a, b, v, g, and n.

8. If repetitions are not permitted, (a) how many four-digit numbers can be formed from the digits 1, 3, 5, 7, and 9? (b) How many of these numbers are less than 5000? (c) How many of these numbers are divisible by 5?

9. If repetitions are permitted, (a) how many four-digit numbers can be formed from the digits 1, 3, 5, 7, and 9? (b) How many of these numbers are less than 5000? (c) How many of these numbers are divisible by 5?

10. An election is to be held to elect 4 people to the local school board. How many election outcomes are possible if 12 names have been placed in nomination?

11. The personnel and budget committee of Ecap University has approved 6 assistant professor positions, 4 associate professor positions, and 2 full professor positions. In how many ways can promotions be granted if the promotion board has received 20 requests for promotion to assistant professor, 12 requests for promotion to associate professor, and 6 requests for promotion to full professor?

12. The mathematics department of City University has an opening at the instructor and assistant professor levels. If 30 applications have been received, in how many ways can these positions be filled?

13. A system of 3 linear equations is to be chosen from 6 given linear equations. In how many ways can this be done?

14. If $P(A) = 0.60$, $P(B) = 0.30$, and $P(A \cap B) = 0.20$, find (a) $P(A \cup B)$, (b) $P(A/B)$, (c) $P(B/A)$.

15. A hand of 6 cards is dealt from a standard deck of 52 cards. Set up a probability model for this process, and find the probability that the hand dealt contains (a) 3 tens, (b) 4 eights, (c) 2 tens and 3 aces, (d) 2 tens or 3 aces, (e) 3 clubs and 2 hearts, (f) 3 clubs or 2 hearts.

16. In connection with tossing a pair of well-balanced dice (one red and one green), are the events the sum showing is 7 and green shows 4 independent?

17. Events A, B, and C are independent, and $P(A) = \frac{1}{5}$, $P(B) = \frac{1}{4}$, and $P(C) = \frac{1}{3}$. Determine (a) $P(A \cap B \cap C)$, (b) $P(A^c \cap B^c \cap C)$, (c) $P(A^c \cap B)$, (d) $P(A \cup B)$, (e) $P(A \cup B^c)$.

18. A pair of dice are tossed 6 times in succession. Find the probability that (a) an even sum shows 4 times, and (b) an even sum shows 4 or more times. (c) What assumption underlies your probability model?

Self-Test 4

NOTE: Combinatorial expressions or products which may arise need not be multiplied out.

1. The Science Department of Ecap University has 80 faculty, 30 of which are biology faculty, 25 of which are chemistry faculty, 15 of which are physics faculty, and 10 of which are geology faculty. A committee of 8 is to be chosen from the Science Department to study curriculum changes.

    (a) Set up a probability model for the selection process.

    (b) State the assumption on which your probability model is based.

    (c) Find the probability that the committee chosen contains 3 biology faculty.

    (d) Find the probability that the committee chosen contains Robert Weiss and Janet Fox of the Science Department.

    (e) Find the probability that the committee chosen contains 3 biology faculty and 2 chemistry faculty.

    (f) Find the probability that the committee chosen contains 3 biology or 2 chemistry faculty.

2. In a certain mass production process the probability that a defective item is produced is 0.01. A run of 10,100 items is produced.

    (a) Set up a probability model for this production process.

    (b) On what assumption is your probability model based?

    (c) Let X denote the number of defectives produced. Use a normal curve approximation to find $P(X \le 105)$. As a point of information note that $\sqrt{99.99} = 10$.

3. The following is a probability model for the process of tossing a certain die: $S = \{1, 2, 3, 4, 5, 6\}$;

$$P(1) = \frac{2}{13}, \ P(2) = \frac{1}{13}, \ P(3) = \frac{3}{13}, \ P(4) = \frac{2}{13}, \ P(5) = \frac{3}{13}, \ P(6) = \frac{2}{13}$$

Let A denote the event than an odd number shows.

(a) Find $P(A)$.

(b) State the relative frequency interpretation of the conclusion reached in (a).

(c) In tossing the die described by the above model 1300 times an odd number was observed to show 795 times. Does this evidence establish that the conclusion reached in (a) is valid? Explain.

(d) Is the conclusion reached in (a), interpreted in relative frequency terms, true? Explain.

(e) Are the events "an odd number shows" and "a number greater than 3 shows" independent? Explain.

(f) Find the probability that a number greater than 3 shows given that an even number shows.

4. Let X denote the number of successes in connection with a Bernoulli trial model with $n = 625$ and $p = 1/5$. Use a normal curve approximation to find $P(124 \leq X \leq 130)$.

5. On the morning news the radio commentator says that "there is a 75 percent probability of rain today." How should this probabilistic statement be interpreted? Explain. What are the odds in favor of this event?

6. Experience indicates that 80 percent of the applicants for a stenographer's position with the Algis Corporation are competent, and 20 percent are not competent. A proficiency test has been developed, and it is estimated that a competent applicant will pass the test with probability 0.9, whereas an incompetent applicant will pass the exam with probability 0.3.

(a) If an applicant passes the test, what is the probability that he is competent?

(b) If an applicant fails the test, what is the probability that he is competent?

7. Vortex Stereo Systems wishes to examine the effectiveness of a major advertising campaign, which announced major reductions on a number of stereo systems. During the 10-day period that followed the conclusion of the campaign, each person who entered the store was asked whether or not he had been aware of the advertised reductions. Records kept show that 70 percent had been aware of the advertised reductions, and 30 percent had not known about the reductions. Of those who had been aware of the reductions, 20 percent bought a system and 80 percent did not; of those who had not known about the reductions, 10 percent bought a system and 90 percent did not. If a person bought a system, what is the probability that he had known about the advertised reductions?

Self-Test 5

1. The Ajax Company has been awarded a contract to build a new superhighway. In one section through which the highway is to pass the soil is either clay (with an estimated probability of 0.80) or rock. A geological test is to be used for which the following data are available. Of the areas tested known to contain rock, 80 percent of the time the test yielded a positive result, and 20 percent of the time it yielded an incorrect result; of the areas tested known to contain clay, 85 percent of the time the test yielded a positive result, and 15 percent of the time it yielded an incorrect result.
   (a) If the test indicates the presence of rock where the highway is to be built, what is the probability that the soil actually is rock?
   (b) If the test indicates the presence of clay, what is the probability that the soil actually is clay?

2. Five percent of the staff of the Aras Publishing Company occupy management-level positions, 60 percent are in marketing, and 35 percent have other miscellaneous jobs. Of those in management, 60 percent earn more then $25,000 per year; 20 percent of the marketing staff and 15 percent of those in the miscellaneous category earn more than $25,000 per year. If a member of the staff selected at random is found to earn more than $25,000, what is the probability that the person is in marketing?

3. A certain system under study is described by an homogeneous Markov chain with the following transition matrix:

$$T_1 = \begin{bmatrix} \frac{1}{3} & \frac{1}{4} & \frac{5}{12} \\ \frac{1}{6} & \frac{5}{12} & \frac{5}{12} \\ \frac{1}{4} & \frac{1}{4} & \frac{1}{2} \end{bmatrix}$$

What is the probability that the system (a) passes from state 1 to state 3 in two steps; (b) passes from state 2 to state 1 in two steps; (c) passes from state 2 to state 1 in three steps; (d) passes from state 3 to state 1 in three steps?

4. The parachutist Ogden Stiles has a probability of 0.40 of landing on a preset target. Consider a sequence of 50 jumps.
   (a) Describe a probability model for this process.
   (b) On what assumption is this model based?
   (c) State, but do not evaluate, the probability that Ogden will land on target 22 times.
   (d) Can a normal curve approximation be used to estimate the probability that Ogden lands on target 22 times? Explain.
   (e) Assuming that the answer to (d) is yes, find a normal curve approximation for the probability that Ogden lands on target 22 times.

5. Consider a particle in random motion along a line under the following conditions.
   1. The particle can occupy any one of the positions 1, 2, 3, 4, and 5, called states (see Figure 1).
   2. Time is measured in discrete units, $t = 1, 2, 3, 4, 5, 6$. At time $t = 1$, the particle is in one of the states 1, 2, 3, 4, 5. If the particle is initially at or subsequently reaches state 1, then at the next time value it is found in state 2. If the particle is initially at or subsequently reaches state 5, then at the next time value it is found in state 4. The states 1 and 5 are called reflecting states.
   3. If a time $t = 1$ the particle is not in reflecting state, then at time 2 the particle is found in either the same state or in a neighboring state that is one space unit to the right or one space unit to the left of where it was at time $t = 1$. This behavior is repeated at times $t = 2, 3, 4, 5$.
   4. The probability that a particle in a nonreflecting state remain in that state is $\frac{1}{6}$. The probability that it jumps to the left is $\frac{2}{6}$, and the probability that it jumps to the right is $\frac{3}{6}$. This probability assignment is independent of time.

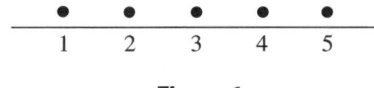

Figure 1

(a) Write the matrix of transition probabilities of the homogeneous Markov chain describing this process.

(b) Determine T_2, the matrix of transition probabilities after two steps.

(c) Find the probability of the particle (1) goes from state 3 to state 3 in two time steps, (2) goes from state 4 to state 5 in two time steps, (3) goes from state 3 to state 1 in three time steps, (4) goes from state 4 to state 4 in three time steps.

6. Let X denote the number of successes in connection with a Bernoulli trial model with $n = 240$ and $p = 0.40$. Find (a) $P(X = 82)$, (b) $P(X \leq 90)$, (c) $P(86 \leq X \leq 94)$, (d) $P(X > 100)$, (e) $P(102 \leq X \leq 114)$.

7. Dapper Dan, the well-known collector of rare dice, has added another rare pair to his collection. With reference to the usual set of 36 outcomes, these dice have the property that outcomes with even sum are three times as likely to show as outcomes with odd sum.

   (a) Define a probability function on S_1 that best reflects the nature of these dice.
   (b) Find the probability that (1) the sum of the numbers showing 7, (2) the sum of the numbers showing is greater than or equal to 8, (3) the sum of the numbers showing is less than 6.

8. The squirrel population seems to have exploded in Bell City and the City Council would like to obtain an estimate of the size of the squirrel population. Darius Consultants were hired to obtain a maximum likelihood estimate of the population

size.

Fifty squirrels were caught, tagged, and released. Shortly thereafter 40 were caught and it was found that 2 had been previously caught.

(a) Based on these results, what is the maximum likelihood estimate of the squirrel population in Bell City?

(b) What assumption(s) underlie this estimate?

(c) Is it possible that the estimate is considerably off the mark? Explain.

9. Population $Q = \{3, 7, 9, 11, 13\}$. Consider the process of drawing a sample of size 2 at random from Q.

(a) Define a probability model for this process.

(b) On what assumption is this model based?

(c) Define the sampling distribution of the mean $\bar{x}$

(d) Find $P(6 \leq \bar{x} \leq 8)$.

CHAPTER 10

Mathematical Models for Conflict Situations

10.1. INTRODUCTION TO GAME THEORY

In the mathematical theory of games there is envisioned two or more parties (called **players**) with conflicting interests who have a certain freedom of choice in the selection of options, but whose control over the situation is partial at best. A player has control over his own actions within the rules that define the game, but does not have control over his opponents' actions, nor does he have control over the element of chance if it is present. The problem is to define, within the rules that govern the game, a concept of optimal strategies for the players and develop methods for determining these optimal strategies. If the number of opponents is n, we speak of an ***n*-person game.** If whatever any player wins at the end of the game is lost by other players, so that the sum of the payoffs to all players is zero (winnings are denoted by positive payoffs and losses by negative payoffs), then the game is called a **zero-sum game.**

Decisive direction to the modern theory of games was provided by the publication in 1944 of *The Theory of Games and Economic Behavior* by the mathematician John von Neumann and the economist Oskar Morgenstern.* In 1994 the Nobel Prize in Economics was awarded to the mathematician John Nash and the economists John Harsanyi and Reinhard Selten for their contributions to the mathematical theory of games and its application to economics. Had they lived, von Neumann and Morgenstern would undoubtedly have shared in the award.

---

*Princeton, N.J.: Princeton University Press, 3rd ed., 1953.

In this introduction to game theory we restrict our attention to a special class of two-person zero-sum games, the games of matrix type. Later in the chapter we shall explore the role of linear programming in solving game-theory problems.

To illustrate a matrix-type two-person zero-sum game, consider the following situation. There are two opponents, Mr. Row and Mr. Column, and a rectangular array of numbers

$$\begin{array}{c} & \begin{array}{ccc} C_1 & C_2 & C_3 \end{array} \\ \begin{array}{c} R_1 \\ R_2 \end{array} & \left[\begin{array}{ccc} 2 & -2 & -3 \\ 3 & 1 & 2 \end{array} \right] \end{array}$$

which contains two rows R_1 and R_2 and three columns C_1, C_2, and C_3. Mr Row is to choose a row and Mr. Column is to choose a column. Neither player knows the other player's choice. When the choices made are revealed, the payoff to each player is specified by the entry in the above array in the intersection of the row and column chosen. Thus if Mr. Row chooses row R_1 and Mr. Column chooses column C_1, the payment of Mr. Column to Mr. Row is 2 units, 2 dollars, let us say. (The units may be monetary or of some other nature.) If Mr. Row chooses row R_1 and Mr. Column chooses column C_2, the payment of Mr. Column to Mr. Row is $-\$2$, which means that Mr. Row pays Mr. Column \$2. Since the game is a zero-sum game, the amount won by one player is equal to the amount lost by the other player. The rectangular array of numbers that specifies the payoffs is called the **payoff matrix to the row player**. The problem is to determine each player's best option so as to guarantee the best possible outcome under the conditions that define the game.

> More generally, a **two-person zero-sum game of matrix type** satisfies the following conditions:
>
> 1. There are two opponents, called the **row and column players.**
> 2. A payoff matrix describing the payoffs from the column player is defined and is known to both players. The amount won by one player is equal to the amount lost by the other player so that the sum of the payoffs is zero.
> 3. Each player makes one move (a row is chosen by the row player and a column is chosen by the column player) without knowing his opponent's choice.
> 4. The payoff from the column player to the row player is described by the value in the payoff matrix that is in the intersection of the row and column chosen.
> 5. It is assumed that both players will do the best they can within the restrictions of the game.

The problem is to define and determine optimal strategies for the row and column players within the context imposed.

Optimal Pure Strategies

Returning to the adventures of Mr. Row and Mr. Column, the question is: what choice should each player elect to guarantee himself the best possible outcome no matter what his opponent does? Looking at the situation from Mr. Row's point of view, we see that if Mr. Row chooses R_1, the worst that can happen is that Mr. Column may pick C_3, in which case Mr. Row loses $3. If Mr. Row chooses R_2, the worst that can happen is that Mr. Column may pick C_2, in which case Mr. Row wins $1. Thus the safe choice for Mr. Row is row R_2, since he is guaranteed to win $1 even if Mr. Column makes his best choice, and he may do better if Mr. Column makes a poor choice. Thus Mr. Row looks at the minimum payoff offered by each row, called the **row minima,** and chooses the row, R_2 in this case, that yields the maximum of these row minima, called **maximin.** Thus maximin $= 1$.

Looking at the situation from Mr. Column's point of view, we see that if Mr. Column chooses C_1, the worst that can happen is that Mr. Row may choose R_2, in which case Mr. Column loses $3. If Mr. Column chooses C_2, the worst that can happen is that Mr. Row may choose R_2, in which case Mr. Column loses $1. If Mr. Column chooses C_3, the worst that can happen is that Mr. Row may choose R_2, in which case Mr. Column loses $2. Thus the safe choice for Mr. Column is C_2, since he is guaranteed that Mr. Row can win no more than $1 even if Mr. Row makes his best choice, and he may do better if Mr. Row makes a poor choice. Thus Mr. Column looks at the maximum payoff in each column, called the **column maxima,** and chooses the column, C_2 in this case, that yields the minimum of these row maxima, called **minimax.** Thus minimax $= 1$. In this situation maximin $=$ minimax $= 1$. The results of this analysis are summarized in Figure 10.1. Row minima are set down next

		Mr. Column			
		C_1	C_2	C_3	Row minima
Mr.	R_1	2	-2	-3	-3
Row	R_2	3	1	2	1
					Maximin $= 1$
Column maxima		3	1	2	Minimax $= 1$

Figure 10.1

to each row, column maxima are set down below each column, and maximin and minimax are recorded as shown.

> More generally, if maximin = minimax, the choice of the row that yields this common value is called an **optimal pure strategy for the row player**, and the choice of the column that yields this common value is called an **optimal pure strategy for the column player**. The common value of maximin and minimax is called the **value of the game**. A **fair game** is one with value 0.

Thus Mr. Row's optimal pure strategy is to choose row R_2, Mr. Column's optimal pure strategy is to choose column C_2, and the value of their game is 1.

Let us note that the value 1 is the smallest number in its row, R_2, and the largest number in its column, C_2; such a number is called a **saddle value**, and its position in the payoff matrix, here row R_2, column C_2, is called a **saddle point** of the payoff matrix. It can be proved that if K is a saddle value, then K is the value of the game, and the optimal pure strategies for the row and column players are given by the saddle point describing the location of K.

EXAMPLE 1

For the two-person zero-sum game with payoff matrix

$$\begin{bmatrix} -2 & 4 & -2 & 1 \\ -1 & 2 & 1 & 3 \\ -2 & 1 & 2 & -3 \end{bmatrix}$$

determine optimal pure strategies for the row and column players and the value of the game.

The row minima and column maxima are summarized in Figure 10.2. Since maximin = minimax = -1, the row player's optimal strategy is R_2, the column player's optimal strategy is C_1, and the value of the game is -1.

	C_1	C_2	C_3	C_4	Row minima
R_1	-2	4	-2	1	-2
R_2	-1	2	1	3	-1
R_3	-2	1	2	-3	-3
					Maximin = -1
Column maxima	-1	4	2	3	Minimax = -1

Figure 10.2

EXAMPLE 2

The Asta and Audre companies are competing for shares in the estimated $5 million minicomputer market. If both companies conduct high-intensity advertising campaigns, they will each get 50 percent of the market. If Asta runs a high-intensity campaign to Audre's low-intensity campaign, Asta will capture 70 percent of the market (to Audre's 30 percent); if Asta runs a low-intensity campaign against Audre's high-intensity blitz, Asta will capture 40 percent of the market (to Audre's 60 percent). If both run low-intensity campaigns, Asta will capture 65 percent of the market (to Audre's 35 percent). With respect to these assumptions, determine each company's optimal strategy.

The outcome resulting from the various options can be expressed in terms of the game defined by the payoff matrix shown in Figure 10.3, where the numbers given express the percentages of the market Audre concedes to Asta. Thus if Asta chooses row 1 (opts for a high-intensity campaign), while Audre chooses column 1 (opts for a low-intensity campaign), Asta will capture 65 percent of the market.

$$
\begin{array}{cc}
 & \text{Audre} \\
\text{Asta} & \begin{array}{c} \text{Low} \\ \text{High} \end{array} \begin{bmatrix} \text{Low} & \text{High} \\ 0.65 & 0.40 \\ 0.70 & 0.50 \end{bmatrix}
\end{array}
$$

Figure 10.3

The row minima and column maxima are summarized in Figure 10.4, from which we see that maximin = minimax = 0.50. Thus it follows as a valid conclusion

		Audre		Row minima
		Low	High	
Asta	Low	0.65	0.40	0.40
	High	0.70	0.50	0.50
Column maxima		0.70	0.50	Maximin = 0.50 Minimax = 0.50

Figure 10.4

that the Asta Company's optimal strategy is to play row 2 (opt for a high-intensity campaign), and the Audre Company's optimal strategy is to play column 2 (opt for a high-intensity campaign). The anticipated outcome is that both companies will obtain an equal share of the market.

But should the companies implement these findings and will the anticipated outcome come to pass? Here too, we can only note that game theory, like any mathematical discipline, guarantees the validity of its conclusions, not their truth. The valid conclusions are true if the underlying assumptions are sufficiently realistic, and thus both companies would be well advised to review the assumptions made. Is it sufficiently realistic to reduce this competitive situation to the game-theory terms that have been used, or are there some basic features that have not been adequately taken into account? This is the key consideration.

EXERCISES

For the games defined by the following payoff matrices, which specify the payoffs from the column player to the row player, determine the optimal strategies for the row and column players and the value.

1. $\begin{bmatrix} 1 & 3 \\ 0 & -2 \end{bmatrix}$ 2. $\begin{bmatrix} 4 & 2 \\ 3 & -3 \end{bmatrix}$ 3. $\begin{bmatrix} 2 & 3 & 4 \\ -1 & 3 & 1 \end{bmatrix}$

4. $\begin{bmatrix} 4 & -1 & 0 \\ 2 & 3 & 1 \\ 3 & 2 & -1 \end{bmatrix}$

5. The Row and Column corporations plan to open fast-food restaurants in a new, still developing shopping center. The shopping center is situated on a rectangular site and various locations are available. If both corporations choose center sites or both choose off-center sites, each will capture 50 percent of the business. If the Row Corporation chooses a center site and the Column Corporation chooses an off-center site, then the Row Corporation will capture 65 percent of the business (to the Column Corporation's 35 percent); if the Row Corporation chooses an off-center site and the Column Corporation chooses a center site, the Row Corporation will capture 40 percent of the business (to the Column Corporation's 60 percent). State the payoff matrix that describes the situation, and determine the optimal strategies for both players with respect to the assumptions made. What is the value of the game, and how do you interpret this value?

Mixed Strategies

Maximin is not equal to minimax for all matrix games, and thus the approach used to define optimal strategies is of limited applicability. To illustrate, consider the following version of the game of matching pennies. Two players each show head or tail without knowing the other player's choice. If they both show heads, the row player pays the column player 4¢, and if they both show tails the row player pays the column player 3¢. If the row player shows heads and the column player shows tails, the column player pays the row player 5¢; if the row player shows tails and the column player shows heads, the column player pays the row player 2¢. The payoff matrix for the row player is shown in Figure 10.5, from which we see that maximin = -3, while minimax = 2.

$$\begin{array}{c c} & \begin{array}{c c} H & T \end{array} \\ \begin{array}{c} H \\ T \end{array} & \begin{bmatrix} -4 & 5 \\ 2 & -3 \end{bmatrix} \end{array} \quad \begin{array}{c} \text{Row minima} \\ -4 \\ -3 \end{array}$$

Column maxima   2   5   Maximin = -3
                            Minimax = 2

Figure 10.5

It's a matter of chance as to who wins on any one play of the game, and the analysis that suffices when maximin equals minimax is insufficient here. Let us imagine the game being played a large number of times, and try to develop an optimal long-run strategy for playing the game. If the row player plays row 1 (shows heads) all the time, the column player would react by playing column 1 (showing heads) and thus winning 4¢ on every play. A similar situation arises if row 2 is played all the time. Obviously, the row player should play row 1 a certain percentage of the time and row 2 the rest of the time, and in a random way so that no pattern can be discerned and taken advantage of. At this point it is useful to introduce the language of probability. To say, for example, that row 1 should be played with probability $\frac{1}{3}$ and row 2 should be played with probability $\frac{2}{3}$ is interpreted to mean that row 1 should be played roughly one third of the time and row 2 should be played two thirds of the time in a random way. More generally, to say that row 1 should be played with probability p and row 2 should be played with probability $1 - p$ is interpreted to mean that row 1 should be played roughly p of the time and that row 2 should be played the rest, or $1 - p$, of the time.* The question is, what is the best value of p? To answer this question, we must introduce the idea of expected value.

The **row player's expected value with respect to column 1 being played by the column player** is the sum of the products of the payoffs in column 1 times the probabilities of these payoffs. Since the payoffs in column 1 are -4 and 2 (see Figure 10.6), and the probabilities with which they occur are the probabilities with which

Play row 1 with probability p $\begin{bmatrix} -4 & 5 \\ 2 & -3 \end{bmatrix}$
Play row 2 with probability $1 - p$

Figure 10.6

---

*These brief remarks on probability are sufficient for our game-theory needs in this chapter. A systematic discussion of probability is presented in Chapters 6–9.

row 1 and row 2 are chosen, that is, p and $1 - p$, respectively, the row player's expected value with respect to column 1 being played by the column player is

$$y_1 = (-4)p + 2(1 - p)$$
$$= -6p + 2, \quad \text{where } 0 \leq p \leq 1$$

Similarly, **the row player's expected value with respect to column 2 being played by the column player** is defined as the sum of the products of the payoffs in column 2 times the probabilities of these payoffs. Since the payoffs in column 2 are 5 and -3, and the probabilities with which they occur are p and $1 - p$, respectively, the row player's expected value with respect to column 2 being played by the column player is

$$y_2 = 5p + (-3)(1 - p)$$
$$= 8p - 3, \quad \text{where } 0 \leq p \leq 1$$

Expected value y_1 can be interpreted as approximating the average amount won per game by the row player with respect to column 1 being played by the column player, assuming a large number of plays of the game, as the following analysis shows. By definition,

$$y_1 = (-4)p + 2(1 - p)$$

Since p is roughly the fraction of the time row 1 is played, and $1 - p$ is the fraction of the time row 2 is played, y_1 is approximately given by

$$y_1 \simeq (-4) \frac{\begin{bmatrix} \text{no. of times} \\ \text{row 1 is played} \end{bmatrix}}{\begin{bmatrix} \text{no. of times} \\ \text{the game is} \\ \text{played} \end{bmatrix}} + (2) \frac{\begin{bmatrix} \text{no. of times} \\ \text{row 2 is played} \end{bmatrix}}{\begin{bmatrix} \text{no. of times} \\ \text{the game is} \\ \text{played} \end{bmatrix}}$$

Since the number of times the game is played is a common denominator, we can add numerators to obtain

$$y_1 \simeq \frac{(-4)\begin{bmatrix} \text{no. of times} \\ \text{row 1 is played} \end{bmatrix} + (2)\begin{bmatrix} \text{no. of times} \\ \text{row 2 is played} \end{bmatrix}}{[\text{no. of times the game is played}]}$$

Since -4 is the amount won by the row player when row 1 is played and 2 is the amount won by the row player when row 2 is played, we have

$$(-4)\begin{bmatrix} \text{no. of times} \\ \text{row 1 is played} \end{bmatrix} = \begin{bmatrix} \text{row player's winnings} \\ \text{when row 1 is played} \end{bmatrix}$$

$$(2)\begin{bmatrix} \text{no. of times} \\ \text{row 2 is played} \end{bmatrix} = \begin{bmatrix} \text{row player's winnings} \\ \text{when row 2 is played} \end{bmatrix}$$

The sum of the winnings with respect to rows 1 and 2 being played is the row player's total winnings. Thus we obtain

$$y_1 \simeq \frac{\text{[row player's total winnings]}}{\text{[no. of times the game is played]}}$$

which is the row player's average winnings per game with respect to the column player playing column 1.

Similarly, the expected value y_2 approximates the row player's average winnings per game with respect to the column player playing column 2, assuming a large number of plays of the game.

We want to choose p so as to maximize, in some sense, the row player's expected average winnings per game. This idea can be precisely expressed in the following way.

> Let
> $$m = \text{minimum of } y_1 \text{ and } y_2$$
> Choose p so as to maximize m. That is, choose p so as to maximize the minimum of the expected values with respect to columns 1 and 2 being played by the column player.

To determine p we turn to the graphs of y_1 and y_2 shown in Figure 10.7. From the graphs of y_1 and y_2 we obtain the graph of $m = $ minimum of y_1 and y_2, shown in Figure 10.8.

Setting y_1 equal to y_2 and solving for p yields

$$-6p + 2 = 8p - 3$$
$$-14p = -5$$
$$p = \tfrac{5}{14}$$

Substituting $\tfrac{5}{14}$ for p in y_1 or y_2 yields $y_1 = y_2 = -\tfrac{1}{7}$. Thus $y_1 = -6p + 2$ intersects $y_2 = 8p - 3$ at the point $(\tfrac{5}{14}, -\tfrac{1}{7})$. As is clear from Figure 10.8, to maximize the minimum of y_1 and y_2, we take p equal to $\tfrac{5}{14}$, the p-value of the intersection point. If $p < \tfrac{5}{14}$, then m, the minimum of y_1 and y_2, is y_2, which is less than $-\tfrac{1}{7}$; if $p > \tfrac{5}{14}$, then m, the minimum of y_1 and y_2, is y_1, which is less than $-\tfrac{1}{7}$. Thus the maximum value of m is $-\tfrac{1}{7}$, which is attained when $p = \tfrac{5}{14}$.

To put this result into operation, we must employ a chance device that selects row 1 roughly $\tfrac{5}{14}$ of the time and row 2 the rest of the time in a random way. One simple chance device can be constructed by putting 5 white marbles and 9 black marbles of the same size into a bag. Reach into the bag and, without looking, pick a marble; if it is white, play row 1 (select heads), and if black, play row 2 (select tails). Over the long run the row player can expect to lose approximately one-seventh of a cent per game, and this is the best he can expect to do.

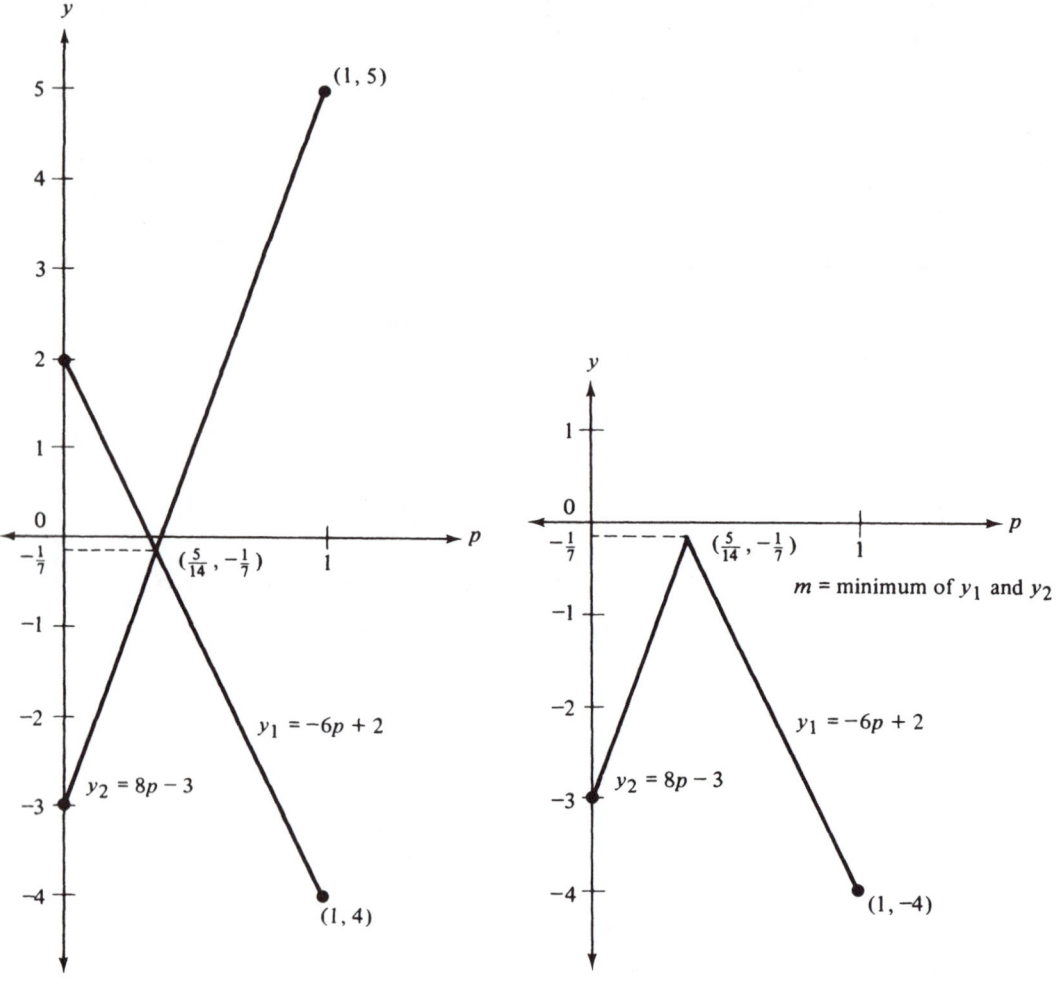

Figure 10.7 **Figure 10.8**

More generally, to maximize **m**, the minimum of the expected values y_1 and y_2 with respect to the column player playing columns 1 and 2, respectively, take for p the p-value of the point of intersection of y_1 and y_2. To obtain this value, called the **row player's optimal mixed strategy**, set y_1 equal to y_2, and solve for p.

The common value of y_1 and y_2 for this optimal value of p, called the **value of the game**, describes how well the row player can expect to do over the long run when rows 1 and 2 are played at random by means of his optimal mixed strategy. Here, too, a game with value zero is called a *fair game*.

EXAMPLE 3

Determine the row player's optimal strategy and the value of the game for the game defined by the payoff matrix

$$\begin{bmatrix} 1 & 3 \\ 4 & 2 \end{bmatrix}$$

First let us note that maximin = 2 while minimax = 3, so that there is no optimal pure strategy for the row player. We must seek an optimal mixed strategy. Let p denote the probability with which row 1 is to be played, and $1 - p$ the probability with which row 2 is to be played. We have

$$\begin{array}{c} \text{probability } p \\ \text{probability } 1 - p \end{array} \begin{bmatrix} 1 & 3 \\ 4 & 2 \end{bmatrix}$$

The row player's expected values with respect to the column player playing columns 1 and 2 are given by

$$y_1 = 1p + 4(1 - p) = -3p + 4$$
$$y_2 = 3p + 2(1 - p) = p + 2$$

Setting y_1 equal to y_2 and solving for p yields

$$-3p + 4 = p + 2$$
$$-4p = -2$$
$$p = \tfrac{1}{2}$$

Thus $1 - p = \tfrac{1}{2}$, and the row player's optimal mixed strategy is to play row 1 with probability $\tfrac{1}{2}$ and row 2 with probability $\tfrac{1}{2}$; that is, play row 1 about one half the time and row 2 the rest of the time in a random manner. This can be done by using a chance device, such as a bag containing two equal-sized marbles of different colors, black and white, for example. Choose a marble without looking; if it's white, play row 1, and if it's black, play row 2.

The common value of y_1 and y_2 for $p = \tfrac{1}{2}$ (obtained by substituting $\tfrac{1}{2}$ for p in y_1 or y_2) is 2.5; thus the value of the game is 2.5.

Let us now turn our attention to the column player, who is probably beginning to feel neglected by now. Again consider the game of matching pennies considered earlier, for which

$$\begin{bmatrix} -4 & 5 \\ 2 & -3 \end{bmatrix}$$

is the payoff matrix for the row player. To obtain the payoff matrix for the column player describing the payoffs of the row player to the column player, we change the sign of the values in this matrix. This yields

$$\begin{bmatrix} 4 & -5 \\ -2 & 3 \end{bmatrix}$$

and the column player's matrix.

Let r denote the probability with which column 1 is to be played, and $1 - r$ the probability with which column 2 is to be played. This yields the situation described by Figure 10.9. Expected-value concepts, analogous to the ones defined for the row

$$\begin{matrix} \text{Played} & \text{Played} \\ \text{with} & \text{with} \\ \text{probability } r & \text{probability } 1 - r \end{matrix}$$
$$\begin{bmatrix} 4 & -5 \\ -2 & 3 \end{bmatrix}$$

Figure 10.9

player, are defined as follows. The **column player's expected value with respect to row 1 being played by the row player** is the sum of the products of the payoffs in row 1 times the probabilities of these payoffs. For this problem, we have

$$z_1 = 4r + (-5)(1 - r)$$
$$= 9r - 5, \quad \text{where } 0 \leq r \leq 1$$

Similarly, the **column player's expected value with respect to row 2 being played by the row player** is the sum of the products of the payoffs in row 2 times the probabilities of these payoffs. This yields

$$z_2 = (-2)r + 3(1 - r)$$
$$= -5r + 3, \quad \text{where } 0 \leq r \leq 1$$

The interpretation of these expected values is entirely analogous to that given for their row player counterparts. We can interpret z_1 as approximating the average amount won per game by the column player with respect to row 1 being played by the row player, assuming that the game is played a large number of times. Similarly, for z_2 replace row 1 by row 2 in the preceding statement.

Let
$$n = \text{minimum of } z_1 \text{ and } z_2$$

The column player's **optimal mixed strategy** is defined as the value of r for which n is maximized. A graphical analysis of z_1 and z_2, similar to the kind given for the row player's expected values, shows that the column player's optimal mixed strategy is the value of r that is the r-value of the point of intersection of z_1 and z_2. To obtain this value, set z_1 equal to z_2 and solve for r.

For the problem at hand, we obtain
$$9r - 5 = -5r + 3$$
$$14r = 8$$
$$r = \tfrac{4}{7}$$

Thus $1 - r = \tfrac{3}{7}$, and the column player's optimal mixed strategy is to play column 1 with probability $\tfrac{4}{7}$ and column 2 with probability $\tfrac{3}{7}$. The common value of z_1 and z_2 for $r = \tfrac{4}{7}$ is $\tfrac{1}{7}$, the negative of the value of the game $-\tfrac{1}{7}$. Thus the column player's anticipated average winnings per play is equal to the row player's average losses per play over the long run, which makes sense.

EXAMPLE 4

Determine the column player's optimal mixed strategy for the game described in Example 3.

Since the row player's payoff matrix is
$$\begin{bmatrix} 1 & 3 \\ 4 & 2 \end{bmatrix}$$
the column player's payoff matrix is
$$\begin{bmatrix} -1 & -3 \\ -4 & -2 \end{bmatrix}$$

Let r denote the probability with which column 1 is to be played and $1 - r$ the probability with which column 2 is to be played. The column player's expected values with respect to the row player playing rows 1 and 2 are given by
$$z_1 = (-1)r + (-3)(1 - r) = 2r - 3$$
$$z_2 = (-4)r + (-2)(1 - r) = -2r - 2$$

Setting z_1 equal to z_2 and solving for r yields
$$2r - 3 = -2r - 2$$
$$4r = 1$$
$$r = \tfrac{1}{4}$$

Thus $1 - r = \tfrac{3}{4}$, and the column player's optimal mixed strategy is to play column 1 with probability $\tfrac{1}{4}$ and column 2 with probability $\tfrac{3}{4}$. The common value of z_1 and z_2 for $r = \tfrac{1}{4}$ is -2.5, the negative of the value of the game. By playing his optimal mixed strategy, the column player can expect to do no worse than lose an average of 2.5 units per play over the long run.

EXERCISES

For the games defined by the following payoff matrices, which specify the payoffs from the column player to the row player, determine the optimal strategies for the row and column players and the value.

6. $\begin{bmatrix} 1 & -1 \\ -1 & 1 \end{bmatrix}$
7. $\begin{bmatrix} 6 & 2 \\ 3 & 4 \end{bmatrix}$
8. $\begin{bmatrix} 4 & 2 \\ 2 & 3 \end{bmatrix}$
9. $\begin{bmatrix} 5 & 2 \\ 2 & 6 \end{bmatrix}$

Show that the optimal mixed strategies for the row and column players are the same for the games defined by the following payoff matrices.

10. $\begin{bmatrix} -4 & 5 \\ 2 & -3 \end{bmatrix}$ and $\begin{bmatrix} 1 & 10 \\ 7 & 2 \end{bmatrix}$

11. $\begin{bmatrix} -4 & 5 \\ 2 & -3 \end{bmatrix}$ and $\begin{bmatrix} -4+c & 5+c \\ 2+c & -3+c \end{bmatrix}$

Exercises 10–11 illustrate the general result that adding the same amount to all entries of a payoff matrix does not change the row and column players' optimal mixed strategies. The value of the game is, however, increased (or decreased) by the amount added.

12. A beverage company can stress taste or low calorie level in its television advertising. Preliminary studies indicate that advertising which focuses on taste is effective on 50 percent of the viewers not over 40 years old who see the ads, and effective on 30 percent of the over-40 group who see the ads, whereas ads based on calorie level are effective on 25 percent of the not-over-40 group, and effective on 60 percent of the over-40 group. Assuming that the same proportion of viewers in the two age groups is exposed to the ads, and viewing the situation as a game with the beverage company and the market as opponents, set up the pay-off matrix, and determine the beverage company's optimal strategy and the value of the game. What does implementation of this optimal strategy call for, and how is the value of the game to be interpreted?

10.2. A POINT OF CONTACT WITH LINEAR PROGRAMMING*

In this section we explore a most remarkable connection between two-person zero-sum games and linear programming. As we shall see,

> The problem of finding optimal mixed strategies for the row and column players can be formulated as linear programming problems that are dual to each other. Both problems can then be solved simultaneously by use of the simplex method.

For simplicity, we shall restrict our attention to games defined by payoff matrices with two rows and two columns.

To begin, let us reconsider the game defined by the payoff matrix

$$\begin{bmatrix} 1 & 3 \\ 4 & 2 \end{bmatrix}$$

(discussed in Section 10.1, Example 3) and the problem of determining the row and column players' optimal mixed strategies and the value of the game. Let p and $1 - p$ denote the probabilities with which rows 1 and 2 are to be played. The row player's expected values y_1 and y_2 with respect to columns 1 and 2, respectively, being played by the column player are

$$y_1 = 1p + 4(1 - p)$$
$$y_2 = 3p + 2(1 - p)$$

---

*Sections 3.10 and 4.3 are prerequisites for the discussion in this section.

Since p and $1 - p$ are nonnegative, and not both are zero, y_1 and y_2 are both positive. (If the graphs of y_1 and y_2 are drawn, both will be seen to lie above the p-axis.) Thus m, the minimum of y_1 and y_2, is positive.

To continue, we introduce variables x and y defined as follows:

$$x = \frac{p}{m}, \quad y = \frac{1-p}{m}$$

Since p and $1 - p$ are nonnegative and m is positive, it follows that x and y are nonnegative; that is,

$$x \geq 0$$
$$y \geq 0$$

Adding x and y yields

$$x + y = \frac{p}{m} + \frac{1-p}{m} = \frac{1}{m}$$

Our problem is to maximize m. But maximizing m is equivalent to minimizing $1/m$, which is equal to $x + y$. Thus the row player wants to

Minimize $R(x, y) = x + y$

subject to

$$x \geq 0$$
$$y \geq 0$$

What other conditions must x and y satisfy? That m is the minimum of $y_1 = 1p + 4(1 - p)$ and $y_2 = 3p + 2(1 - p)$ means that y_1 and y_2 are greater than or equal to m. That is,

$$1p + 4(1 - p) \geq m \tag{10.1}$$
$$3p + 2(1 - p) \geq m \tag{10.2}$$

Since x and y are defined by $x = p/m$ and $y = (1 - p)/m$, it follows that

$$p = xm, \quad 1 - p = ym$$

Substituting xm for p and ym for $1 - p$ in relations (10.1) and (10.2) yields

$$1xm + 4ym \geq m$$
$$3xm + 2ym \geq m$$

Since m is positive, dividing both sides of these relations by m yields the following additional conditions that must be satisfied by x and y:

$$x + 4y \geq 1$$
$$3x + 2y \geq 1$$

In summary, then, the row problem of determining p so as to maximize m, the minimum of y_1 and y_2, leads to the following linear program:

$$\text{Minimize } R(x, y) = x + y$$

subject to

$$x \geq 0$$
$$y \geq 0$$
$$x + 4y \geq 1$$
$$3x + 2y \geq 1$$

where $p = xm$, $1 - p = ym$, and $x + y = 1/m$.

Turning to the column player, let us first note that the column player's payoff matrix is

$$\begin{bmatrix} -1 & -3 \\ -4 & -2 \end{bmatrix}$$

Let r and $1 - r$ denote the probabilities with which columns 1 and 2 are to be played. The column player's expected values with respect to rows 1 and 2 being played by the row player are

$$z_1 = (-1)r + (-3)(1 - r)$$
$$z_2 = (-4)r + (-2)(1 - r)$$

Since r and $1 - r$ are nonnegative and are being multiplied by negative numbers to form z_1 and z_2, z_1 and z_2 are both negative. (If the graphs of z_1 and z_2 are drawn, both will be seen to lie below the r-axis.) Thus n, the minimum of z_1 and z_2, is negative.

We next define variables u and w as follows:

$$u = \frac{r}{n}, \qquad w = \frac{1 - r}{n}$$

Since r and $1 - r$ are nonnegative and n is negative, it follows that u and w are negative or zero; that is,

$$u \leq 0$$
$$w \leq 0$$

Adding u and w yields

$$u + w = \frac{r}{n} + \frac{1 - r}{n} = \frac{1}{n}$$

The column player wants to maximize n. Since maximizing n is equivalent to minimizing $1/n$, which equals $u + w$, the column player needs to

$$\text{Minimize } u + w$$
subject to
$$u \leq 0$$
$$w \leq 0$$

To obtain other conditions that must be satisfied, let us note that n is the minimum of $z_1 = (-1)r + (-3)(1-r)$ and $z_2 = (-4)r + (-2)(1-r)$ means that z_1 and z_2 are greater than or equal to n; that is,

$$(-1)r + (-3)(1-r) \geq n \qquad (10.3)$$
$$(-4)r + (-2)(1-r) \geq n \qquad (10.4)$$

Since u and w are defined by $u = r/n$ and $w = (1-r)/n$, it follows that

$$r = un, \qquad 1 - r = wn$$

Substituting un for r and wn for $1 - r$ in relations (10.3) and (10.4) yields

$$(-1)un + (-3)wn \geq n$$
$$(-4)un + (-2)wn \geq n$$

Since n is negative, dividing both sides of the preceding inequalities by n reverses their sense and yields the following conditions:

$$(-1)u + (-3)w \leq 1$$
$$(-4)u + (-2)w \leq 1$$

Thus the column player's problem of maximizing n, the minimum of z_1 and z_2, reduces to the following:

$$\text{Minimize } u + w$$
subject to
$$u \leq 0$$
$$w \leq 0$$
$$(-1)u + (-3)w \leq 1$$
$$(-4)u + (-2)w \leq 1$$

where $r = un$, $1 - r = wn$, and $u + w = 1/n$.

To take this one step further, we introduce variables s and t defined by

$$s = -u, \qquad t = -w$$

By substituting s and t for $-u$ and $-w$, and $-s$ and $-t$ for u and w, we obtain

$$\text{Minimize } -s - t = -(s + t)$$

subject to

$$-s \leq 0$$
$$-t \leq 0$$
$$s + 3t \leq 1$$
$$4s + 2t \leq 1$$

where $r = -sn$, $1 - r = -tn$, and $s + t = -1/n$. To minimize $-(s + t)$ is equivalent to maximizing $s + t$, which we shall call $C(s, t)$. Multiplying $-s \leq 0$ and $-t \leq 0$ by -1 yields $s \geq 0$, $t \geq 0$. Thus the column player's problem can be stated in linear-programming terms as follows:

$$\text{Maximize } C(s, t) = s + t$$

subject to

$$s \geq 0$$
$$t \geq 0$$
$$s + 3t \leq 1$$
$$4s + 2t \leq 1$$

where $r = -sn$, $1 - r = -tn$, and $s + t = -1/n$.

In summary, then, for the game defined by the row player's payoff matrix

$$\begin{bmatrix} 1 & 3 \\ 4 & 2 \end{bmatrix}$$

the problem of determining optimal mixed strategies for the row and column players reduces to solving the following linear programs:

Minimize $R(x, y) = x + y$	Maximize $C(s, t) = s + t$
subject to	subject to
$x \geq 0$	$s \geq 0$
$y \geq 0$	$t \geq 0$
$x + 4y \geq 1$	$s + 3t \leq 1$
$3x + 2y \geq 1$	$4s + 2t \leq 1$
where $p = xm$	where $r = -sn$
$1 - p = ym$	$1 - r = -tn$
$x + y = 1/m$	$s + t = -1/n$

It is worthwhile to note that the constraints $x + 4y \geq 1$ and $3x + 2y \geq 1$ of the row player's linear program come from columns 1 and 2 of the payoff matrix, respectively.

The tabular arrays of these linear programs are shown in Figures 10.10(a) and (b), from which we see that these programs are duals.

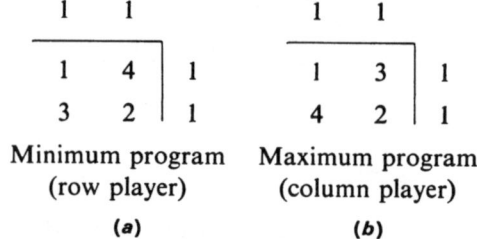

Minimum program (row player)   Maximum program (column player)

(a)   (b)

Figure 10.10

Thus both can be solved at the same time by the simplex method. The simplex tableaus are shown in Figure 10.11, from which we see that

$$s = \tfrac{1}{10} \quad \text{and} \quad t = \tfrac{3}{10}$$

1	3	1	0	1	ST_1
④	2	0	1	1	
−1	−1	0	0	0	
0	⑤/₂	1	−¼	¾	ST_2
1	½	0	¼	¼	
0	−½	0	¼	¼	
0	1	⅖	−1/10	3/10	ST_3
1	0	−⅕	3/10	1/10	
0	½	⅕	⅕	⅖	

Figure 10.11

Thus

$$-\frac{1}{n} = s + t = \frac{2}{5} \quad \text{and} \quad n = -\frac{5}{2}$$

This yields

$$r = -sn = -\tfrac{1}{10} \cdot (-\tfrac{5}{2}) = \tfrac{1}{4}$$

$$1 - r = -tn = -\tfrac{3}{10} \cdot (-\tfrac{5}{2}) = \tfrac{3}{4}$$

The column player's optimal mixed strategy is to play column 1 with probability $\frac{1}{4}$ and column 2 with probability $\frac{3}{4}$. This result was obtained in Section 10.1, Example 4.

From the indicator row of tableau ST_3, we have

$$x = \tfrac{1}{5} \quad \text{and} \quad y = \tfrac{1}{5}$$

Thus

$$\frac{1}{m} = x + y = \frac{2}{5} \quad \text{and} \quad m = \frac{5}{2}$$

This yields

$$p = xm = \tfrac{1}{5} \cdot \tfrac{5}{2} = \tfrac{1}{2}$$

$$1 - p = ym = \tfrac{1}{5} \cdot \tfrac{5}{2} = \tfrac{1}{2}$$

The row player's optimal mixed strategy is to play row 1 with probability $\frac{1}{2}$ and row 2 and probability $\frac{1}{2}$. The value of the game is $m = \frac{5}{2}$. These results were obtained in Section 10.1, Example 3.

EXAMPLE 1

The payoff matrix for the game of matching pennies considered in Section 10.1 (p. 316) is

$$\begin{bmatrix} -4 & 5 \\ 2 & -3 \end{bmatrix}$$

Formulate the problem of finding optimal mixed strategies for the row and column players in linear-programming terms, and determine these optimal strategies and the value of the game.

If we carefully study the analysis that led to formulating the problem of finding the row player's optimal strategy in linear-programming terms, it will be seen that the positivity of y_1 and y_2, the row player's expected values, played a central role. The positivity of y_1 and y_2 was guaranteed by the positivity of the values in the payoff matrix. The problem here is that not all the values in the given payoff matrix are positive, and thus the positivity of y_1 and y_2 cannot be guaranteed. This obstacle can be overcome by using a theorem of game theory which says that if a two-person zero-sum matrix game is modified by adding a constant c to all entries of its payoff matrix, then the value of the game is increased (or decreased) by the amount c, but the optimal strategies of the row and column players are not changed (see Section 10.1, Exercises 10 and 11).

Thus if we add 5 to each entry of our payoff matrix, thereby obtaining

$$\begin{bmatrix} 1 & 10 \\ 7 & 2 \end{bmatrix}$$

the value of the game is increased by 5, but the optimal strategies for the row and column players remain the same. Since all entries are positive, we are assured that the expected values y_1 and y_2 of the row player are positive.

From this payoff matrix we obtain the row player's linear program:

$$\text{Minimize } R(x, y) = x + y$$

subject to

$$x \geq 0$$
$$y \geq 0$$
$$x + 7y \geq 1$$
$$10x + 2y \geq 1$$

where $p = xm$, $1 - p = ym$, and $x + y = 1/m$. The last two constraints, $x + 7y \geq 1$ and $10x + 2y \geq 1$, are obtained from columns 1 and 2, respectively, of our payoff matrix.

The tabular array for this linear program is shown in Figure 10.12(a), from which we obtain the tabular array of its dual, the column player's linear program [Figure 10.12(b)].

$$\begin{array}{cc|c} 1 & 1 & \\ \hline 1 & 7 & 1 \\ 10 & 2 & 1 \end{array} \qquad \begin{array}{cc|c} 1 & 1 & \\ \hline 1 & 10 & 1 \\ 7 & 2 & 1 \end{array}$$

Minimum program (row player)   Maximum program (column player)

(a)   (b)

Figure 10.12

From Figure 10.12(b), we obtain the column player's linear program:

$$\text{Maximize } C(s, t) = s + t$$

subject to

$$s \geq 0$$
$$t \geq 0$$
$$s + 10t \leq 1$$
$$7s + 2t \leq 1$$

where $r = -sn$, $1 - r = -tn$, and $s + t = -1/n$. The simplex method solutions of these linear programs are shown in Figure 10.13, from which we have

$$s = \frac{2}{17} \quad \text{and} \quad t = \frac{3}{34}, \quad -\frac{1}{n} = \frac{7}{34} \quad \text{and} \quad n = -\frac{34}{7}$$

This yields

$$r = -sn = -\frac{2}{17} \cdot \left(-\frac{34}{7}\right) = \frac{4}{7}$$

$$1 - r = -tn = -\frac{3}{34} \cdot \left(-\frac{34}{7}\right) = \frac{3}{7}$$

The column player's optimal strategy is to play column 1 with probability $\frac{4}{7}$ and column 2 with probability $\frac{3}{7}$.

From the indicator row of tableau ST_3, we have

$$x = \frac{5}{68} \quad \text{and} \quad y = \frac{9}{68}, \quad \frac{1}{m} = \frac{7}{34} \quad \text{and} \quad m = \frac{34}{7}$$

$$\begin{array}{|cc:cc:c|l}
\hline
1 & \circled{10} & 1 & 0 & 1 & ST_1 \\
7 & 2 & 0 & 1 & 1 & \\
\hline
-1 & -1 & 0 & 0 & 0 & \\
\hline
\frac{1}{10} & 1 & \frac{1}{10} & 0 & \frac{1}{10} & ST_2 \\
\circled{\frac{34}{5}} & 0 & -\frac{1}{5} & 1 & \frac{4}{5} & \\
\hline
-\frac{9}{10} & 0 & \frac{1}{10} & 0 & \frac{1}{10} & \\
\hline
0 & 1 & \frac{7}{68} & -\frac{1}{68} & \frac{3}{34} & ST_3 \\
1 & 0 & -\frac{1}{34} & \frac{5}{34} & \frac{2}{17} & \\
\hline
0 & 0 & \frac{5}{68} & \frac{9}{68} & \frac{7}{34} & \\
\hline
\end{array}$$

Figure 10.13

This yields

$$p = xm = \frac{5}{68} \cdot \frac{34}{7} = \frac{5}{14}$$

$$1 - p = ym = \frac{9}{68} \cdot \frac{34}{7} = \frac{9}{14}$$

The row player's optimal strategy is to play row 1 with probability $\frac{5}{14}$ and row 2 with probability $\frac{9}{14}$. Since 5 was added to each entry of the original payoff matrix to obtain the modified payoff matrix that we worked in terms of, we must subtract 5 from the value of the modified game to obtain the value of the original game. This yields

$$m = \frac{34}{7} - 5 = -\frac{1}{7}$$

as the value of the original game.

EXERCISES

For the games defined by the following payoff matrices, which state the payoffs from the column player to the row player, formulate the problems of determining mixed optimal strategies for the row and column players in linear-programming terms and solve.

1. $\begin{bmatrix} 6 & 2 \\ 3 & 4 \end{bmatrix}$
2. $\begin{bmatrix} 4 & 2 \\ 2 & 3 \end{bmatrix}$
3. $\begin{bmatrix} 5 & 2 \\ 2 & 6 \end{bmatrix}$
4. $\begin{bmatrix} 6 & 4 \\ 3 & 5 \end{bmatrix}$
5. $\begin{bmatrix} 1 & -1 \\ -1 & 1 \end{bmatrix}$
6. $\begin{bmatrix} 4 & -2 \\ -2 & 5 \end{bmatrix}$

7. By following in the footsteps of the analysis presented in this section, formulate the problem of determining mixed optimal strategies for the row and column players in linear-programming terms for the game defined by the following payoff matrix and solve.

$$\begin{bmatrix} 4 & 3 & 2 \\ 2 & 1 & 4 \\ 3 & 4 & 2 \end{bmatrix}$$

10.3. SELF-TESTS FOR CHAPTER 10

Allow 50 or so minutes for each self-test. Go over the first one before proceeding to the second.

Self-Test 1

1. For the games defined by the following payoff matrices, determine the optimal strategies for the row and column players and the value.

(a) $\begin{bmatrix} 1 & -1 & 2 \\ 4 & 3 & 5 \\ -1 & 2 & 1 \end{bmatrix}$
(b) $\begin{bmatrix} -1 & 3 & 0 & -2 \\ 3 & 4 & 2 & 5 \\ 1 & -2 & 1 & 3 \end{bmatrix}$

(c) $\begin{bmatrix} 4 & -2 \\ -2 & 3 \end{bmatrix}$ and $\begin{bmatrix} 4+c & -2+c \\ -2+c & 3+c \end{bmatrix}$

2. For the games defined by the following payoff matrices, formulate the problems of determining mixed optimal strategies for the row and column players in linear programming terms and solve.

(a) $\begin{bmatrix} 2 & -3 \\ -1 & 2 \end{bmatrix}$
(b) $\begin{bmatrix} 3 & -1 \\ -1 & 2 \end{bmatrix}$

Self-Test 2

1. For the games defined by the following payoff matrices, determine the optimal strategies for the row and column players and the value.

   (a) $\begin{bmatrix} 6 & 1 \\ 3 & 2 \end{bmatrix}$

   (b) $\begin{bmatrix} 4 & 3 \\ 3 & 5 \end{bmatrix}$

   (c) $\begin{bmatrix} 5 & 1 \\ 2 & 3 \end{bmatrix}$

   (d) $\begin{bmatrix} -1 & -2 \\ -2 & -3 \end{bmatrix}$

2. For the games defined by the following payoff matrices, formulate the problems of determining mixed optimal strategies for the row and column players in linear programming terms and solve.

   (a) $\begin{bmatrix} 2 & 1 \\ 1 & 3 \end{bmatrix}$

   (b) $\begin{bmatrix} 4 & -1 \\ -1 & 2 \end{bmatrix}$

Structure: Validity Versus Truth

11.1. DEDUCTIVE REASONING, VALIDITY, AND TRUTH

To simply illustrate the nature of a valid conclusion consider the two statements

(1) All x's are y's.

(2) All y's are z's

and the statement

(3) All x's are z's.

Suppose we take statements (1) and (2) as a starting point for the purpose of seeing what conclusions follow as a logical consequence. In general, a collection of statements set down for such a purpose is called an **hypothesis**. Each statement in an hypothesis is referred to, variously, as an **assumption, premise, postulate,** or **axiom,** depending on context.

The **proof** or **argument** which establishes that a purported conclusion does indeed follow as an inescapable consequence of the postulates is said to be a **valid proof** or **valid argument** and the conclusion of a valid proof is said to be **valid with respect to the postulates,** or **hypothesis** made up of the postulates. Valid conclusions of an hypothesis are called **theorems**.

These are basic definitions which require flushing out if we are to get a secure hold on them. To begin this process we return to our mini-system consisting of (1), (2), and (3).

Postulate *P*1:	All *x*'s are *y*'s.
Postulate *P*2:	All *y*'s are *z*'s.
Conclusion *C*1:	All *x*'s are *z*'s.

These statements assert relationships between various classes of objects identified as *x*'s, *y*'s, and *z*'s. *P*1 forces the class of *x*'s within the class of *y*'s, with *P*2 forcing the class of *y*'s within the still more inclusive class of *z*'s. It follows as an inescapable consequence that the *x*'s, like it or not, are forced within the *z*'s, which is the content of *C*1. Conclusion *C*1 is valid with respect to postulates *P*1 and *P*2, or put another way, *C*1 is valid with respect to the hypothesis consisting of *P*1 and *P*2; it is a theorem in this system and we may upgrade it from *C*1 to *T*1, for theorem.

A convenient way to picture these relationships is by means of diagrams of the following sort. Represent each class by points inside a closed curve. Thus, *P*1 is represented by placing the class of *x*'s within the class of *y*'s as shown in Figure 11.1(a). *P*2 is represented by placing the class of *y*'s within the class of *z*'s as shown in Figure 11.1(b). The diagrams help us to get a better grip on the assumed relationships between the *x*'s, *y*'s, and *z*'s.

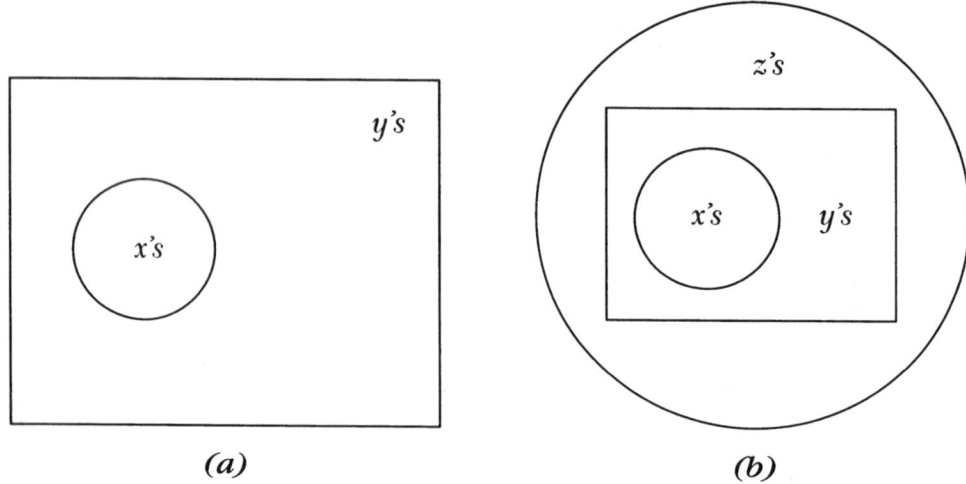

Figure 11.1

The *x*'s, *y*'s, and *z*'s are "abstract" entities in the fullest sense of the term and it is important to note that *T*1 is a valid consequence of *P*1 and *P*2, irrespective of the nature of the *x*'s, *y*'s, and *z*'s. The validity of *T*1 from *P*1 and *P*2 is a structural condition, irrespective of the content assigned to the *x*'s, *y*'s and *z*'s. It's somewhat like having a person in the flesh—this is, the real person—who we may dress up in many ways—business suit, beach wear, evening attire, what have you. The person's appearance changes, sometimes radically, but it is fundamentally the same person. So it is with valid conclusions. We may color-in x, y, and z in many ways, but doing so changes the appearance of $P1$, $P2$ and $T1$ and not the validity of $T1$ on the basis of $P1$ and $P2$.

Suppose, for example, that we add color to x, y, and z as follows. Let:

$$x = \text{apple}, \quad y = \text{people}, \quad z = \text{animal}$$

This gives us the hypothesis:

$P1a$: All apples are people.

$P2a$: All people are animals.

and the valid conclusion:

$T1a$: All apples are animals.

But $T1a$ is wrong, you say. Well, yes and no; we must be very careful to distinguish the sense in which $T1a$ is wrong from the sense in which it is correct. $T1a$ is correct in the sense of being valid with respect to $P1a$ and $P2a$; it is wrong, that is, false, as a statement about the relationship between apples and animals. Adding the coloring $x = \text{apple}$, $y = \text{people}$, $z = \text{animal}$ to the scene does not change anything about the structure of the argument which determines validity, but it does introduce a truth/falsity dimension into the scene which, of course, complicates it.

A system made up of postulates and theorems is called, as one would expect, a **postulate system**. The process of obtaining valid conclusions from the postulates of the system is called **deduction** or **deductive reasoning**. A postulate system I obtained from an "abstract" postulate system APS by assigning representations to its abstract terms (x, y, z or equivalents) is called an **interpretation** of the APS. Validity relationships are maintained in passing from an "abstract" postulate system APS (such as $P1$, $P2$, and $T1$) to any interpretation of APS (such as I_1 consisting of $P1a$, $P2a$, $T1a$).

If we let

$$x = \text{dog}, \quad y = \text{mammal}, \quad z = \text{animal}$$

we obtain interpretation I_2 with postulates

$P1b$: All dogs are mammals.

$P2b$: All mammals are animals.

and theorem

$T1b$: All dogs are animals.

If we let

$$x = \text{apple}, \quad y = \text{cat}, \quad z = \text{fruit}$$

we obtain interpretation I_3 with postulates

$P1c$:   All apples are cats.

$P2c$:   All cats are fruit.

and theorem

$T1c$:   All apples are fruit.

The interpretations I_1, I_2, and I_3 illustrate the following relationships between the validity of a statement and its truth.

> **1.** If T is a theorem in an interpretation I of an APS and the hypothesis of I is true, that is, its postulates are true, then T is true. Valid conclusions deduced from true postulates are true.

This is illustrated by I_2. It makes sense since in obtaining a valid conclusion T of an hypothesis H we do not go beyond H. If H is true and we do not go beyond it to obtain T, then it is not surprising that T is also true.

> **2.** If the theorems of an interpretation I of an APS are true, then we cannot conclude that the hypothesis H of I is true. H might be true; some of the postulates of H might be true and others false; it might be false in its entirety.

I_3 illustrates an interpretation with a true theorem arising from false postulates.

> **3.** If a theorem T of an interpretation I of an APS is false; then some of the postulates of I must be false.

This third property follows from the fact that if the postulates of I were true, then we could not obtain from them a false theorem T. I_1 illustrates an interpretation with a false theorem ($T1a$: All apples are animals) arising from a system with a false postulate ($P1a$: All apples are people).

In summary then, we have:

1. If the postulates are true, then the theorems must be true.
2. If the theorems are true, then the postulates may or may not be true.
3. If a theorem is false, then some of the postulates must be false.

Invalid Arguments

Consider the following structure.

Hypothesis. P1: All x's are y's.
P2: All z's are y's.
Conclusion. C1: All x's are z's.

A diagrammatic representation of P1 and P2 is shown in Figure 11.2. The argument is not valid, or **invalid**, because C1 is not forced by the hypothesis. We are forced by P1 to place

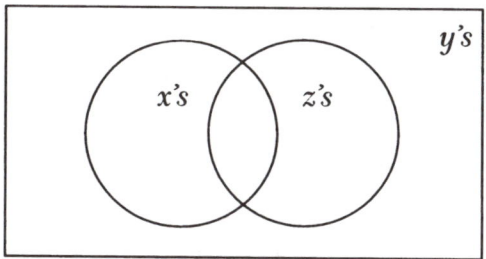

Figure 11.2

the class of x's within the class of y's and by P2 to place the class of z's within the class of y's, but this does not force us to place the x's within the z's. It may be that the x's are contained within the z's, but "may be" isn't strong enough for the argument to be valid.

The mini-systems considered here enable us to initiate a discussion of validity versus truth. For an adequate perspective we must consider a richer framework, which is taken up in the next section.

EXERCISES

Determine the validity of the arguments for the proposed conclusions in 1-7. Explain the basis for your answer.

1. Hypothesis. P1: All mammals are frogs.
P2: Joe Warren is a mammal.
Conclusion. C1: Joe Warren is a frog.

2. Hypothesis. P1: All oranges are blueberries.
   P2: All blueberries are fruit.
   Conclusion. C1: All oranges are fruit.

3. Hypothesis. P1: Some college students are geniuses.
   P2: All freshmen are geniuses.
   Conclusions. C1: All freshmen are college students.
   C2: No freshmen are college students.
   C3: Some freshmen are college students.
   C4: Some college students are freshmen.

4. Hypothesis. P1: No college students are geniuses.
   P2: All freshmen are college students.
   Conclusion. C1: No freshmen are geniuses.

5. Hypothesis. P1: No math professors are bores.
   P2: Some math professors are human.
   Conclusion. C1: Some humans are not bores.

6. Hypothesis. $P1$:   All babies are beautiful.
                 $P2$:   Amy is beautiful.
   Conclusion. $C1$:   Amy is a baby.

7. Hypothesis. $P1$:   Some x's are y's.
                 $P2$:   All x's are z's.
   Conclusions. $C1$:   Some z's are y's.
                 $C2$:   Some z's are not y's.

8. What is it about the "wisdom" stated in the cartoon that prompted the reply, "What have you guys been drinking?"

11.2. DUBIOUS DEDUCTIONS

"What did you learn in geometry?" asked Jeff Arnold of his daughter Jenny. "I don't remember," replied Jenny. "Well, then, what did you do in geometry?" countered Mr. Arnold. "Most of the time we were doing proofs," answered Jenny.

Deductive proofs, which define the backbone of mathematics, are first encountered by most of us in a course in geometry. The development of geometry, called Euclidean geometry after the Greek geometer Euclid of Alexandria (c. 300 B.C.), begins with a statement of some basic postulates. Its theorems are then logically deduced from the postulates and, as theorems begin to accumulate, from a mixture of postulates and previously proved theorems. In developing a proof each statement in the sequence of statements defining the proof is justified by a statement within the system itself (postulate, theorem, construction, definition). Diagrams play a fundamental role in guiding us to the conclusion we seek to reach.

But sometimes the diagrams take on a life of their own and our orderly sequence of logical inferences is broken by employing visual evidence which, strictly speaking, goes beyond the legitimate means of justification that we have available to us in the system. The following two examples illustrate the difficulties that arise.

> **Purported theorem:** There exists a triangle with two right angles.

Proof. Consider two circles that intersect at two points which we shall term A and B (see Figure 11.3). Let $\overline{AC}, \overline{AD}$ denote their respective diameters from A. Let $\overline{CD}$ meet

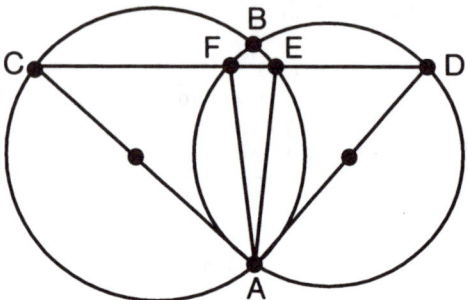

Figure 11.3

the respective circles in E, F. Thus angle AEC is a right angle, since it is inscribed in semicircle AEC. Similarly, angle AFD is a right angle. Thus triangle AEF has two right angles.

The argument seems airtight, but something is wrong since a triangle, with angle sum 180 degrees, cannot have two right angles. The diagram is compelling, but a very carefully drawn diagram might suggest that $\overline{CD}$ passes through B, so that AEF is not a triangle at all but a line segment. While a very carefully drawn diagram might suggest this, there is nothing in the postulates or theorems of Euclidean geometry which allows us to argue that $\overline{CD}$ passes through B.

The following is a well known theorem of Euclidean geometry, but let us look carefully at the proof that is usually presented.

Theorem. The base angles of an isosceles triangle are equal.

Given: Triangle ABC with $\overline{AC} = \overline{BC}$.
To prove: Angle A = Angle B.

Proof.

Statement	Justification
1. Draw the bisector of angle C.	1. Every angle has a bisector.
2. Extend it to meet AB at D.	2. A line may be extended.

3. In triangles *ACD* and *BCD*, $\overline{AC} = \overline{BC}$.
4. Angle 1 = Angle 2.
5. $\overline{CD} = \overline{CD}$
6. Triangle *ACD* is congruent to triangle *BCD*.
7. Angle *A* equals angle *B*.

3. Hypothesis.
4. Definition of angle bisector.
5. Identity.
6. Side-Angle-Side.
7. Corresponding parts of congruent triangles are equal.

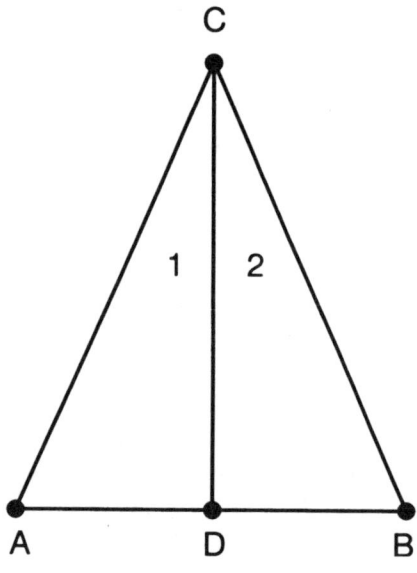

Figure 11.4

The proof and diagram are convincing, but is it, by itself, a valid argument? Strictly speaking, we would have to say no. A difficulty arises in Step 2. The justification, a line may be extended, does not say that a line may be extended to meet another line, *AB* in this case; the lines might be parallel, and we need something in the system itself that would rule this out.

Let us suppose that we can get around this difficulty and conclude that the bisector of angle *C* does intersect line *AB* at *D*. But where is *D*? For the proof to hold up we need *D* to fall between *A* and *B*, but there is nothing in the system that says it must. Figure 11.4 is so convincing on this point that our first reaction might be to say that it's "obvious" that *D* is between *A* and *B*; where else could it be? "Obvious" is not the same as deductive proof and postulates and theorems on "betweeness" are needed to support such conclusions.

What is the place of diagrams, then? Should we abandon them? The answer to this last question is a resounding NO, NO, NO. Diagrams are invaluable for suggesting ideas and providing us with a sense of what we want to do and what we are obtaining. They are essential allies; but their use should not be equated to formal deductive proof, which must come out of the underlying system itself.

EXERCISES

Pinpoint the gaps in the following proofs.

1. Playfair's form of the parallel postulate:

If given a line L and point P not on L, then there is one and only one line which passes through P and is parallel to L. (see Figure 11.5)

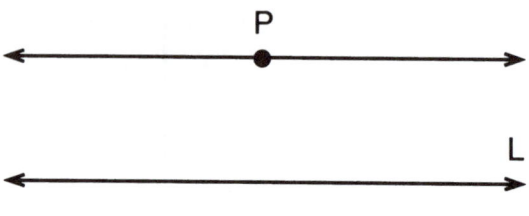

Figure 11.5

Proof

1. Drop a perpendicular from point P to line L (see Figure 11.6). To this perpendicular erect a perpendicular PE from the point P. This second perpendicular is parallel to line L by the theorem that two perpendiculars to the same line are parallel. Since it is possible to drop only one perpendicular from a given point to a given line, and it is possible to erect only one perpendicular to a line from a point lying on it, the parallel line PE is unique.

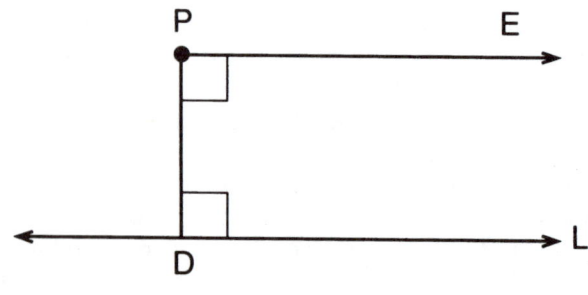

Figure 11.6

2. Return to Exercise 1 for a "proof" by the Greek geometer Proclus (410-485).

From Exercise 1 we have that there exists a line M passing through P parallel to L. (see Figure 11.7).

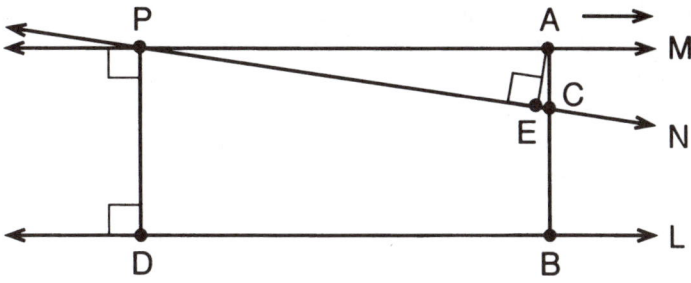

Figure 11.7

The problem is to show that M is unique. Suppose there is another line N through P parallel to L. Then N makes an acute angle with PD that lies on one side or the other of PD. Suppose it's the right side of PD. The part of N to the right of P is thus contained in the region bounded by L, M, and PD.

Let A denote any point of M to the right of P; let AB denote the perpendicular to L at B, and let C denote the point at which AB intersects N. Then $\overline{AB} > \overline{AC}$. Let A recede on M; then $\overline{AC}$ increases without bound since $\overline{AC} \geq \overline{AE}$, the perpendicular from A to line N. Thus AB, which is at least as large as $\overline{AC}$, increases without bound.

But the distance between two parallel lines must be bounded. Thus we have a contradiction to the result that $\overline{AC}$ increases without bound, which means that our supposition that there is another line N through P parallel to L is untenable.

11.3. MORE ON PROOF AND POSTULATE SYSTEMS

To obtain a more complete perspective on proof and postulate systems in mathematics we examine some mini-systems that are more substantial than those considered in Section 11.1.

First, we turn our attention to the system I_a founded on the following postulates.

P1a: Every line is a collection points which contains at least two points.
P2a: For any two points there is at least one line containing them.
P3a: For any line there is a point not contained by it.
P4a: There is at least one line.

We have not defined the basic terms point and line and at first sight it might seem unnecessary to do so. "We all know what point and line indicate," you might say. Yes, most of us think of point as indicating position in space and line as being a path produced by a straight edge of indefinite length, but, as we saw in the preceding section, these associations have the power to cloud our minds when it comes to constructing valid proofs.

To get around this difficulty let us use uncharged terms—zog and glob, for example, keeping in mind that one interpretation that we may give to these terms is point and line. We may also use this envisioned interpretation to draw pictures to help guide our thinking, being careful to keep them at a distance, so-to-speak, so that they do not dominate our thinking to the extent that they close our minds to other possible interpretations.

The abstract postulate system that emerges, call it **Glob Theory**, is based on the following postulates.

> P1: Every glob is a collection of zogs which contains at least two zogs.
> P2: For any two zogs there is at least one glob containing them.
> P3: For any glob there is a zog not contained by it.
> P4: There is at least one glob.

What are globs and zogs? They are undefined. Every postulate system begins with undefined terms, undefined in the sense that no unique characterization in terms of more basic entities is given. There is no way around this. If we were to define glob and zog in terms of mumbo and jumbo, let us say, then mumbo and jumbo would be our basic undefined terms. If we sought to define mumbo and jumbo in terms of more basic elements, then they would be our undefined terms, and on it goes. Postulates P1 through P4 do not uniquely define zog and glob, but state some relationships between them.

If we interpret zog and glob as point and line in the usual sense, we emerge with interpretation I_a consisting of postulates $P1a$ through $P4a$. On the other hand, suppose we interpret zog and glob as follows:

> zog: The points given by the coordinates (1,1), (5,1), (3,5).
> glob: The circular arcs joining these points of the circle determined by them (see Figure 11.8).

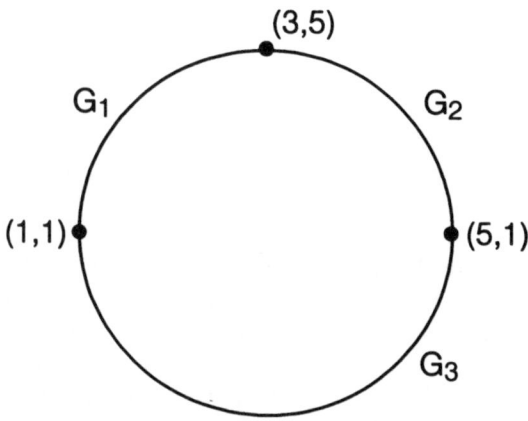

Figure 11.8

There are three globs in this interpretation: G_1, G_2, and G_3.

The postulates in this interpretation of Glob Theory, call it I_b, take the following form:

P1b: Every circular arc is a collection of points which contains at least two points (that is, at least two of (1,1), (5,1), (3,5)).
P2b: For any two points there is at least one circular arc containing them.
P3b: For any circular arc there is a point not contained by it.
P4b: There is at least one circular arc.

Suppose we interpret zog as an ordered triple of numbers which is a solution common to at least two of the following three equations, which we interpret as globs.

$$x - 2y + z = 3 \quad (11.1)$$
$$2x - 3y - z = 7 \quad (11.2)$$
$$5x - 8y - z = 20 \quad (11.3)$$

Glob Theory's postulates in terms of this interpretation take the following form:

P1c: Every equation determines a solution set which contains at least two members.
P2c: For any two solutions (of a pair of equations) there is at least one equation satisfied by them.
P3c: For any equation there is a solution (of a pair of equations) which does not satisfy it.
P4c: There is at least one equation.

We term this interpretation I_c.

For I_a, I_b, and I_c the conditions stated in the postulates of its parent Glob Theory are satisfied. Such interpretations of an abstract postulate system are of particular interest; they are called **models** of the parent *APS*. There are two types of models—concrete and ideal—which we should distinguish between. A model of an abstract postulate system is said to be **concrete** if the interpretation assigned to its undefined terms are objects and relations adopted from the real world; the model is said to be **ideal** if the interpretation assigned to its undefined terms are objects and relations adopted from some other postulate system. I_a and I_b are ideal models of Glob Theory.

> A fundamental property of models is that theorems established for the parent abstract postulate system describe relations which hold for any model as well. Since validity is determined by structure and not by how undefined terms are interpreted, theorems proved in the abstract hold for models of an *APS* as well.

We now turn to proving some theorems to illustrate proof in this sort of setting.

The models I_a, I_b, and I_c of Glob Theory contain at least three zogs, which suggests the possibility that Glob Theory itself contains at least three zogs. Let us undertake to develop a formal proof of this suggested result.

Possibility: Glob Theory contains at least three zogs.

Idea: Get a glob into play since a glob contains at least two zogs. As a visual aid we might refer "informally" to something like Figure 11.9(a). This gives us two zogs, p and q, and we are well under way.

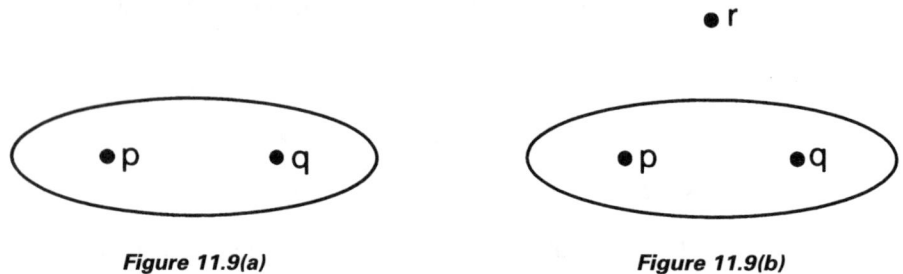

Figure 11.9(a)   Figure 11.9(b)

We can nail down our third zog r by appealing to $P3$. As a visual aid we have Figure 11.9(b). These figures are just that, helpful visual aids; they do not comprise the formal proof.

At this point, with the ideas in hand, we are ready to write down the formal proof.

Theorem 1. Glob Theory contains at least three zogs.

Proof

Statement	Justification
1. Let L denote a glob.	1. $P4$
2. L contains at least two zogs, p and q.	2. $P1$
3. There is a zog r not in L.	3. $P3$
4. p, q, and r are distinct zogs.	4. Summary of preceding results.

The observation that the zogs are distinct, with justification, is most important. For convenience we have introduced letters to denote the zogs, but the fact that different letters are being used does not by itself guarantee that the zogs are different.

What about the globs? How many can we count on in general? The interpretations I_a, I_b and I_c of Glob Theory suggest that there are at least three. Let us see if we can develop a package of ideas to prove this result. A possibility of this sort which is suggested by consideration of special cases is called a **conjecture.**

Conjecture. Glob Theory contains at least three globs.

Idea: we can generate an initial glob, call it L_1, by appealing to $P4$ (see Figure 11.10(a)). We can get other globs into play by introducing zogs and then appealing to $P2$ (see Figure 11.10(b)).

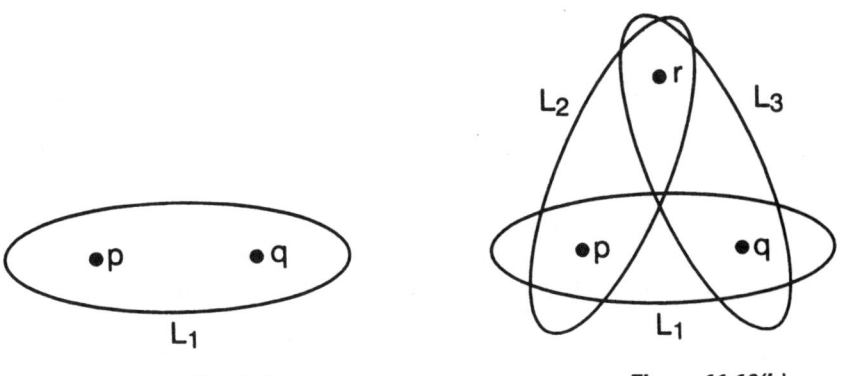

Figure 11.10(a)

Figure 11.10(b)

We are now ready to write down a formal proof.

Theorem 2. Glob Theory contains at least three globs.

Proof.

Statement	Justification
1. Let L_1 denote a glob.	1. P4
2. L_1 contains at least two zogs, p and q.	2. P1
3. There is a zog r not in L_1.	3. P3
4. There is a glob $L_2 \neq L_1$ which contains zogs p and r.	4. P2; $L_2 \neq L_1$ since L_2 contains r which is not in L_1.

Now we must be very careful. Figure 11.10(b) is helpful in providing us with a start and initial direction, but it is also potentially misleading. The suggestion conveyed is that we follow up by asserting the existence of glob $L_3 \neq L_1$ which contains r and q and exhibiting globs L_1, L_2, and L_3 to conclude the proof. But how can we be sure that L_3 is not the same as L_2? We can get around this difficulty by introducing L_3 as containing r and q and considering two cases.

5. Case 1. L_3 is not the same as L_1 and L_2, in which case our work is done.

6. Case 2. L_3 is the same as L_2, in which case p, q, and r are contained by the same glob, $L_3 \neq L_2$, visually illustrated by Figure 11.11(a).

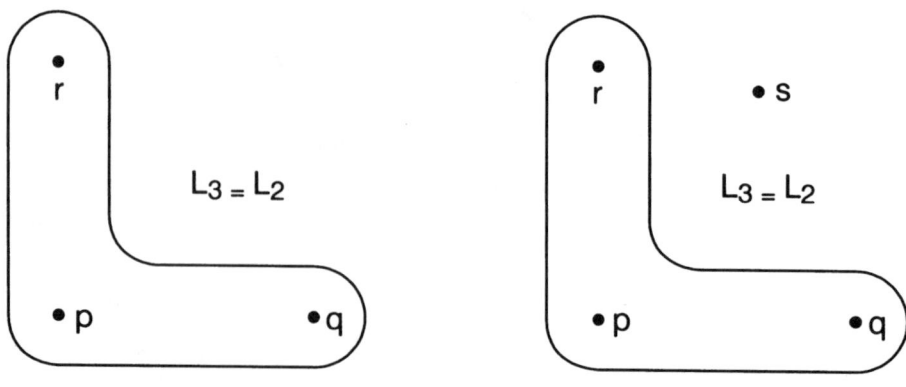

Figure 11.11(a)   **Figure 11.11(b)**

Then, by P3, there is a zog s not contained by $L_3 = L_2$, visually illustrated by Figure 11.11(b). By P_2 there is a glob $L_4 \neq L_2$ which contains r and s. $L_4 \neq L_1$ contains r which is not contained by $L_1 \cdot L_1$, L_2 and L_4 are distinct globs.

To this point in its development Glob theory consists of two undefined terms, four postulates, and two theorems. In undertaking to prove other theorems we may now employ theorems 1 and 2 to justify assertions that we make.

The theorems, proved in the abstract for Glob Theory, hold for all models of Glob Theory as well. It's just a matter of shifting from the langauge of zogs and globs to the terms of the model. For I_a we have:

$T1a$: I_a contains at least three points.
$T2a$: I_a contains at least three lines.

For I_b we have:

$T1b$: I_b contains at least three points (of the form (1,1), (5,1), (3,5)).
$T2b$: I_b contains at least three circular arcs.

EXERCISES

1. Specify two ideal models for Glob Theory. How do Theorems 1 and 2 of Glob Theory read in terms of these models?

2. Specify a concrete model for Glob Theory. How do Theorems 1 and 2 of Glob Theory read in terms of this model?

3. Having established that there exists at least three zogs in Glob Theory, we might attempt to climb the mathematical ladder another rung and establish that Glob Theory has at least four zogs.

   Is the following argument to prove this result valid? Explain

   Conjecture 1:   Glob Theory contains at least four zogs.

   Proof.

Statement	Justification
1. Let L denote a glob.	1. P4
2. L contains at least two zogs, p and q.	2. P1
3. Let r denote a zog not on L.	3. P3
4. Let K denote a glob containing r.	4. T2
5. Let s denote another zog in K.	5. P1
6. p, q, r, and s are four zogs in Glob Theory.	6. Summary of preceding steps.

4. Michael Vlasik, an "expert" on Glob Theory, conjectured that this postulate system contains at least six zogs. Is the following argument that Michael developed to prove this conjecture valid? Explain.

Conjecture 2: Glob Theory contains at least six zogs.
Proof.

Statement	Justification
1. Let L, K, and M denote globs.	1. T2
2. Let a and b denote zogs in L, c and d zogs in K, and e and f zogs in M.	2. P1
3. a, b, c, d, e, and f are six zogs in Glob Theory.	3. Summary of preceding steps.

5. William Schneider, another student of Glob Theory, argued that since the proof given in 3 to prove that Glob Theory contains at least four zogs was invalid, this conjecture is not a theorem of Glob Theory. Would you agree? Explain.

6. Melinda Hu, another student at Glob Theory, expressed the view that Conjectures 1 and 2 are not theorems of this system. Is she justified in this view? Explain.

7. **Neighborhood Theory**, with undefined terms blob and neighborhood, is based on the following three postulates.

   $P1.$ There are at least three blobs.

   $P2.$ For any two blobs there is at least one neighborhood containing them.

   $P3.$ For any two blobs p and q there is a neighborhood P containing p and a neighborhood Q containing q where P and Q have no blobs in common.

   (a) State two models for Neighborhood Theory. The conjectures stated in (b) through (d) have been proposed concerning Neighborhood Theory. Consider each and prove it a theorem or show that it is not a theorem of Neighborhood Theory.

   (b) $C1$: There are at least three neighborhoods.

   (c) $C2$: Not all blobs are contained in any one neighborhood.

   (d) $C3$: No two neighborhoods have a blob in common.

8. **Mumbo-Jumbo Theory**, with undefined terms mumbo and jumbo, is based on the following five postulates.

   $P1.$ For any two jumbos there is exactly one mumbo containing them.

   $P2.$ Not all jumbos are contained by the same mumbo.

   $P3.$ There is at least one mumbo.

   $P4.$ Every mumbo contains at least three jumbos.

   $P5.$ For any jumbo, there are exactly two mumbos containing it.

   (a) Try to construct a model for Mumbo-Jumbo Theory. What seems to be the difficulty in doing this?

The conjectures stated in (b) and (c) have been proposed concerning Mumbo-Jumbo Theory. Consider each and prove it a theorem or show that it is not a theorem of Mumbo-Jumbo Theory.

(b) C1: There are at least four jumbos.

(c) C2: There is a jumbo which is contained by three mumbos.

(d) If C2 is a theorem, what are its implications for Mumbo-Jumbo Theory?

9. Show that interpretation I_c of Glob Theory (p. 348) is a model.

10. Consider the following interpretation I_d of Glob Theory. zog: an ordered pair of numbers which is a solution common to at least two of the following four equations, which we interpret as globs.

$$-x + y = 2, \quad -x + y = -2$$
$$x + y = 2, \quad x + y = -2$$

(a) What are the zogs in I_d.

(b) Is I_d a model of Glob Theory?

11. Consider the following interpretation I_e of Glob Theory. zog: an ordered triple of numbers which is a solution common to at least two of the following three equations, which we interpret as globs.

$$x - 2y + 2z = -1$$
$$3x + 2y + 2z = 9$$
$$2x - 3y - 3z = 6$$

Is I_e a model of Glob Theory? Explain

11.4. CONSISTENCY AND INDEPENDENCE

A postulate system is said to be **consistent** if it does not contain contradictory statements (two postulates, postulate and theorem, or two theorems). This is the most fundamental property required of a postulate system. A postulate system that is not consistent, called **inconsistent,** is worthless. Mumbo-Jumbo Theory, stated in Exercise 8 to the previous section, is an example of an inconsistent postulate system. Conjecture C2, There is a jumbo which is contained by three mumbos, is a theorem of Mumbo-Jumbo Theory which contradicts a postulate P5: For any jumbo there are exactly two mumbos containing it.

The consistency of a postulate system may take two forms, absolute and relative. The **absolute consistency** of a postulate system is established by providing a concrete model for it, which establishes that the system is as consistent as the real world. The **relative consistency** of a postulate system G is established by providing an ideal model I for it, which establishes that G is as consistent as I. Providing an ideal model I for G shifts the burden of the consistency of G to I. It shows that if G is inconsistent, then so is I.

To illustrate these ideas we return to Glob Theory, considered in the previous section, based on the following postulates.

P1: Every glob is a collection of zogs which contains at least two zogs.
P2: For any two zogs there is at least one glob containing them.
P3: For any glob there is a zog not contained by it.
P4: There is at least one glob.

Interpretations I_a and I_b discussed in the previous section are ideal models of Glob Theory which show that Glob Theory is as consistent as Euclidean geometry and the real number system, respectively. A concrete model I_c of Glob Theory is suggested by Figure 11.10(b) itself, which is reproduced in Figure 11.12.

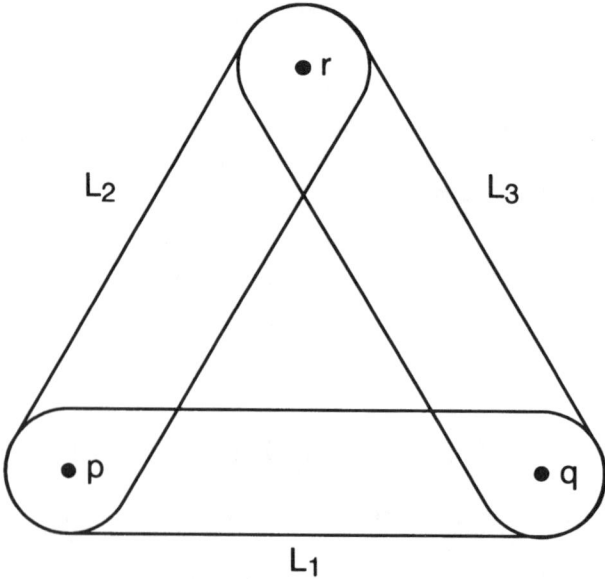

Figure 11.12

Let p, q, and r denote any three concrete objects (3 books; 3 bees in a hive; 3 teachers in a department; whatever); we take these as our zogs. For our globs we take the subsets of two of p, q, and r. We have:

$$L_1 = \{p,q\}, \quad L_2 = \{p,r\}, \quad L_3 = \{q,r\}$$

The postulates of Glob Theory take the following form in terms of this interpretation of zog and glob:

*P*1c: Every set of two objects of the three contains at least two objects.
*P*2c: For any two objects there is at least one set of two containing them.
*P*3c: For any set of two objects there is an object not contained by it.
*P*4c: There is at least one set of two objects.

These conditions are satisfied and I_c is a concrete model of Glob Theory. This establishes the absolute consistency of Glob Theory.

> Let us observe that analytic geometry sets up an ideal model for Euclidean geometry which establishes that Euclidean geometry is as consistent as the real-number system. By establishing ideal models the consistency of many branches of mathematics can be reduced to the consistency of a basic branch.

A postulate of a postulate system is said to be **independent** in the system if it is not a valid consequence of the other postulates of the system. If one desires a statement *P* within a postulate system (either as a postulate or theorem) and *P* is independent in the system, then the only way to have it in the system is to take it as a postulate since it cannot be had as a theorem.

One approach to showing that a postulate *P* is independent in a postulate system *G* is to exhibit an interpretation for the terms of *G* for which the other postulates of *G* are satisfied, but *P* is not satisfied. Having done this, it follows that *P* could not be a theorem of *G* since, if it were, if would also have to be a theorem in the aforenoted interpretation, which it is not.

EXAMPLE 1

Postulate *P*3 of Glob Theory, (for any glob there is a zog not contained by it) is independent.

Consider two books b_1 and b_2. We take these as our zogs and set $L_1 = \{b_1, b_2\}$ as our glob. Postulates *P*1, *P*2, and *P*4 are satisfied for this interpretation, but *P*3 is not. Thus, *P*3 is independent; if it were a theorem of postulates *P*1, *P*2, and *P*4, then the statement, for any set of two books, there is a book not contained by it, would be satisfied in this interpretation, which is not the case.

EXERCISES

1. Is Mumbo-Jumbo Theory, defined in Exercise 8 of Section 11.3, consistent? Explain.

2. An **Abelian Group** is a postulate system which consists of a set G of objects, called **elements of G**, and an operation denoted by •, subject to the following postulates:

    P1:   To every pair of elements x and y of G, given in the stated order there corresponds an unique element of G, denoted by $x \bullet y$. (Closure postulate for the operation •.)

    P2:   If x, y, and z are any elements of G, then $x \bullet (y \bullet z) = (x \bullet y) \bullet z$. (Associative postulate for the operation •.)

    P3:   There is an unique element e of G, called the identity element, having the property that if x is any element of G, $x \bullet e = e \bullet x = x$.

    P4:   To each element x of G there corresponds an unique element x', called the inverse of x, having the property that $x \bullet x' = x' \bullet x = e$.

    P5:   If x and y are any elements of G, then $x \bullet y = y \bullet x$. (Commutative postulate for the operation •.)

    (a) Is Abelian Group Theory consistent? Explain.

    (b) Show that postulate *P5* is independent.

3. Victor P. Neighborhood, son of the founder of Neighborhood Theory, proposed an extension of Neighborhood Theory (Sec. 11.3, Ex. 7, p. 354) by adding the following postulate to *P1–P3*. *P4*: There exist at most two neighborhoods. Is Victor's Extended Neighborhood Theory consistent? Explain.

4. Janice Neighborhood, on the other hand, proposed an extension by adding the following postulate to *P1–P3*. *P4(a)*: There exist at most three neighborhoods.

    (a) Is Janice's Extended Neighborhood Theory consistent? Explain.

    (b) Is *P4(a)* independent? Explain.

11.5. A CLASSICAL PROBLEM OF INDEPENDENCE AND ITS AFTERMATH

Euclidean Geometry

    Geometry, and more generally mathematics, as a system of postulates, definitions, and theorems or propositions deduced from the postulates, is the great contribution of the ancient Greek mathematicians collectively. To any such system they gave the name "Elements". Although not the first of its kind, the most successful Elements were those compiled by Euclid of Alexandria around 300 B.C. Euclid set himself the task of taking the mathematics known at the time and organizing it into a deductive system. This involved deciding on which statements were to serve as the basic assumptions of the system (postulates), which were to be theorems, and providing proofs for the theorems, original ones when necessary. Euclid also had to decide how definitions were to be formulated. It was an enormous undertaking and, considering the decisive influence which his work has had, a brilliantly successful one.

Euclid took five geometric postulates as a basis for his deductive treatment of the geometry we now call Euclidean geometry and introduced them in the following manner.

Let the following be postulated:

I. To draw a straight line from any point to any point.
II. To produce a finite straight line continuously in a straight line.
III. To describe a circle with any center and distance.
IV. That all right angles are equal to one another.
V. That, if a straight line falling on two straight lines makes the interior angles on the same side less than two right angles, the two straight lines, if produced indefinitely, meet on that side on which the angles are less than two right angles. That is, in terms of Figure 11.13, if the sum of angles A and B is less than 180°, then lines N and M will intersect at some point on the same side as angles A and B.

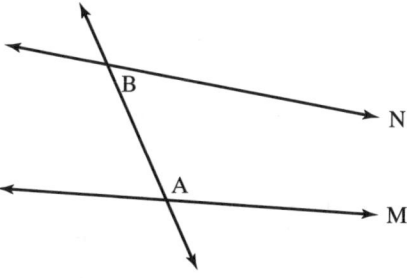

Figure 11.13

In more modern terminology, postulates I and II say: Through any two points, one straight line can be drawn; a line segment determines a line which is indefinite in extent. Postulate III says that a circle is determined when its center and radius are prescribed. Postulate IV says that all right angles are congruent. Postulates I-IV seem to express obvious truths in a simple fashion.

Postulate V, in contrast, in much more complex and as such stands in violation of the accepted criteria for postulates of the time, that they must express self-evident truths about spatial relations in a simple way. Euclid was aware of this, but part of his extraordinary accomplishment was to recognize that such a statement was needed to support the most complex of his geometric deductions and to boldly take it as a postulate after attempts to deduce it from simpler statements were unsuccessful. Euclid's fifth postulate is known as his **parallel postulate**, although the term parallel does not occur in it. He defines parallel lines as lines which being in the same plane and being produced indefinitely in both directions do not meet. The label parallel postulate is appropriate because it is equivalent to the following statement which involves the term parallel, equivalent in the sense that it plus the other four Euclidean postulates imply Euclid's fifth postulate, and vice versa.

Plairfair's form of Euclid's parallel postulate:

> If given a line *L* and point *P* not on *L*, then there is one and only one line which passes through *P* and is parallel to *L* (Figure 11.14).

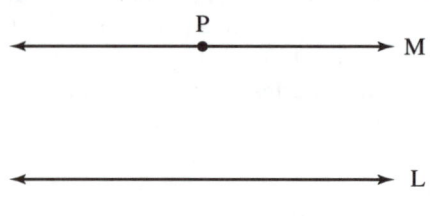

Figure 11.14

This equivalent to Euclid's parallel postulate is the one usually cited as Euclid's parallel postulate in textbook expositions of Euclidean geometry. It is known as Plairfair's postulate, after the Scottish physicist and mathematician John Plairfair (1748-1819) who popularized it in a very successful textbook that he wrote on Euclid's *Elements*.

For the ancient Greek mathematicians and philosophers and their successors in mathematics, philosophy and science Euclidean geometry served as a description of the space in which we live and as an intellectual discipline. It was considered the essence of physical space.

A Non-Euclidean Challenger

Euclid's contemporaries and successors greatly admired the organization of geometry which he had achieved in his *Elements,* but they were also dissatisfied with the price that had been paid in the form of a fifth postulate which could neither be considered simple nor self-evident. The parallel postulate problem that arose was to free Euclid from this blemish by either deducing the fifth postulate from Euclid's other postulates, or replacing it with an equivalent postulate which was simple and self-evident. The problem attracted many scholars from many lands. Some attempts to solve it were ingenious, but all were unsuccessful. In a departure from earlier approaches to the parallel postulate problem, which were direct, three mathematicians, working independently, brought a *reducio ad absurdum* or indirect approach to the problem. Each tried to show that Euclid's fifth was a consequence of his other postulates by showing that if the fifth or its equivalent is replaced by its negation, the amended set of postulates has implications that are contradictory. None were successful. The results they obtained were in contradiction to the nature of space as it was then perceived, but they were not in contradiction with each other.

Chapter 11 / Structure: Validity Versus Truth   361

In the early nineteenth century the parallel postulate problem was taken up by three men who reached startingly different conclusions from their predecessors. Nicolai Ivanovich Lobachevsky (1792-1856), of the then recently established Kazan University in southern Russia, was first to publicly announce and publish his results. Lobachevsky took the contradiction of Plairfair's equivalent of Euclid's fifth in the following form:

> **Lobachevsky's parallel postulate:** If given a line L and a point P not on L, there are at least two lines which pass through P and are parallal to L (see Figure 11.15).

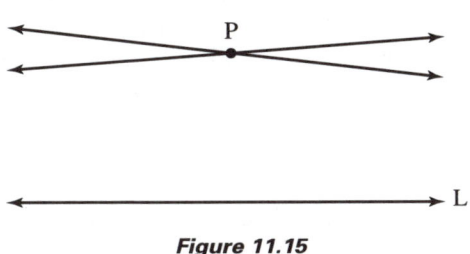

Figure 11.15

On the face of it this statement, which we shall term Lobachevsky's parallel postulate, seems absurd. In deducing the consequences of the amended system consisting of Euclid's postulates, with his fifth replaced by Lobachevsky's parallel postulate, Lobachevsky concluded that his system, now called non-Euclidean or Lobachevskian geometry, forms a consistent whole, although it was strikingly at variance with reality as it was then understood to be. Lobachevsky first outlined his ideas in a paper which he presented at a meeting of the mathematics and physics division of Kazan University held on February 26, 1826. Three years later he elaborated on his ideas in his paper "On the Principles of Geometry", published in the Kazan Messenger. In the years 1835-1855 Lobachevsky further developed his non-Euclidean Geometry in a series of works. He worked in virtual isolation.

He was appreciated by his colleagues as an outstanding teacher and administrator (having served as Rector of Kazan University from 1827-1846, and Assistant Guardian of the Kazan Educational District, (1846-1855)), but his ideas on geometry were incomprehensible to them and were treated with tolerance at best and derision and ridicule at worst. One of Lobachevsky's papers came to the attention of Carl Friedrich Gauss (1777-1855), who appreciated its worth and had Lobachevsky elected a member of the Gottingen Scientific Society in 1842. In letters to friends Gauss expressed the highest praise for Lobachevsky's work, but he never gave it public support. By the early 1820's Gauss had satisfied himself that a system based on a denial of Euclid's fifth together with Euclid's other postulates is consistent but, apart from sharing his ideas in letters with trusted friends, kept his views to himself. Gauss was a private person who shunned controversy. As he put it in a letter to one of his friends, he feared the "clamor of the Boeotians".

The basic idea of a non-Euclidean geometry arose from still another source. A Hungarian friend of Gauss, Farkas Bolyai (1775-1856), had spent much time trying to prove the parallel postulate from Euclid's other postulates. The problem attracted his son Janos Bolyai (1802-1860), who concluded that the system based on a denial of Euclid's fifth in the form, through a point not on a given line there are infinitely many lines passing through the point parallel to the given line, was free of contradiction. This system Bolyai called *The Science of Absolute Space*. It was published in 1832 as an appendix to his father's text on geometry. Gauss' reaction to this work of the younger Bolyai was similar to the case of Lobachevsky, private approval but no public support. *The Science of Absolute Space* was Bolyai's one and only work on non-Euclidean geometry. Gauss' lack of public support and his discovery that Lobachevsky had anticipated him in publication left him so embittered that he did not further extend his ideas.

Some of the differences between Euclidean and Lobachevskian geometry due to the change in the parallel postulate are summarized in the following comparison Table 11.1.

Lobachevsky's far reaching investigations into what we now call Lobachevskian geometry convinced him that this system is mathematically legitimate in the sense of being consistent, that is, free of contradictory statements. While convincing, Lobachevsky's analysis fell short of being conclusive on this point. In 1868 Eugenio

Table 11.1

	Euclidean	Lobachevskian	
Given line L and point P not on L, there exist	one and only one line	infinitely many lines	through P parallel to L
Parallel lines	are equidistant	are never equidistant	
If a line intersects one of two parallel lines, it	must	may or may not	intersect the other
If two lines intersect in one point, there is	no line	one line	parallel to both
Two lines parallel to the same line	must	may or may not	be parallel
A line which meets one of two parallels	must	may or may not	meet the other
If two lines are parallel there are	infinitely many lines	at most one line	perpendicular to both
The number of circles passing through three noncolinear points is	exactly one	at most one	
The angle sum of a triangle is	equal to	less than	180°
The area of a triangle is	independent of	proportional to 180° minus	its angle sum
Rectangles	exist	do not exist	

Beltrami (1835-1900) developed an ideal model for Lobachevskian geometry within the Euclidean framework and in doing so showed that if an inconsistency existed in Lobachevskian geometry, then an inconsistency must also exist in Euclidean geometry. A mathematical Samson seeking to topple Lobachevskian geometry on the basis of inconsistency would also, if successful, topple Euclidean geometry; Lobachevskian geometry is thus as mathematically legitimate as Euclidean geometry.

Showing that a less familiar, controversial structure is as structually legitimate as a more familiar one brings the less familiar one closer to us. But it also raises another question. Is the more familiar structure, or perhaps we should say, seemingly more familiar structure, as consistent as the less familiar one? Models of Euclidean geometry have been given within the framework of Lobachevskian geometry, so that in the final analysis we can say that each is as mathematically legitimate as the other.

It was firmly believed for 2200 years that Euclid's parallal postulate was not independent of his other postulates. The development of a non-Euclidean geometry by Lobachevsky based on a contradiction to Euclid's parallel postulate and the proof of its relative consistency in terms of Euclidean geometry not only showed the independence of Euclid's parallel postulate, but was responsible for shaping attitudes which underlie the nature of modern mathematics.

EXERCISES

Carefully consider the following statements. If you agree with a statement, explain the basis for your agreement; if you disagree with a statement, explain the basis for your disagreement.

1. The parallel postulate problem was to show that Euclid's parallel postulate is true.

2. Lobachevsky sought to prove that Euclidean geometry is inconsistent.

3. Since Lobachevskian geometry contains statements which contradict statements in Euclidean geometry, Lobachevskian geometry cannot be considered a realistic description of physical space.

4. If Euclidean geometry is not a true description of space, then the proofs of some Euclidean theorems must be in error.

5. Eugenio Beltrami showed that if Lobachevskian geometry is consistent, then Euclidean geometry must be inconsistent.

6. Janos Bolyai's main achievement was to show that Euclid's parallel postulate is false.

7. A postulate is a statement which is true.

8. "Karl Friedrich Gauss, as a youth, sought to prove that parallel lines will never meet. His failure to do so led him to suspect that there is something imprecise about the geometry of Euclid that is taught in school. It seemed that if parallel lines were extended far enough, they might curve enough to meet." ("Finding of Blue Galaxies Backs Big Bang Theory," *The New York Times,* June 13, 1965).

11.6. MATHEMATICAL MODELS FOR REAL WORLD PHENOMENA

As we have seen, the other sense in which the term **mathematical model,** often shortened to **model,** is used is in connection with developing a mathematical portrait of a real world phenomenon of interest. Just as a person's portrait may be sketched in many ways, so too may a mathematical portrait of a real world phenomenon be developed in many ways, depending on its features of interest to be captured and the ingenuity and sensitivity of the mathematical artist developing the model. A mathematical portrait consists of assumptions, or equivalently, postulates, that the mathematical artist is willing to make, postulates which lead to valid conclusions—the theorems of the model.

How do we determine how "successful" is a mathematical portrait of our subject of interest? If its postulates are "reasonably" accurate, as determined by experimentation and observation, then the model is judged a success. The problem is that in general a model's postulates do not lend themselves to a truth/falsity determination by experimentation and observation. This being the case, close attention must be paid to the realism of the postulates of the model. In this situation we are like the artist sketching a portrait in dim light. We would like to turn a bright light on the subject, but it is not available and so we must proceed carefully with the dim light we do have.

Another source of light is to be found in the model's theorems. If their truth/falsity can be established, then doing so will throw light, indirect light, to be sure, on the postulates. If the theorems are shown to be true, or to allow for an important shade of gray, realistic, then the postulates, as we saw illustrated in Section 11.1, may be realistic or possibly unrealistic. Realistic theorems may arise from unrealistic postulates, so that establishing the realism of a model's theorems does not definitively establish the realism of its postulates. But it does serve to strengthen our confidence in the postulates as providing a good working portrait of the phenomenon under study, with the understood qualification that there may be important aspects of the portrait which are incomplete and possibly inaccurate. On the other hand, if a theorem of the model is show to be false, then this sends us an unequivocal message that the portrait provided by its postulates is in need of revision. The revision called for may be relatively minor or very drastic.

Suppose, for example, that a production model has been developed for a company that manufactures stereo systems, product models RA-5 and RA-9, let us say. A theorem of this model says that for the company to maximize its monthly profit on

RA-5 and RA-9 it should set its production schedule to make 10,000 RA-5 and 12,000 RA-9 units per month for a projected maximum profit of $750,000. This valid conclusion of the model may sound good, but is it realistic? The acid test entails setting the production schedule at these levels and seeing if the actual monthly profit realized is in the neighborhood of the projected $750,000. If it turns out not to be the case, that is, we have a valid but unrealistic conclusion, this acid test could prove rather costly to the company. The question is, what options are available prior to going ahead with the acid test and implementing the derived production schedule? Only one, and that is to pay particularly close attention to the realism of the postulates of the model. This is the decisive juncture. If there are reservations about the realism of the assumptions being made, then such should be resolved before implementing the production schedule obtained as a valid conclusion. This may require that the model, that is, its postulates, be refined and that a new production schedule be derived from the refined model.

EXERCISES

1. Consider the following statements; for each one state, with explanation, whether you agree or disagree.
   (a) If some of a model's theorems have been confirmed by experimentation/observation, then the model's postulates must be realistic.
   (b) If some of a model's postulates are unrealistic, then some of its theorems must be false.
   (c) If a model's postulates are unrealistic, then some of its theorems may be true.
   (d) If a model's postulates are true, then some of its theorems may be false.
   (e) Postulates of a model, by their very meaning, are true statements.
   (f) If a theorem is false, then it cannot be a valid statement.
   (g) If a conclusion obtained from a model is true, then it must be a theorem of the model.
   (h) A conclusion obtained from a model can be shown to be a theorem by showing that it is true.

2. Raymond J. Ellis of Ecap University developed a mathematical model to describe the nature and behavior of gravity waves. The realism of the postulates of Ellis's model cannot be determined by directly subjecting them to experimental verification.
   (a) What can be done to access the realism of Ellis's model? Explain.
   (b) Over a period of twenty years ten valid conclusions of Ellis's model were subjected to experimental tests.
      (i) Suppose all ten of these conclusions were shown to be realistic. Would this establish that Ellis's model is as close to a perfect portrait of gravity waves that one could hope to obtain? Explain.

(ii) Suppose the first nine of Ellis's conclusions were shown to be realistic, but that the tenth was shown to be unrealistic. What implications would this have for the Ellis model?

3. A team of economists headed by Janet Valdez developed a mathematical model for foreign trade between a group of countries in Central and South America. One of the valid consequences of this model holds that the gross domestic product of each of the participating countries would rise by at least 5% per year for ten years if all tariffs between the countries were eliminated. Before implementing this conclusion by eliminating tariffs a commission made up of the economic ministers of the countries involved undertook to review the Valdez model. What should be the focus of the review? How so?

4. Jupiter Motors is in the enviable position of having a shipping schedule problem. Its Jupiter 500 sport car is very popular and it has become difficult to keep up with the demand. J.M. is obligated to supply its distributors in New York, Atlanta, Dallas, and Seattle 6100, 8200, 6500, and 9000 Jupiter 500's per month, respectively. The cars are to be shipped from plants in Arkin, Hastings, and Freelawn. The problem is to satisfy these commitments at minimum total cost.

Operations research groups, working independently, set up two mathematical shipping models, designated by SM-1 and SM-2, for this problem. A valid consequence of SM-1 is that the shipping schedule shown in Table 11.2 should be implemented to obtain a minimum monthly shipping cost of $650,000. A valid consequence of SM-2 is that the shipping schedule shown in Table 11.3 should be implemented to obtain a minimum monthly shipping cost of $725,000.

Table 11.2 Shipping Schedule Based on SM-1

	New York	Atlanta	Dallas	Seattle
Arkin	2000	4500	3000	1500
Hastings	1100	2500	1500	3000
Freelawn	3000	1200	2000	4500

Table 11.3 Shipping Schedule Based on SM-2

	New York	Atlanta	Dallas	Seattle
Arkin	1000	3000	4000	3000
Hastings	3200	1500	1000	2400
Freelawn	1900	3700	1500	3600

(a) Would you implement SM-1's shipping schedule because its projected total minimum cost, $650,000 per month, is less than the $725,000 projected total minimum cost of implementing SM-2? Explain.

(b) If your answer to (a) is no, on what basis would you implement one of the aforenoted shipping schedules?

(c) Is it possible that you might not implement either of the aforenoted shipping schedules? How so?

(d) Does the lower projected cost of SM-1 imply that SM-2's projected cost is not valid? How so?

11.7. WHICH GEOMETRY IS RIGHT FOR SPACE?

Since each is as consistent as the other, Euclidean and Lobachevskian geometries can peacefully coexist as mathematical structures very nicely. But can they coexist as rival models of physical space? If you believe in physical space as being synonymous with a unique set of principles, Euclidean principles, then the answer is no; from such a point of view there is only one way to do it, the Euclidean way, and that's that. The relative consistency of Lobachevskian geometry in terms of its Euclidean counterpart makes it arbitrary, and thus indefensible, to choose one as a unique model for space on the basis of intellectual comfort. An examination of the postulates or theorems of each geometry against the behavior of physical space in terms of experimentation and observation emerges as the only satisfactory means for settling the issue of which geometry is the more realistic description of physical space.

The program that emerges entails assigning physical meanings to point and line and, in terms of these physical representations, empirically testing the realism of some key theorems of both geometries. An important property of Lobachevskian geometry is that for "small regions" it differs little from Euclidean geometry. For "sufficiently large" regions the differences become more substantial, but how large "sufficiently large" must be in order for significant differences to be revealed by experimental results is not clear. One key result in both geometries concerns the interior angle sum of a triangle. In Euclidean geometry this is 180 degrees. A possible approach suggested by this is to interpret point as the position of a celestial body, line segment as a beam of light projected from one such body to another, and determine the interior angle sum of large celestial triangles determined by three such celestial points. This is what Lobachevsky did in taking as his celestial points the Earth, Sun and star Sirius. He concluded that in this triangle the interior angle sum cannot differ from 180 degrees by more than 0.00000372″. The result is inconclusive because we cannot answer the question of whether the distances involved in this celestial triangle are sufficiently large to reveal deviations from Euclidicity. Another complication, which Lobachevsky recognized, is that the geometry used to describe space goes hand in hand with properties of matter in space. It might be that a discrepancy observed in the interior angle sum of a triangle from 180 degrees could be explained by retaining the assumptions of Euclidean geometry but at the same time modifying some physical assumptions involving mechanics or optics. By the same token the absence of a discrepancy might be compatible with the assumptions of non-Euclidean geometry and suitable adjustments in our assumptions about the behavior of matter in space.

As to the application of geometry to practical measurements, engineering, surveying, and the like, "in the small", as we say, Euclidean geometry, suitably interpreted, is confirmed to a high degree of approximation as a theory of physical space. Lobachevskian geometry "in the small" approximates Euclidean geometry. Many of its principles, though not all, may be considered physically confirmed to a high degree of approximation. Engineers, surveyors and scientists who use Euclidean geometry in practice could use Lobachevskian geometry instead, but it is simpler to use Euclidean geometry since its formulas tend to be simpler.

11.8. SELF-TESTS FOR CHAPTER 11

Allow 100 or so minutes for each self-test. Go over the first before proceeding to the second.

Self-Test 1

1. Determine the validity of the arguments for the proposed conclusions. Explain the basis for your answer.

Hypothesis.	$P1$:	All dogs are mammals.
	$P2$:	Some mammals are frogs.
	$P3$:	Some mammals are white.
Conclusions.	$C1$:	Some dogs are frogs.
	$C2$:	Some frogs are white.
	$C3$:	Some dogs are white.

2. Pinpoint the gap(s) in the "proof" of the following statement.

If a line L intersects two parallel lines M and N, the sum of the interior angles lying on the same side of L is 180 degrees (see Figure 11.16).

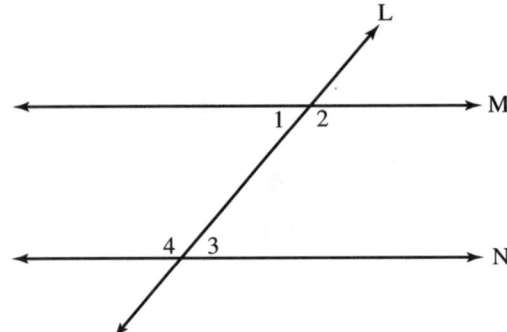

Figure 11.16

Proof.

Three cases are possible:

1. The sum of the interior angles on the same side of L exceeds 180 degrees.
2. The sum of the interior angles on the same side of L is less than 180 degrees.
3. The sum of the interior angles on the same side of L equals 180 degrees.

In the first case we have:

$$\text{angle } 1 + \text{angle } 4 > 180°, \quad \text{angle } 2 + \text{angle } 3 > 180°$$

Thus:

The angle sum of 1 through 4 exceeds 360°

But the sum of the four interior angles is equal to two straight angles, 360 degrees. This contradiction shows that Case 1 is untenable. The same reasoning shows that Case 2 is untenable. Thus we are left with Case 3, which is what is to be proved.

3. For the postulates of Glob Theory see Section 11.3 (p. 348). Jan Hansen, another "expert" on Glob Theory, proposed the following proof of the conjecture that this postulate system contains at least six globs. Is her argument valid? Explain.

Conjecture 3: Glob Theory contains at least six globs.

Proof.

Statement	Justification
1. Let p, q, r, and s denote four zogs.	1. Conjecture 1

2. Let G_1 denote the glob containing p and q, G_2 the glob containing p and r, G_3 the glob containing p and s, G_4 the glob containing q and r, G_5 the glob containing q and s, and G_6 the glob containing r and s.

2. P2

3. G_1 through G_6 number six globs.

3. Summary of preceding steps.

4. Robert Glob, brother of the founder of Glob Theory (Sec. 11.3, p. 348), proposed an extension of Glob Theory by adding the following postulate to $P1-P4$. P5: There are at least four zogs.

   (a) Is Robert's Extended Glob Theory consistent? Explain.

   (b) Is P5 an independent postulate? Explain.

   (c) Is Jane Hansen's proof of conjecture 3 (see Question 3) valid in Robert's extension? Explain.

5. Gladys Glob proposed an extension of Glob Theory by adding the following postulate to $P1-P4$. P5(a): There are at least three globs.

   (a) Is Gladys' extension consistent? Explain.

   (b) Is $P5(a)$ independent? Explain.

   (c) Gladys decided to go one step further and replace her $P5(a)$ by $P5(b)$: There are at least five globs. Is $P5(b)$ independent? Explain.

6. A mathematician proposed a geometry based on the postulates of Euclidean geometry with two replacements:

Euclidean postulate	Replacement
1. Parallel postulate: Given a line L and point P not on it, there is one line passing through P parallel to L.	1. There are no parallel lines.
2. Given two distinct points, there is one line containing them	2. Given two distinct points, there is at least one line (possibly more than one) containing them.

   One argument brought against this proposed geometry is that it must contain contradictory statements (i.e., be inconsistent) because its parallel postulate (there are no parallel lines) and its second statement (two distinct points do not determine one and only one line) are contrary to the nature of physical space. Would you agree with this point of view?

7. "The *reducio ad absurdum* or indirect approach to the parallel postulate problem was to show that the negation of Euclid's parallel postulate is false." Do you agree? Explain.

8. "In geometry truth and validity mean the same thing." Do you agree? Explain.

9. "The modern concept of geometry is the same as that advanced by the ancient Greek mathematicians." Do you agree? Explain.

10. "Japanese scientists have reported that small gyroscopes lose weight when spun under cetain conditions. . . . Dr. Park of The University of Maryland said the finding, if proved true, would almost certainly be explained by general relativity, Albert Einstein's geometric theory of gravitation. . . . Dr. Forword, who aids the Air Force in its propulsion work, said the sheer volume of anti-gravity claims threw doubt on the validity of the new finding." (*The New York Times*, December 28, 1989, p. A17.)

    (a) In what sense might general relativity explain the gyroscope phenomenon?
    (b) If the gyroscope phenomenon holds up and it cannot be explained by general relativity, would implications would this have for general relativity?
    (c) In what sense does the "sheer volume of anti-gravity claims" throw doubt on the validity of the new gyroscope finding?

11. ". . . theorists calculated that if a certain class of atoms could be chilled to temperatures below any that exist in nature, the atoms would merge with each other to become hugh "superatoms:" . . . The creation of a Bose-Einstein condensate, as this hypothetical superatomic state of matter is called, . . . would not only demonstrate the validity of some outlandish predictions of quantum theory, but would create a form of matter that may never have existed anywhere before." (M. Browne, "Physicists Get Warmer in Search for Weird Matter Close to Absolute Zero," *The New York Times,* Aug. 23, 1994, C1.)

    (a) Would the creation of a Bose-Einstein condensate demonstrate the validity of some outlandish predictions of quantum theory? Explain.
    (b) What would the creation of a Bose-Einstein condensate demonstrate? Explain.

Self-Test 2

1. Determine the validity of the arguments for the proposed conclusions. Explain the basis for your answer.

Hypothesis.	P1:	All x's are y's.
	P2:	Some x's are z's.
	P3:	Some w's are x's.
Conclusions.	C1:	Some y's are z's.
	C2:	Some y's are w's.
	C3:	Some z's are w's
	C4:	Some z's are not w's.
	C5:	Some z's are not x's.
	C6:	Some z's are not y's.

2. Pinpoint the gap(s) in the "proof" of the following statement.

   Theorem: The diagonals of a parallelogram bisect each other.
   Given: Parallelogram *ABCD* with diagonals $\overline{AC}$ and $\overline{BD}$.
   To Prove: $\overline{AE} = \overline{EC}$, $\overline{BE} = \overline{ED}$

   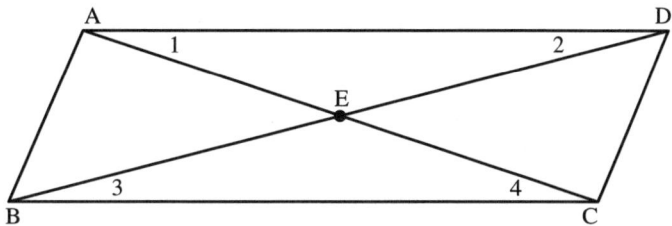

   Figure 11.17

   Proof.

Statement	Justification
1. *AD* is parallel to *BC*; *AB* is parallel to *DC*.	1. By hypothesis *ABCD* is a parallelogram.
2. Angle 1 = Angle 4. Angle 2 = Angle 3.	2. Alternate interior angles of parallel lines are equal.
3. $\overline{AD} = \overline{BC}$.	3. Opposite sides of a parallelogram are equal.
4. Triangles *AED* and *CEB* are congruent.	4. Angle-Side-Angle.
5. $\overline{AE} = \overline{EC}$, $\overline{BE} = \overline{ED}$.	5. Corresponding parts of congruent triangles are equal.

3. **Yuk Theory,** with undefined terms luk and yuk, is based on the following five postulates.

   *P*1. Every yuk is a collection of luks containing at least one luk.

   *P*2. There are at least two luks.

   *P*3. For any two luks, there is one and only one yuk containing them.

   *P*4. For any yuk there is a luk not contained by it.

   *D*1. Two yuks are called **parallel** if there is no luk contained by both.

   *P*5. If *L* is a yuk and *p* is a luk not contained by it, there is exactly one yuk containing *p* which is parallel to *L*.

   (a) State two models for Yuk Theory.

   The conjectures stated in (b) through (h) have been proposed concerning Yuk Theory. Consider each and prove it a theorem or show that it is not a theorem of Yuk Theory.

(b) *C*1: Every luk is contained by at least two yuks.

(c) *C*2: Every yuk contains at least two luks.

(d) *C*3: There are at least three luks.

(e) *C*4: There are at least four luks.

(f) *C*5: There are at least five yuks.

(g) *C*6: There are at least six yuks.

(h) *C*7: There are at least seven yuks.

4. Is Yuk Theory, defined in Question 3, consistent? Explain.

5. Marcel Yuk formulated modified Yuk Theory (MYT) by adding the following postulate P6 to the five postulates of Yuk Theory. P6: There are at most five yuks. Is MYT consistent? Explain.

6. Consider the surface of a sphere in Euclidean space (see Figure 11.18).

(a) Concerning Yuk Theory, interpret luk as meaning a point on the surface of this sphere and yuk as meaning a great circle on the surface of this sphere (that is, a circle which breaks the sphere into hemispheres). Is this interpretation a model of Yuk Theory? Explain.

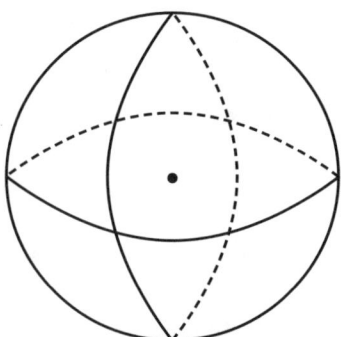

Figure 11.18

(b) Concerning Glob Theory (p. 348), interpret zog and glob in the same way. Is this interpretation a model of Glob Theory? Explain.

7. Is postulate *P*5 of Yuk Theory independent?

8. "A refined radar technique that may settle the current debate over the validity of Einstein's general theory of relativity has been successfully tested." (*The New York Times,* Feb. 28, 1968, p. 20.)

(a) Could such a technique be used to settle a question of validity?

(b) What issue might such a technique help to resolve?

9. "The most accurate long-distance measurements ever made, by means of radio signals between the Viking spacecraft on Mars and antennas on Earth, have produced new confirmation of Einstein's theory of relativity, a Viking project scientist reported today." (*The New York Times,* Jan. 7, 1977, p. A8.) In what sense was the theory of relativity confirmed?

10. "Evidence reported by physicists last fall suggested that the particles [protons], the basic building blocks of matter, had long, but finite lifetimes. But a recent report by researchers who participated in an Ohio study says that the proton lifetimes may be even longer than the billions of years previously estimated. They also say this could mean theories predicting such decay are invalid." (*The New York Times,* Jan. 23, 1983).

   (a) Would this evidence render the conclusions of such theories of proton decay invalid?

   (b) What effect would this evidence have on theories of proton decay?

Appendix on Tables

Table A: Standard Normal Curve Areas

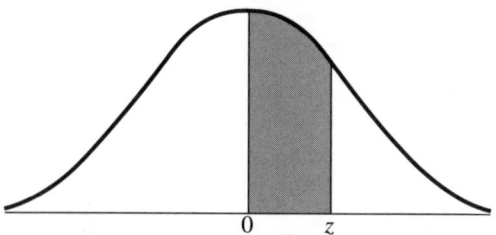

Area of the region under the standard normal curve from 0 to z, shown shaded.

z	0	1	2	3	4	5	6	7	8	9
0.0	.0000	.0040	.0080	.0120	.0160	.0199	.0239	.0279	.0319	.0359
0.1	.0398	.0438	.0478	.0517	.0557	.0596	.0636	.0675	.0714	.0753
0.2	.0793	.0832	.0871	.0910	.0948	.0987	.1026	.1064	.1103	.1141
0.3	.1179	.1217	.1255	.1293	.1331	.1368	.1406	.1443	.1480	.1517
0.4	.1554	.1591	.1628	.1664	.1700	.1736	.1772	.1808	.1844	.1879
0.5	.1915	.1950	.1985	.2019	.2054	.2088	.2123	.2157	.2190	.2224
0.6	.2257	.2291	.2324	.2357	.2389	.2422	.2454	.2486	.2517	.2549
0.7	.2580	.2611	.2642	.2673	.2704	.2734	.2764	.2794	.2823	.2852
0.8	.2881	.2910	.2939	.2967	.2995	.3023	.3051	.3078	.3106	.3133
0.9	.3159	.3186	.3212	.3238	.3264	.3289	.3315	.3340	.3365	.3389
1.0	.3413	.3438	.3461	.3485	.3508	.3531	.3554	.3577	.3599	.3621
1.1	.3643	.3665	.3686	.3708	.3729	.3749	.3770	.3790	.3810	.3830
1.2	.3849	.3869	.3888	.3907	.3925	.3944	.3962	.3980	.3997	.4015
1.3	.4032	.4049	.4066	.4082	.4099	.4115	.4131	.4147	.4162	.4177
1.4	.4192	.4207	.4222	.4236	.4251	.4265	.4279	.4292	.4306	.4319
1.5	.4332	.4345	.4357	.4370	.4382	.4394	.4406	.4418	.4429	.4441
1.6	.4452	.4463	.4474	.4484	.4495	.4505	.4515	.4525	.4535	.4545
1.7	.4554	.4564	.4573	.4582	.4591	.4599	.4608	.4616	.4625	.4633
1.8	.4641	.4649	.4656	.4664	.4671	.4678	.4686	.4693	.4699	.4706
1.9	.4713	.4719	.4726	.4732	.4738	.4744	.4750	.4756	.4761	.4767
2.0	.4772	.4778	.4783	.4788	.4793	.4798	.4803	.4808	.4812	.4817
2.1	.4821	.4826	.4830	.4834	.4838	.4842	.4846	.4850	.4854	.4857
2.2	.4861	.4864	.4868	.4871	.4875	.4878	.4881	.4884	.4887	.4890
2.3	.4893	.4896	.4898	.4901	.4904	.4906	.4909	.4911	.4913	.4916
2.4	.4918	.4920	.4922	.4925	.4927	.4929	.4931	.4932	.4934	.4936
2.5	.4938	.4940	.4941	.4943	.4945	.4946	.4948	.4949	.4951	.4952
2.6	.4953	.4955	.4956	.4957	.4959	.4960	.4961	.4962	.4963	.4964
2.7	.4965	.4966	.4967	.4968	.4969	.4970	.4971	.4972	.4973	.4974
2.8	.4974	.4975	.4976	.4977	.4977	.4978	.4979	.4979	.4980	.4981
2.9	.4981	.4982	.4982	.4983	.4984	.4984	.4985	.4985	.4986	.4986
3.0	.4987	.4987	.4987	.4988	.4988	.4989	.4989	.4989	.4990	.4990

Answers to Odd-Numbered Exercises and Self-Tests

Section 1.1 (p. 6)

1. $y = 2x + 2$

3. $y = -x + 3$

5. $y = -2x - 1$

7. $3y = x + 2$

9. $x = 1$

Section 1.2 (p. 11)

1. (5, 4)

3. (1, 4)

5. (2, 3)

7. $(\frac{20}{3}, \frac{29}{3})$

9. (50, 40)

11. (45, 10)

13. $(\frac{211}{16}, \frac{155}{16})$

Section 1.3 (p. 19)

1. (a) $y = 4.5x - 1.7$ **(b)** For $x = 5$, $y = 20.8$; if the president of the company is complimented five times per week for his intellectual keenness, then the predicted average annual increment is 20.8 hundred dollars.

3. (a) $y = 13.946x - 32.704$ **(b)** For $x = 30$, $y = 385.68$; for sales personnel who obtain a score of 30 on the aptitude test, the predicted average first-year sales volume is $385.68 thousand.

5. (a) $y = 0.0534x + 1.145$ **(b)** For $x = 35$, $y = 3.0$; for students who are 35 years old at the time of graduation, the predicted average grade-point average is 3.0.

Section 1.4 (pp. 30 and 34)

1. $(5, 4)$
3. $(1, 4)$
5. $(\frac{20}{3}, \frac{29}{3})$
7. $(2, 3)$
9. $y = 2x + 5$, x may be chosen at will (arbitrarily)

11. $(150{,}000; 100{,}000)$
13. $(1, -1, -1)$
15. $(-2, 3, -1)$
17. $(-1/2, -1, 2, 2)$

Section 1.5 (p. 40)

1. (a) Let x, y, and z denote the total costs of service departments S_1, S_2, and S_3, respectively. The condition that the total cost of each service department must equal the overhead of the department (for March) plus the charges for services provided by the other service departments leads to the following relations:

$$x = 40{,}000 + 0.1y + 0.2z$$
$$y = 30{,}000 + 0.1x + 0.05z$$
$$z = 20{,}000 + 0.05x + 0.1y$$

By transposing terms we obtain the following system of equations:

$$x - 0.1y - 0.2z = 40{,}000$$
$$-0.1x + y - 0.05z = 30{,}000$$
$$-0.05x - 0.1y + z = 20{,}000$$

(b) To the nearest dollar, $x = \$48{,}831$, $y = \$36{,}186$, $z = \$26{,}060$.
(c) $P_1 = \$43{,}128$, $P_2 = \$46{,}872$

3.
$$x + 0.095y = 2{,}589.415$$
$$0.01x + 1.010y + 0.01z = 1{,}529.7$$
$$0.48x + 0.480y + z = 12{,}253.24$$
$$x = 2{,}457.63, \quad y = 1{,}387.18, \quad z = 10{,}407.73$$

Section 1.6 (p. 47)

1. $x = 5 - z$, $y = z$, z arbitrary
3. $y = -6 + 2z$, $x = 10 - z$, z arbitrary
5. $x = -8 + w$, $y = -2$, $z = 2$, w arbitrary
7. We introduce variables s, t, u, v, w, and x as shown in the network diagram. Variable x denotes the number of cars passing between points B and C per hour, v the number of cars passing between points C and E per hour, and so on. For equilibrium at the points A, B, C, D and E, the number of cars entering each of these points per hour must equal the number of cars leaving each of these points per hour. This leads to the following conditions:

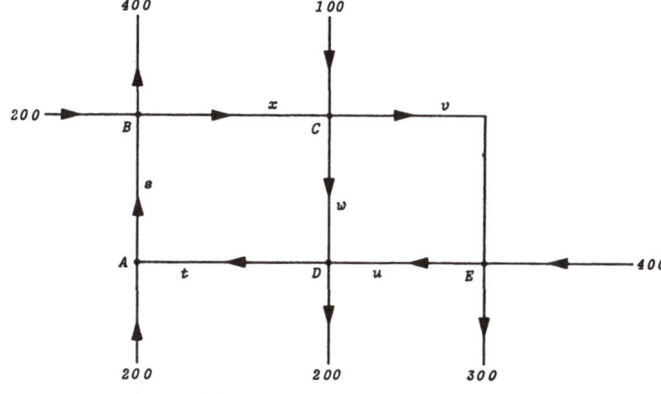

$$A:\ t + 200 = s$$
$$B:\ s + 200 = x + 400$$
$$C:\ x + 100 = v + w$$
$$D:\ u + w = t + 200$$
$$E:\ v + 400 = u + 300$$

Transposing and rewriting terms yields the following system of equations:

$$s - t = 200$$
$$s - x = 200$$
$$v + w - x = 100$$
$$-t + u + w = 200$$
$$u - v = 100$$

Solving this system yields:

$$s = x + 200$$
$$t = x$$
$$u = -w + x + 200$$
$$v = -w + x + 100$$

w and x are arbitrary

For this solution to make sense in terms of the traffic background in question, the values given to the underlying variables must, at a minimum, be restricted to non-negative integers. From $s = x + 200$, we obtain

$$x = s - 200$$

$x \geq 0$ yields

$$s - 200 \geq 0$$
$$s \geq 200$$

Thus traffic equilibrium cannot be maintained if fewer than 200 cars per hour pass between points A and B, which is clear from the network structure.

The condition $t = x$ tells us that traffic equilibrium cannot be maintained if the number of cars passing between D and A per hour is not equal to the number of cars passing between B and C.

From $u \geq 0$ and $u = -w + x + 200$, we have

$$-w + x + 200 \geq 0$$
$$x + 200 \geq w$$
$$w \leq x + 200$$

This tells us that traffic equilibrium cannot be maintained if the number of cars passing between C and D per hour exceeds 200 plus the number of cars passing between B and C per hour.

Replacing x in $w \leq x + 200$ by $s - 200$ yields

$$w \leq s,$$

which tells us that traffic equilibrium cannot be maintained if the number of cars passing between C and D per hour exceeds the number of cars passing between A and B per hour.

Section 1.7 (p. 49)

1. $(0, 2)$, $(3, -2)$, $(-4, 0)$
3. $(2, 3)$, $(6, 2)$
5. $(3, -2)$, $(-4, 0)$, $(2, 3)$, $(6, 2)$
7. $(3, -1, 4)$, $(\frac{1}{2}, \frac{1}{3}, -3)$, $(-2, 3, 1)$
9. $(2, 3, 1)$, $(1, 3, -1)$, $(4, 2, 1)$, $(\frac{1}{2}, \frac{1}{3}, -3)$
11. $(2, 3, 1)$, $(3, -1, 4)$, $(\frac{1}{2}, \frac{1}{3}, -3)$, $(-2, 3, 1)$

Section 1.7 (p. 51)

21. $x > -\frac{2}{3}$

23. $x \leq 2$

25. $y < 2$

27. $x < -15$

29. $x > \frac{5}{2}$

31. $y \geq 2$

Section 1.7 (p. 56)

33.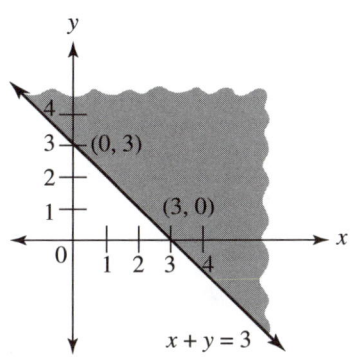

37.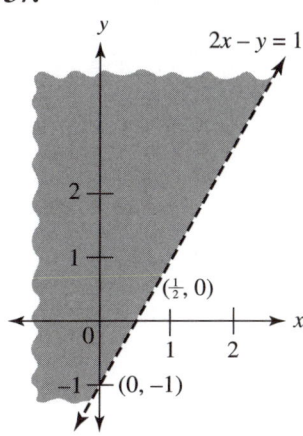

35.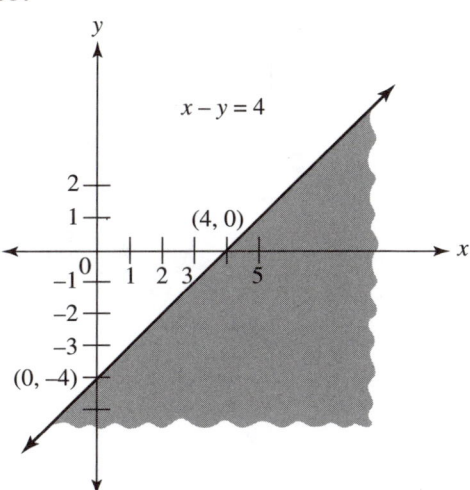

39.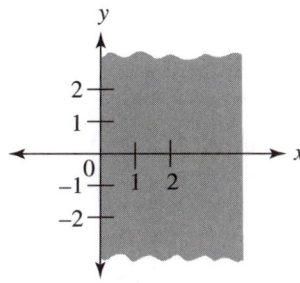

41.

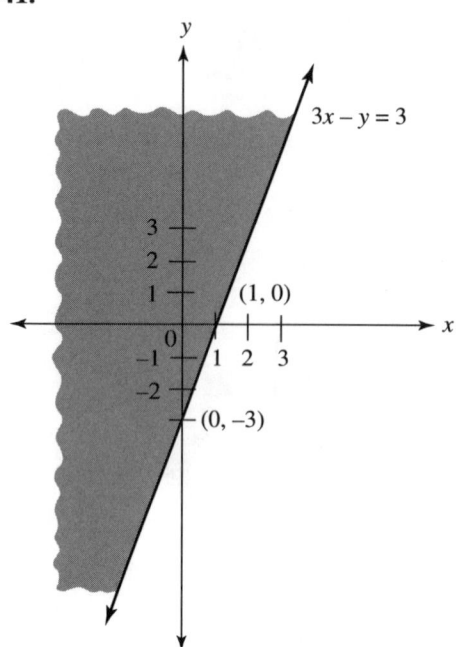

45.

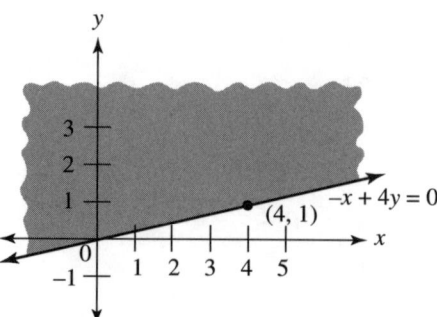

43.

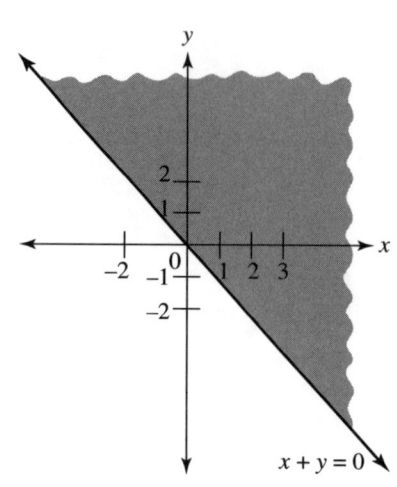

47.

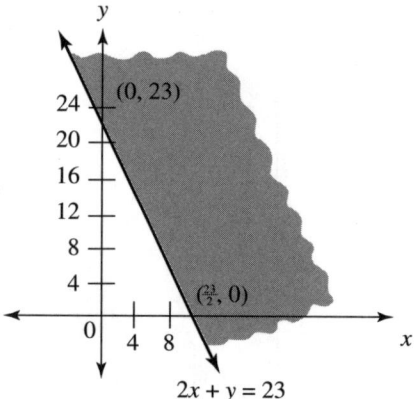

Section 1.7 (pp. 58 and 62)

49. (8,2), (8,17), (20,5), (9,12), (14,7)

51. (1,1,2), (0,1,3), (2,1,1), (4,1,0), (2,1,2)

53.

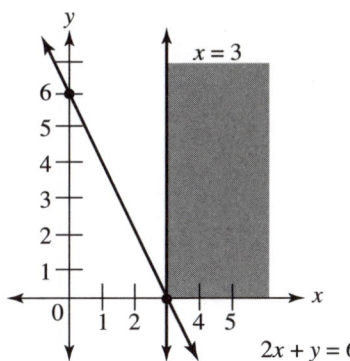

59.

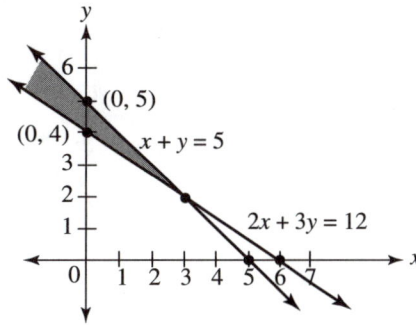

55.

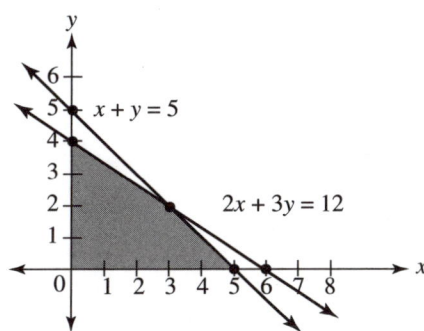

61.

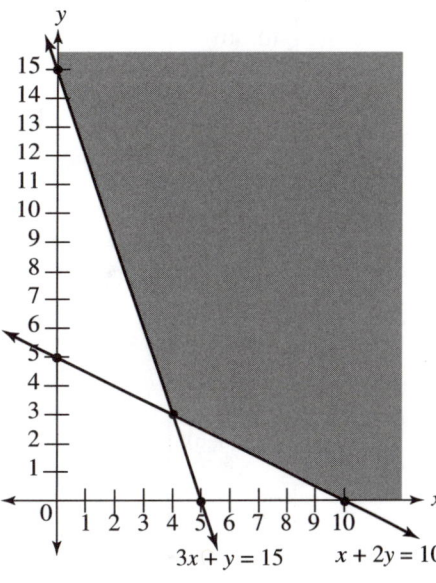

57.

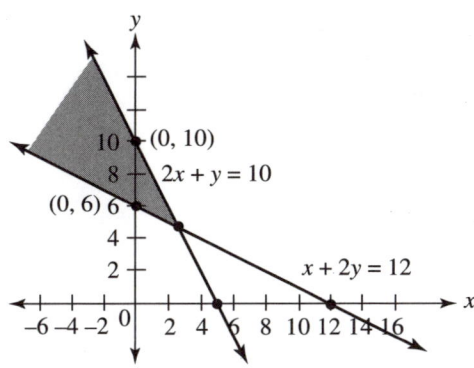

63.

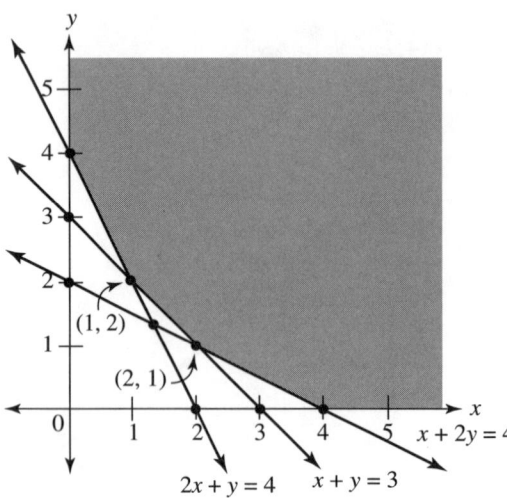

65.

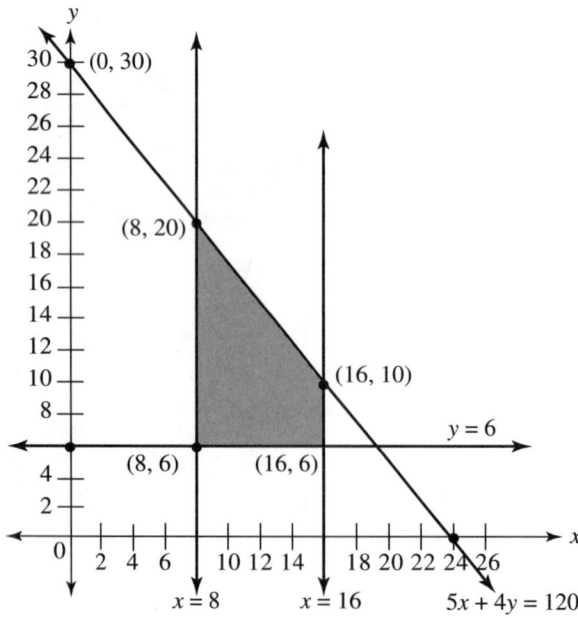

67.

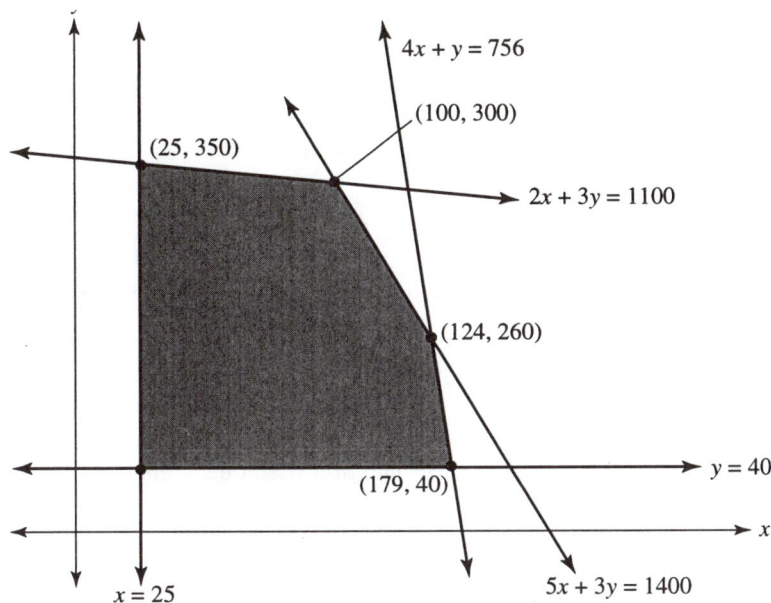

Section 1.8 (p. 64)

1. not linear
3. linear
5. $C(0, 15)$ is not defined, $C(0, 23) = 92$, $C(\frac{23}{2}, 0)$ is not defined, $C(\frac{20}{3}, \frac{29}{3}) = 72$, $C(\frac{75}{4}, 0) = \frac{375}{4}$
7. $I(8, 2) = 0.84$, $I(25, 0)$ is not defined, $I(8, 17) = 1.74$, $I(20, 5) = 2.1$

Section 2.2 (p. 72)

1. **(a)** and **(b)** Mathematics is precise in the sense that its methods yield valid conclusions with respect to the assumptions made. Sometimes different assumptions are made about a situation under study, and these assumptions lead to different conclusions (solutions), each of which is valid (inescapable consequence) with respect to its assumptions. Different valid conclusions are not in conflict insofar as validity is concerned because they come from different assumptions. They are in conflict insofar as truth is concerned.

   (c) No; although both values are mathematically correct (that is, valid with respect to their respective assumptions), we are not assured that the larger value is more realistic than the smaller one, and that the true maximum profit is the predicted 230,000 value.

   (d) The assumptions of the M1 model were highly unrealistic.

Section 3.2 (p. 81)

1. The basic data are summarized in Table 1.

Table 1

Product	No. made	Profit per unit	Construction time per unit (hours)	Finishing time per unit (hours)
DT-1	x	140	8	3
DT-2	y	150	5	2
			≤ 2210	≤ 860

The assumptions underlying these data and the conditions that at least 50 DT-1 units and 50 DT-2 units must be made per week lead to LP-2.

Section 3.3 (p. 88)

1. (6,0); 30
3. (1,2); 2.7
5. (45,15); 5.7
7. $(\frac{50}{3}, \frac{65}{3})$; 3350
9. (50, 40); 13,800
11. (75,0) and, more generally, (75,y), where y is an integer between 0 and 25, inclusive; 12,420.

Section 3.4 (p. 91)

1. **(a)** The corner-point method cannot by itself guarantee that profit will be maximized when 300 ZKB-47 and 250 ZKB-82 units are made daily and sold. Whether these output levels or other ones will maximize profit is a question of truth, and the corner-point method, as a mathematical technique, can only ensure the validity of the predicted output levels with respect to the linear programming model that was set up for the profit maximization problem in question. If the assumptions that the model reflects are realistic, then the aforementioned output levels, obtained as a valid conclusion of these assumptions, will maximize profit; if these assumptions are not realistic, then it might well happen that other output levels would yield a higher profit than that projected for 300 ZKB-47 and 250 ZKB-82 units.

    (b) The basis for implementing the conclusion that 300 ZKB-47 and 250 ZKB-82 units be made and sold daily is the belief, based on an analysis of the company's operations and the market, that the assumptions made are realistic.

Section 3.6 (p. 106)

1. Maximize $P = 180x + 120y$
   subject to
   $$x \geq 0, y \geq 0$$
   $$4x + 3y \leq 320$$
   $$5x + 2y \leq 330$$

   x and y are the number of $K15$ and $K31$ units, respectively, to be made per week.

   Sol. $(50, 40)$; max. value $13{,}800$ (see Exercise 9 of Section 3.3 (p. 88)).

3. Minimize $C = 45x + 50y$
   subject to
   $$x \geq 0, y \geq 0$$
   $$3x + 5y \geq 12$$
   $$x + y \geq 3$$

   x and y are the number of pounds of pork and beef, respectively, to be used in putting together a can of meat.

   Sol. $(\frac{3}{2}, \frac{3}{2})$; minimum value 142.5.

5. Maximize $I = 0.09x + 0.06y$
   subject to
   $$x \geq 0, y \geq 0$$
   $$x + y \leq 25$$
   $$-x + 4y \geq 0$$
   $$x \geq 8$$

   x and y are the amounts (in millions of dollars) allocated for loans and securities, respectively.

   Sol. $(20, 5)$; maximum value 2.1.

7. Two linear programs emerge, one for the faculty council and the other for the administration. The constraints are the same for both linear programs since the underlying conditions are the same.

   Max. $F = x + y$
   subject to
   $$x \geq 0, \ y \geq 0$$
   $$x \leq 22, y \geq 3$$
   $$x + 2y \leq 30$$
   $$-x + 4y \leq 0$$

   Sol. $(22, 4)$; maximum value 26.

   Min. $C = 500x + 1000y$
   subject to
   (same constraints)

   x and y denote the number of associate and full-professor slots, respectively, to be established.

   Sol. $(12, 3)$; minimum value 9000.

9. Min. $C = 1.2x + 1.8y$

    subject to

    $x \geq 0, y \geq 0$

    $x + y = 2{,}000{,}000$

    $30x + 32y \geq 62{,}400{,}000$

    x and y denote the number of tons of steel produced subject to the F14 and F24 filter systems, respectively.

    Sol. (800,000; 1,200,000); minimum value 3,120,000.

11. Max. $P = 24x + 30y + 26z$

    subject to

    $x \geq 0, y \geq 0, z \geq 0$

    $2x + 3y + z \leq 16$

    $x + y + 3z \leq 17$

    $2x + y + z \leq 12$

    x, y, and z denote the number of C15, C24, and C51 minicomputer units, respectively, to be made.

13. Max. $I = 0.1x + 0.18y$

    subject to

    $x \geq 0, y \geq 0$

    $x \leq 41{,}000; y \geq 5{,}000$

    $x + y \leq 50{,}000$

    $-x + 4y \leq 0$

    x and y denote the amounts to be invested in bonds and stock, respectively.

    Sol. (40,000; 10,000); Max. value 5800.

15. Min. $C = 0.8x + 1.25y$

    subject to

    $x \geq 0, y \geq 0$

    $4x + 3y \geq 30$

    $2x + 3y \geq 24$

    x and y denote the number of AA1 and AA2 freight cars, respectively, to be rented.

    Sol. (12, 0); Min. value 9.6.

17. Max. $P = 10x + 12y$

    subject to

    $x \geq 0, y \geq 0$

    $x + 2y \leq 24{,}000$

    $2x + y \leq 30{,}000$

    $x \geq 5000, y \geq 5000$

    x and y denote the target sales volume (in product units) for hospitals and medical supply houses, respectively.

In all of these situations we are translating assumptions made into a linear program, omitting nothing that was explicitly stated and adding nothing of our own. Our concern is that the assumptions made are realistic and that the factors that were left out because they were viewed as negligible may be realistically treated as such.

Section 3.9 (pp. 124 and 128)

1. (a) Min. $C = x + 1.5y$

   subject to

   $x \geq 0, y \geq 0$

   $x + 2y \geq 9$

   $3x + 2y \geq 12$

   Sol. (3/2, 15/4); min. value 7.125.

   (b) $x + y \geq 6$

   (c) (3, 3)

   x and y denote the number of pounds of bonemeal and processed vegetable matter, respectively, to be used to make up a can of Martins Miracle.

3. $2x + y \leq 5$; sol. (1, 3), max value 34.

5. (a) $x + y \leq 33$

   (b) Max. $I = 40x + 44y$

   subject to

   $x \geq 0, y \geq 0$

   $4x + 5y \leq 150$

   $6x + 7y \leq 228$

   $5x + 4y \leq 150$

   $x + y \leq 33$

   x and y denote the number of desks and bookcases, respectively, to be made. Sol. (15, 18); max. value 1392.

7. The variables defined in Table 2 denote the number of tons of Bauxite to be shipped from the sources to the destinations.

Table 2

		Destination		
		Baltimore (D_1)	Cincinnati (D_2)	Pittsburgh (D_3)
Source	Turin (S_1)	X_{11}	X_{12}	X_{13}
	Johnston (S_2)	X_{21}	X_{22}	X_{23}

The following linear program emerges:

$$\text{Minimize } C = 2X_{11} + 3X_{12} + 2X_{13} + 3X_{21} + 4X_{22} + 2X_{23}$$

subject to

$X_{11} \geq 0, X_{12} \geq 0,$ etc.

$$\left.\begin{array}{l} X_{11} + X_{21} \geq 8000 \\ X_{12} + X_{22} \geq 10{,}000 \\ X_{13} + X_{23} \geq 12{,}000 \end{array}\right\} \begin{array}{l} \text{At least the} \\ \text{amts required} \\ \text{will be received.} \end{array}$$

$$\left.\begin{array}{l} X_{11} + X_{12} + X_{13} \leq 15{,}000 \\ X_{21} + X_{22} + X_{23} \leq 20{,}000 \end{array}\right\} \begin{array}{l} \text{The capacity of the} \\ \text{sources cannot be} \\ \text{exceeded.} \end{array}$$

9. To relate candidates 1 (Jones), 2 (Johnson), and 3 (Marks) to jobs 1 (Supreme Court Judge) and 2 (Civil Court Judge) we introduce variables X_{11}, X_{21}, and more generally X_{ij}, to relate candidate i to job j, as summarized in Table 3. X_{ij} can assume one of two values, 0 if candidate i is not assigned job j, 1 if candidate i is assigned job j.

Table 3

Candidate		Job	
		Supreme Court Judge (job 1)	Civil Court Judge (job 2)
Jones	(candidate 1)	X_{11}	X_{12}
Johnson	(candidate 2)	X_{21}	X_{22}
Marks	(candidate 3)	X_{31}	X_{32}

The problem of filling the positions so that the total potential rating is maximized in such a way that each candidate is assigned to at most one job and each job is filled by at most one person is expressed by the following 0-1 integer program:

$$\text{Max. } P = 9X_{11} + 8X_{21} + 10X_{31} + 8X_{12} + 9X_{22} + 8X_{32}$$

subject to

$$\left.\begin{array}{l} X_{11} + X_{12} \leq 1 \\ X_{21} + X_{22} \leq 1 \\ X_{31} + X_{32} \leq 1 \end{array}\right\} \begin{array}{l} \text{Each candidate is} \\ \text{assigned to at} \\ \text{most one job} \end{array}$$

$$\left.\begin{array}{l} X_{11} + X_{21} + X_{31} \leq 1 \\ X_{12} + X_{22} + X_{32} \leq 1 \end{array}\right\} \begin{array}{l} \text{Each job is filled} \\ \text{by at most one} \\ \text{candidate} \end{array}$$

11. Max. $P = 9X_{11} + 10X_{21} + 8X_{31} + 6X_{12} + 10X_{22} + 3X_{32}$
subject to
$$X_{11} \geq 0, X_{21} \geq 0, \text{ etc.}$$
$$X_{11} + X_{12} \leq 1$$
$$X_{21} + X_{22} \leq 1$$
$$X_{31} + X_{32} \leq 1$$
$$X_{11} + X_{21} + X_{31} \leq 1$$
$$X_{12} + X_{22} + X_{32} \leq 1$$

Section 3.10 (p. 137)

1. Min. $F = 320s + 330t$
subject to
$$s \geq 0, t \geq 0$$
$$4s + 5t \geq 18$$
$$3s + 2t \geq 12$$

3. Min. $F = 11u + 10v + 10w$
subject to
$$u \geq 0, v \geq 0, w \geq 0$$
$$2u + v + 2w \geq 22$$
$$3u + v + 2w \geq 30$$
$$u + 3v + w \geq 20$$

5. Min. $H = -s + t$
subject to
$$s \geq 0, t \geq 0$$
$$-s + t \geq 1$$
$$-s - t \geq 1$$

Section 4.1 (p. 145)

1. (10, 12, 0); 50
3. Pivot on $\frac{1}{2}$ in col. 1.
5. Pivot on 1 in col. 2.
7. Pivot on 1 in col. 2.
9. Pivot on 3 in col. 1.
11. Pivot on 2 in col. 3.

Section 4.1 (p. 151)

13. (8, 0, 12); 59
15. (0, 15, 18); 90
17. (18, 16, 0, 0); (18, 0, 16, 0) (18, 0, 0, 16); 230
19. Underlying. L. P. has no sol.
21. (0, 10); 28
23. Pivot on $\frac{1}{6}$ or $\frac{1}{5}$ in col. 2.
25. (80, 0); 480
27. (80, 0); 1600
29. No solution
31. (12, 0, 6); 84

33. Max. $P = 45x + 30y + 35z$
subject to
$$x \geq 0, y \geq 0, z \geq 0$$
$$2x + 2y + z \leq 10$$
$$2x + 3y + z \leq 11$$
$$x + y + 3z \leq 10,$$

where x, y, and z denote the number of units of copper, lead, and zinc, respectively, to be produced by the flotation process. (4, 0, 2) is the solution, and 250 is the maximum value.

Section 4.3 (p. 160)

1. $(\frac{9}{2}, 0); 45$
3. $(\frac{16}{5}, 0); 32$
5. $(6, 5); 208$
7. $(\frac{16}{3}, \frac{4}{3}, 0); \frac{208}{3}$

Section 5.1 (p. 163)

1. $A + B = \begin{bmatrix} 6 & 1 & 2 \\ 6 & 4 & 6 \end{bmatrix},$

   $A - B = \begin{bmatrix} -2 & -3 & 4 \\ -4 & 0 & -6 \end{bmatrix},$

   $B - A = \begin{bmatrix} 2 & 3 & -4 \\ 4 & 0 & 6 \end{bmatrix}$

3. $A + B = \begin{bmatrix} 10 & -8 \\ 9 & -6 \end{bmatrix},$

   $A - B = \begin{bmatrix} -6 & 6 \\ 3 & 2 \end{bmatrix}$

5. $\begin{bmatrix} -2 & -2 & 1 \\ -4 & -5 & -2 \\ -3 & 0 & 0 \end{bmatrix}$

7. $A + B = \begin{bmatrix} 5 & 1 & -1 \end{bmatrix}$

   $A - B = \begin{bmatrix} 1 & 3 & 9 \end{bmatrix}$

9. (a) 4, (b) -6, (c) 17, (d) -6

Section 5.1 (p. 167)

11. $AB = \begin{bmatrix} 5 & 12 & -5 \\ -2 & 0 & -1 \\ -3 & 4 & -4 \end{bmatrix}$

$A^2 = \begin{bmatrix} -6 & 5 & 11 \\ -3 & -2 & 1 \\ -9 & -3 & 6 \end{bmatrix}$

$BA = \begin{bmatrix} 3 & 5 & 2 \\ -2 & 3 & 5 \\ 3 & -2 & -5 \end{bmatrix}$

$B^2 = \begin{bmatrix} 5 & 4 & 0 \\ 5 & 10 & -1 \\ 1 & 1 & 4 \end{bmatrix}$

13. $\begin{bmatrix} 0 & 0 \\ 0 & 0 \end{bmatrix}$

15. $\begin{bmatrix} 0 & 0 & 0 \\ 0 & 0 & 0 \\ 0 & 0 & 0 \end{bmatrix}$

17. $\begin{bmatrix} a & b & c \\ d & e & f \\ g & h & i \end{bmatrix}$

19. (a) $A(BC) = (AB)C = \begin{bmatrix} 10 & -5 \\ -5 & 5 \end{bmatrix}$

(b) $A(B + C) = AB + AC = \begin{bmatrix} 16 & -7 \\ 9 & -8 \end{bmatrix}$

(c) $(B + C)A = BA + BC = \begin{bmatrix} -1 & 7 \\ 8 & 9 \end{bmatrix}$

21. (a) $BC = \begin{bmatrix} ej + fm & ek + fn \\ gj + hm & gk + hn \end{bmatrix}$,

$AB = \begin{bmatrix} ae + bg & af + bh \\ ce + dg & cf + dh \end{bmatrix}$,

$AC = \begin{bmatrix} aj + bm & ak + bn \\ cj + dm & ck + dn \end{bmatrix}$

(b) $A(BC) = \begin{bmatrix} a(ej + fm) + b(gj + hm) & a(ek + fn) + b(gk + hn) \\ c(ej + fm) + d(gj + hm) & c(ek + fn) + d(gk + hn) \end{bmatrix}$

$(AB)C = \begin{bmatrix} (ae + bg)j + (af + bh)m & (ae + bg)k + (af + bh)n \\ (ce + dg)j + (cf + dh)m & (ce + dg)k + (cf + dh)n \end{bmatrix}$

(c) $A(B + C) = \begin{bmatrix} a(e + j) + b(g + m) & a(f + k) + b(h + n) \\ c(e + j) + d(g + m) & c(f + k) + d(h + n) \end{bmatrix}$

$AB + AC = \begin{bmatrix} ae + bg + aj + bm & af + bh + ak + bn \\ ce + dg + cj + dm & cf + dh + ck + dn \end{bmatrix}$

23.
$$BE = \begin{bmatrix} \text{plant } P_1 & \text{plant } P_2 \\ 13{,}300 & 13{,}900 \\ 17{,}200 & 18{,}350 \end{bmatrix} \begin{array}{l} \text{CMM steel} \\ \text{CST steel} \end{array}$$

The entries of BE describe the cost of pollution control in making the two types of steel at the two plants: \$13,300 is the cost of pollution control in producing CMM steel at plant P_1; \$13,900 is the cost of pollution control in producing CMM steel at plant P_2; etc.

Section 5.2 (p. 173)

1. $\bar{0} = \begin{bmatrix} 0 & 0 \\ 0 & 0 \\ 0 & 0 \\ 0 & 0 \end{bmatrix}$

3. $I_4 = \begin{bmatrix} 1 & 0 & 0 & 0 \\ 0 & 1 & 0 & 0 \\ 0 & 0 & 1 & 0 \\ 0 & 0 & 0 & 1 \end{bmatrix}$

5. no, since $\begin{bmatrix} 2 & \frac{1}{2} \\ \frac{1}{3} & -1 \end{bmatrix} \cdot \begin{bmatrix} 4 & 1 \\ 2 & 3 \end{bmatrix} = \begin{bmatrix} 9 & \frac{7}{2} \\ -\frac{2}{3} & -\frac{8}{3} \end{bmatrix} \neq \begin{bmatrix} 1 & 0 \\ 0 & 1 \end{bmatrix}$

Section 5.3 (p. 177)

1. $\begin{bmatrix} \frac{1}{13} & \frac{3}{13} \\ \frac{4}{13} & -\frac{1}{13} \end{bmatrix}$

3. $\begin{bmatrix} -\frac{1}{10} & \frac{3}{10} \\ \frac{4}{10} & -\frac{2}{10} \end{bmatrix}$

5. $\begin{bmatrix} 1 & 0 & 0 \\ -1 & \frac{1}{2} & 0 \\ 0 & -\frac{1}{2} & \frac{1}{5} \end{bmatrix}$

7. no inverse

9. $\begin{bmatrix} -\frac{3}{4} & -\frac{1}{4} & \frac{1}{2} \\ \frac{5}{4} & \frac{5}{4} & -1 \\ -\frac{1}{4} & -\frac{3}{4} & \frac{1}{2} \end{bmatrix}$

11. $\begin{bmatrix} \frac{7}{42} & \frac{7}{42} & \frac{1}{6} \\ \frac{11}{42} & -\frac{1}{42} & -\frac{1}{6} \\ \frac{5}{42} & \frac{11}{42} & -\frac{1}{6} \end{bmatrix}$

Section 5.4 (p. 179)

1. $(\frac{16}{7}, \frac{22}{7})$

3. $(6, -2)$

5. $(3, 2, 1)$

Section 5.5 (p. 182)

1. (a) $X = \begin{bmatrix} 25,900.89 \\ 22,199.01 \\ 84,710.89 \\ 16,809.99 \end{bmatrix}$ **(b)** $X = \begin{bmatrix} 28,378.37 \\ 24,496.31 \\ 99,665.84 \\ 19,287.46 \end{bmatrix}$

3. (a) In matrix terms we have $EX = B$, where

$$E = \begin{bmatrix} 1 & -0.6 & -0.6 \\ 0 & 1 & -0.3 \\ -0.1 & -0.1 & 1 \end{bmatrix}, \quad B = \begin{bmatrix} 80,000 \\ 60,000 \\ 50,000 \end{bmatrix}, \quad X = \begin{bmatrix} x \\ y \\ z \end{bmatrix}.$$

(b) $E^{-1} = \begin{bmatrix} 1.0874439 & 0.7399102 & 0.8744394 \\ 0.0336322 & 1.0538116 & 0.3363228 \\ 0.1121076 & 0.1793721 & 1.1210762 \end{bmatrix}$

From $X = E^{-1}B$, we obtain $X = \begin{bmatrix} 175,112 \\ 82,735 \\ 75,784 \end{bmatrix}$.

(c) For $B = \begin{bmatrix} 90,000 \\ 70,000 \\ 70,000 \end{bmatrix}$, $X = E^{-1}B$ yields $X = \begin{bmatrix} 210,874 \\ 100,336 \\ 101,121 \end{bmatrix}$.

Section 5.6 (p. 187)

1. (a) $A = \begin{bmatrix} 0.2 & -0.3 \\ 0.3 & 0.1 \end{bmatrix}$, $D = \begin{bmatrix} d_1 \\ d_2 \end{bmatrix}$, and $X = \begin{bmatrix} x_1 \\ x_2 \end{bmatrix}$ are the input-coefficient, final-demand, and output matrices, respectively. We have:

$$I_2 - A = \begin{bmatrix} 0.8 & -0.3 \\ -0.3 & 0.9 \end{bmatrix}, \quad (I_2 - A)^{-1} = \begin{bmatrix} \frac{90}{63} & \frac{30}{63} \\ \frac{30}{63} & \frac{80}{63} \end{bmatrix}$$

$$X = (I_2 - A)^{-1}D = \begin{bmatrix} \frac{90}{63}d_1 + \frac{30}{63}d_2 \\ \frac{30}{63}d_1 + \frac{80}{63}d_2 \end{bmatrix}$$

(b) $5100 worth of commodity 1 and $5200 worth of commodity 2; $1020 worth of commodity 1 is needed to produce commodity 1, and $1560 worth of commodity 1 is needed to produce commodity 2; $1530 worth of commodity 2 is needed to produce commodity 1, and $520 worth of commodity 2 is needed to produce commodity 2.

(c) $6750 worth of commodity 1 and $6450 worth of commodity 2; $1350 worth of commodity 1 is needed to produce commodity 1, and $1935 worth of commodity 1 is needed to produce commodity 2; $2025 worth of commodity 2 is needed to produce commodity 1, and $645 worth of commodity 2 is needed to produce commodity 2.

SELF-TESTS FOR CHAPTERS 1-5

SELF-TEST 1 (p. 188)

1. $y = \frac{1}{2}x + 4$
2. $x + 4y = 7$
3. $(16, 12)$
4. $(-2, 4)$
5. (a) $y = 0.4384x - 0.7371$; (b) for $x = 45$, $y = 19$; the predicted average income for employees of age 45 in this corporation is $19,000.
6. $(0, -8, -4)$
7. $x = -\frac{19}{7} + 2z$, $y = \frac{2}{7} - z$, z arbitrary
8. no; the 2^x term destroys linearity
9.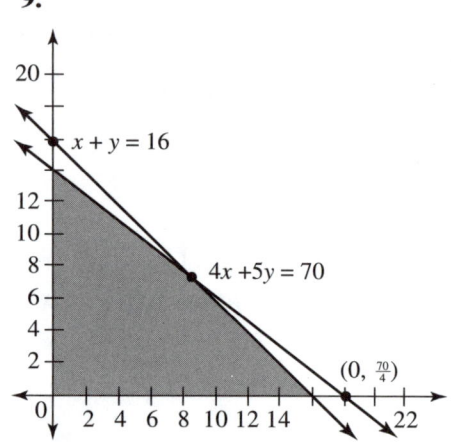
10. (a) No; $F = 3x + y^2$ is not linear. (b) Yes. (c) No; we seek constraints that reflect reality, but there is no guarantee of success. (d) no; the validity of a solution means that it follows from the linear program, which may or may not be realistic. (e) No; see (d).

SELF-TEST 2 (p. 189)

1. Let x and y denote the amounts (in millions of dollars) invested in loans and securities, respectively. We have the following linear program:

   Maximize $I(x,y) = 0.10x + 0.08y$

   subject to

   $$x \geq 0, y \geq 0$$
   $$x + y \leq 60$$
   $$-x + 3y \geq 0$$
   $$x \geq 15$$

   Sol. (45,15); Max. value 5.7

2. $(-2, 1, 3)$

4. Max $P = 2x + 3y$
subject to
$$x \geq 0, y \geq 0$$
$$50x + 60y \leq 12{,}000$$
$$3x + 4y \leq 760$$

3. $(1, 1); 5$
where x and y are the number of K17 and K24 units, respectively, to be stocked. Solution: $(0, 190)$; value 570.

5. **(a)** Different solutions were obtained on the basis of different assumptions. The mathematics is precise in the sense that its methods yield conclusions that are valid with respect to the underlying assumptions. The fact that the simplex method was used in one situation whereas the corner-point method was used in the other instance is irrelevant as far as the validity of the conclusions obtained is concerned. Both procedures, as mathematical methods, yield valid conclusions.

(b) Both solutions are correct in the sense of being valid with respect to their underlying hypotheses. Clearly both cannot be true since the maximum profit cannot be $3000 per week and at the same time $2500 per week.

(c) The basic question is, which solution actually maximizes profit, or to allow a shade of gray, which solution comes closest to approximating the maximum profit? This is the solution that we wish to implement. The assumptions from which these solutions follow as valid consequences should be carefully reviewed as to their realism. If the assumptions that lead to the solution (450,300), with a predicted maximum profit of $2500 per week, are more realistic than the assumptions that lead to the solution (500,280), with a predicted maximum profit of $3000 per week, then the (450,300) solution should be implemented since the predicted $2500 profit value will be closer to reality than the $3000 value. Without question, the predicted $3000 profit is more appealing, but what is more appealing is not necessarily what is realistic.

6. **(a)** $2x + y \geq 6, \quad x + 2y \geq 6$

(b) $(4, 1); 21$

7. No; if a linear program has a feasible point, then its dual may or may not have a solution. von Neumann's theorem says that if both a linear program and its dual have a feasible point, then both have solutions and the same value.

SELF-TEST 3 (p. 191)

1. (a) (i)
Min. $G = 13s + 8t + 12u$
subject to
$$s \geq 0, t \geq 0, u \geq 0$$
$$3s + 2t + 3u \geq 5$$
$$2s + t \geq 6$$
$$s + 2t + 2u \geq 8$$

(ii)
Max. $G = 4s + 3t + 2u + 3v$
subject to
$$s \geq 0, t \geq 0, u \geq 0, v \geq 0$$
$$-2s + 2t + 4u + 3v \leq 10$$
$$s - 4t + u + v \leq 4$$
$$-3s + 8t + 3u + 4v \leq 8$$

(b) (i) (0, 0, 0) and (10, 12, 10), for example, are feasible points for the max. and min. dual linear programs, respectively. Thus, both have solutions and the same value.

(ii) The min. linear program has no feasible point since $-2x + y - 3z \geq 4$ and $2x - 4y + 8z \geq 3$ cannot both be satisfied by nonnegative values of x, y, and z. Neither it nor its dual have solutions, according to von Neumann's theorem.

(c) (i) Max. linear program: (0, 6, 1); 44. Min. dual program: (4/3, 10/3, 0); 44.

2. $AB = \begin{bmatrix} -10 & -2 \\ 8 & 17 \end{bmatrix}$, $BA = \begin{bmatrix} -2 & 8 \\ 17 & 9 \end{bmatrix}$, $A^2 = \begin{bmatrix} 14 & -14 \\ 7 & 7 \end{bmatrix}$, $B^2 = \begin{bmatrix} 7 & 8 \\ 12 & 31 \end{bmatrix}$

3. Yes; since $AB = BA = I_2$.

4.
$0.99x - 0.02y - 0.02z = 30,000$
$-0.08x + y - 0.01z = 20,000$
$-0.02x - 0.15y + z = 95,000$

x, y, and z denote the total costs of the accounting, shipping, and marketing departments, respectively, for October.

5. False; see Problem 2 of Self-Test 2.

6. False; $AB = \bar{0}$ for $A = \begin{bmatrix} 1 & 2 \\ 2 & 4 \end{bmatrix}$, $B = \begin{bmatrix} 2 & 4 \\ -1 & -2 \end{bmatrix}$.

7. False; $AX = AY$ for $A = \begin{bmatrix} 2 & -1 \\ 4 & -2 \end{bmatrix}$, $X = \begin{bmatrix} 0 & 1 \\ 3 & 2 \end{bmatrix}$, $Y = \begin{bmatrix} -1 & -1 \\ 1 & -2 \end{bmatrix}$, but $X \neq Y$.

8. True

9. True

10. False; $A = \begin{bmatrix} 1 & 2 \\ 2 & 4 \end{bmatrix}$ has no multiplicative inverse; see Section 5.3, Example 3 (p. 177).

11. Min. $C = 0.1x - 0.05y + 10,550$

subject to

$x \geq 0$, $y \geq 0$
$x \leq 4000$, $y \leq 5000$
$x + y \leq 5000$
$x + y \geq 1500$

x and y denote the number of tons of coal to be shipped from Brooks and Scranton respectively, to Smithtown. Sol. (0, 5000); min. value 10,300.

Send 0, 5000, and 0 tons of coal from Brooks, Scranton, and Kearn, respectively, to Smithtown; send 4000, 0, and 3500 tons of coal from these respective sources to Warren.

SELF-TEST 4 (p. 192)

1. No inverse
3. (1, 2)
4. $(29, -18, -3)$

2. $F^{-1} = \begin{bmatrix} \frac{5}{16} & \frac{7}{16} & \frac{3}{16} \\ -\frac{1}{16} & \frac{5}{16} & \frac{9}{16} \\ \frac{3}{16} & \frac{1}{16} & \frac{5}{16} \end{bmatrix}$

5. (a) Min. $G = 30x + 28y$
   subject to
   $$x \geq 0, y \geq 0$$
   $$3x + 4y \geq 9$$
   $$5x + y \geq 10$$
   $$x + 2y \geq 16$$

   (b) (0, 0, 0) and (5, 6), for example, are feasible points for the max. linear program and its min. dual, respectively. Thus, both have solutions and the same value.

   (c) (0, 32/9, 110/9); 231.1

6. Max. $C = 72{,}000\,x + 80{,}000y$
   subject to
   $$x \geq 0, y \geq 0$$
   $$x + y \geq 20$$
   $$4x + 5y \leq 70$$

   x and y denote the number of BM23 and BM30 barges, respectively, to be purchased. No solution.

7. (a) $A = \begin{bmatrix} 0.5 & 0.2 \\ 0.2 & 0.5 \end{bmatrix}$, $D = \begin{bmatrix} d_1 \\ d_2 \end{bmatrix}$, $X = \begin{bmatrix} x_1 \\ x_2 \end{bmatrix}$ are the input-coefficient, final-demand, and output matrices, respectively. We have:

   $I_2 - A = \begin{bmatrix} 0.5 & -0.2 \\ -0.2 & 0.5 \end{bmatrix}$ $(I_2 - A)^{-1} = \begin{bmatrix} \frac{50}{21} & \frac{20}{21} \\ \frac{20}{21} & \frac{50}{21} \end{bmatrix}$

   Thus
   $$X = (I_2 - A)^{-1}D = \begin{bmatrix} \frac{50}{21}d_1 + \frac{20}{21}d_2 \\ \frac{20}{21}d_1 + \frac{50}{21}d_2 \end{bmatrix}$$

8. Disagree. Mathematical methods have the advantage of certitude in the sense of validity, not truth. The statement confuses truth with validity. Replace truth where it appears with validity and we have a correct assertion.

Section 6.1 (p. 198)

1. (a) {1, 3, 5, 6}, (b) {1, 3, 4, 5}, (c) {3, 4 ,5, 6}, (d) M, (e) {2, 4, 6}, (f) {1, 2, 4}, (g) {1, 2, 3, 6}, (h) {3, 5}, (i) {5}, (j) {5}, (k) {6}, (l) {1}, (m) {1, 2, 3, 4, 5}, (n) {2, 4}, (o) {1, 3 ,4, 5, 6}, (p) {5}, (q) {1, 2, 4, 5, 6}, (r) {2}

3. A^c is the set of all Presidents of the United States, past and present, who were not Republicans.

Section 6.2 (p. 203)

3. $S = \{A, B, C\}$ is a sample space for the process, because whenever the dice are tossed exactly one of the events A, B, or C occurs.

5. $S_1 = \{C_1, C_2, \ldots, C_{52}\}$, where C_1 is the event that card C_1 is dealt, C_2 is the event that card C_2 is dealt, and so on. $S_2 = \{B, R\}$, where B is the event that a black card is dealt, and R is the event that a red card is dealt. $S_3 = \{P, N\}$, where P is the event that a picture card is dealt, and N is the event that a nonpicture card is dealt.

7. $S_1 = \{(C_1, C_2, C_3, C_4, C_5), (C_1, C_2, C_3, C_4, C_6), \ldots, (C_{48}, C_{49}, C_{50}, C_{51}, C_{52})\}$, where $(C_1, C_2, C_3, C_4, C_5)$ is the event that cards $C_1, C_2, C_3, C_4,$ and C_5 are dealt in any order whatever, etc. S_1 is the collection of events described by all unordered hands of five cards that can be dealt from a standard deck. $S_2\{(C_1C_2C_3C_4C_5), (C_1C_2C_3C_4C_5), (C_{48}C_{49}C_{50}C_{51}C_{52})\}$, where $(C_1C_2C_3C_4C_5)$ is the event that cards $C_1, C_2, C_3, C_4,$ and C_5 are dealt in the order C_1, followed by C_2, followed by C_3, followed by C_4, followed by C_5, etc. S_1 is the collection of events described by all ordered hands of five cards that can be dealt from a standard deck. $S_3 = \{R, B, M\}$, where R is the event that the hand dealt contains red cards only, B is the event that the hand dealt contains black cards only, and M is the event that the hand dealt contains a mixture of black and red cards. $S_4 = \{0, 1, 2, 3, 4\}$, where 0 is the event that the hand dealt contains no aces, 1 is the event that the hand dealt contains 1 ace, etc.

9. (a) $S_1 = \{(S_1, S_2), (S_1, S_3), (S_1, S_4), (S_2, S_3), (S_2, S_4), (S_3, S_4)\}$, where (S_1, S_2) is the event that students S_1 and S_2 are chosen, irrespective of order, etc. $S_2 = \{(S_1S_2), (S_2S_1), (S_1S_3), (S_3S_1), (S_1S_4), (S_4S_1), (S_2S_3), (S_3S_2), (S_2S_4), (S_4S_2), (S_3S_4), (S_4S_3)\}$, where S_1S_2 is the event that students S_1 and S_2 are chosen in the order S_1 followed by S_2, etc.

(b) No; it is not the case that when a sample of 2 is selected only one event in S occurs.

Section 6.4 (p. 208)

1. (a) $P(E) = P(2) + P(4) + P(6) = 3/4 = 0.75$. (b) If a die whose behavior is described by model Y3 is tossed a large number of times, an even number will show approximately 75% of the time. (c) The tossing of the die a large number of times and observing how often an even number shows is irrelevant to the validity issue. The validity of $P(E) = 0.75$ was established by adding up the probabilities with which 2, 4 and 6 show in model Y3 and obtaining 0.75. (c) From tossing the die a large number of times, we have that the relative frequency with which an

even number showed is 0.665, which is markedly at variance with the predicted relative frequency of approximately 0.75 in (b). This establishes that $P(E) = 0.75$, interpreted in relative frequency terms, is false. **(e)** Model Y3 is not realistic for the die in question since a valid conclusion of the model has been shown to be false.

Section 6.5 (p. 214)

1. (a) Jack's assumption, that the sample points in his sample space are equally likely to occur, is not an accurate reflection of the fairness of the dice. Consider, for example, the sample points $\{1,1\}$ and $\{1,2\}$. $\{1,1\}$ can only occur in one way, both dice must fall with 1 showing. $\{1,2\}$ can occur in two ways, the first die shows 1 and the second die shows 2, the first die shows 2 and the second die shows 1. Thus for fair dice it is reasonable to expect that over the long run the event $\{1, 2\}$ will occur approximately twice as often as the event $\{1,1\}$. Thus it would be more reasonable to assign $\{1,2\}$ a probability which is twice that assigned to $\{1,1\}$, instead of assigning both the same probability as Jack has done.

(b) Let x denote the probability to be assigned to $\{1,1\}$. The sample points $\{2,2\}$, $\{3,3\}$, $\{4,4\}$, $\{5,5\}$, $\{6,6\}$ should all be assigned the same probability x since they all can only occur in one way. The remaining sample points, $\{1,2\}$, $\{1,3\}$, $\{2,3\}$, and so on, should all be assigned a probability $2x$ since each can occur in two ways. We have six sample points which will be assigned a probability value x, yielding a sum of $6x$, and fifteen sample points which will be assigned a probability value $2x$, yielding a sum of $30x$. The total sum of the probabilities of the sample points is thus $36x$. Since this total sum must be equal to 1 we have

$$36x = 1, \text{ which yields } x = \tfrac{1}{36}.$$

Thus we are led to define P on Jack's sample space by $P(\{1,1\}) = P(\{2,2\}) = P(\{3,3\}) = P(\{4,4\}) = P(\{5,5\}) = P(\{6,6\}) = 1/36$, and the probability of each of the remaining sample points—$\{1,2\}$, $\{1,3\}$, $\{2,3\}$, and so on—is $2/36$.

3. (a) (1) Mark's probability function is not realistic because it does not take into account the different proportions of good bulbs made by plant $P1$, good bulbs made by plant $P2$, etc. From the data given the proportion of good bulbs (of the total output of 8000 bulbs) made by $P1$ is 4950/8000, the proportion of defective bulbs made by $P1$ is 50/8000, etc. **(2)** Mark's conclusion is valid with respect to his model, but not realistic in terms of the process of choosing a bulb at random from the day's output with the given proportions of good and defective bulbs.

(b) Bob's model is a open to the same sort of criticism leveled at his brother's model.

(c) Let us think of the bulbs as identified 1, 2,..., 8000. We take as our sample space $S = \{b_1, b_2, \ldots, b_{8000}\}$, where b_1 is the event bulb 1 is chosen, b_2 is the event bulb 2 is chosen, etc. Since we envision the bulb as being chosen at random, without bias, from the day's output, we take as our probability function P the one which assigns 1/8000 to each sample point in S.

(d) Yes, as follows: $P(GP1) = 4950/8000$, $P(GP2) = 2985/8000$, $P(DP1) = 50/8000$, $P(DP2) = 15/8000$.

(e) Yes: $P(G) = 7935/8000$, $P(D) = 65/8000$.

5. (a) Consider the sample point (i,j). If $i + j$ is even (such as (1,1), (1,3), etc.) $P(i,j) = 2/54$; if $i + j$ is odd (such as (1,2), (1,4), etc.), $P(i,j) = 1/54$.

(b) (i) P (even sum) $= 36/54$; (ii) $P(i + j = 7) = 6/54$; (iii) $P(i + j < 6) = 14/54$.

7. For convenience let us think of the cards as numbered 1–52. As our sample space we take

$$S = \{C_1, C_2, \ldots, C_{52}\}$$

where C_1 is the event that card 1 is dealt, C_2 is the event that card 2 is dealt, and so on. Let us assume that the deck is well-shuffled, so that the likelihood of any one card being dealt is the same as the likelihood of any other card being dealt. The probability function P that best reflects this assumption is the one that assigns the same value, 1/52, to all 52 sample points in S. We have:

$$P(C_1) = P(C_2) = \cdots = P(C_{52}) = \tfrac{1}{52}$$

In terms of this probability model, we have

$$P(\text{club is dealt}) = \tfrac{13}{52} = 0.25$$

The relative-frequency interpretation of this result is that if a card is dealt from a well-shuffled standard deck of 52 cards a large number of times, a club will be dealt approximately 25% of the time.

$$P(\text{picture card is dealt}) = \tfrac{12}{52} = 0.231$$

If a card is dealt from a well-shuffled standard deck of 52 cards a large number of times, a picture card will be dealt approximately 23.1% of the time.

8. (a) $S_1 = \{b_1, b_2, \ldots, b_{5000}\}$, where b_1 is the event that blade 1 is chosen, and so on.

$S_2 = \{G, D\}$, where G is the event that a good blade is chosen, D is the event that a defective blade is chosen,

$S_3 = \{G-M1, D-M1, G-M2, D-M2\}$, where $G-M1$ is the event that a good blade made by machine $M1$ is chosen, $D-M1$ is the event that a defective blade made by machine $M1$ is chosen, and so on.

(b) & (c) To say that a blade is chosen at random means that the method of choosing the blade is free from bias favoring certain blades over others; in other words, all blades have the same likelihood of being chosen under the selection mechanism employed. The sample space that is most suitable for expressing this point of view is S_1, and the probability function that best reflects the equally-likelihood outcome assumption is P that assigns the same value, 1/5000, to each sample point in S_1. We have then

$$P(b_1) = P(b_2) = \cdots = P(b_{5000}) = \frac{1}{5000}$$

(d) P(blade made by $M1$ is selected) $= \frac{3000}{5000} = 0.60$

P(defective blade made by $M2$ is selected) $= \frac{20}{5000} = 0.004$

P(defective blade is selected) $= \frac{35}{5000} = 0.007$

Section 6.6 (p. 220)

1. (a) $\frac{5}{6}$, (b) $\frac{4}{6}$
3. (a) $\frac{19}{36}$, (b) $\frac{21}{36}$, (c) $\frac{30}{36}$, (d) $\frac{11}{36}$, (e) $\frac{34}{36}$

Section 6.7 (p. 225)

1. Relative-frequency interpretation: if the underlying process is repeated a large number of times, event E will occur in the neighborhood of 80% of the time. Subjective interpretation: 0.80 is a numerical expression of some individual's degree of belief in the occurrence of event E in connection with the underlying process. The relative-frequency interpretation presupposes that the process can be repeated a large number of times and has an objective content that is independent of the observer. The subjective interpretation can be entertained in connection with once-and-only situations. The subjective probability assigned to an event often depends very much on the observer.
3. Both probabilistic statements can only be interpreted in subjective terms. Both are numerical expressions of degree of belief connected with once-and-only situations.
5. Subjective probability terms; 0.95 is a numerical expression of Eric's degree of belief that he will get an A in Sociology this semester.

Section 6.8 (p. 226)

1. 7/9
3. 9 to 1
5. 0.952; 0.968

Section 7.1 (p. 231)

1. $5 \cdot 4 = 20$
3. $5 \cdot 3 = 15$
5. $3 \cdot 4 \cdot 4 \cdot 3 = 144$
7. $9 \cdot 10 \cdot 10 \cdot 10 = 9000$
9. $2^8 = 256$
11. $n(n-1) \cdots 1$
13. $2^3 + 2^4 = 24$
15. (a) $10 \cdot 9 \cdot 8 \cdot 7 \cdot 6 \cdot 5 \cdot 4 = 604{,}800$
    (b) 10^7

Section 7.1 (p. 238)

17. $P(7, 3) = 210$, $P(12, 4) = 11{,}880$, $C(52, 4) = 270{,}725$
    $P(6, 1) = 6$, $C(16, 3) = 560$, $C(18, 2) = 153$,

19. From the point of view that guests desire particular rooms so that order is important, there are $12 \cdot 11 \cdot 10 \cdot 9 \cdot 8 \cdot 7 = 665{,}280$ ways; from the point of view that any room will do, location is not important, there are $C(12, 6)$ ways.

21. $C(5, 3) \cdot C(6, 4)$
23. $26!$
25. (a) $C(50, 3)$, assuming that an unordered sample is to be chosen. (b) $C(46, 3)$
    (c) $C(4, 1) \cdot C(46, 2)$ (d) $C(4, 2) \cdot C(46, 1)$ (e) $C(4, 3)$.

Section 7.2 (p. 249)

1. (a) $\dfrac{C(4, 3) \cdot C(48, 2)}{C(52, 5)} = 0.001736,$

   (b) $\dfrac{C(4, 3) \cdot C(4, 1) \cdot C(44, 1)}{C(52, 5)} = 0.0002708,$

   (c) $\dfrac{C(4, 3) \cdot C(48, 2)}{C(52, 5)} + \dfrac{C(4, 1) \cdot C(48, 4)}{C(52, 5)} -$

   $\dfrac{C(4, 3) \cdot C(4, 1) \cdot C(44, 1)}{C(52, 5)} = 0.0763,$

   (d) $\dfrac{C(4, 3) \cdot C(48, 2)}{C(52, 5)} + \dfrac{C(4, 2) \cdot C(48, 3)}{C(52, 5)} - \dfrac{C(4, 3) \cdot C(4, 2)}{C(52, 5)} = 0.0417,$

(e) $\dfrac{C(13, 2) \cdot C(39, 3)}{C(52, 5)} = 0.2743,$

(f) $\dfrac{C(13, 2) \cdot C(13, 3)}{C(52, 5)} = 0.00858,$

(g) $\dfrac{C(13, 2) \cdot C(39, 3)}{C(52, 5)} + \dfrac{C(13, 3) \cdot C(39, 2)}{C(52, 5)} - \dfrac{C(13, 2) \cdot C(13, 3)}{C(52, 5)} = 0.3473$

3. (a) $\dfrac{C(6, 1) \cdot C(114, 2)}{C(120, 3)} = 0.1376,$

(b) $\dfrac{C(6, 2) \cdot C(114, 1)}{C(120, 3)} = 0.0061,$

(c) $\dfrac{C(6, 3)}{C(120, 3)} = 0.000071,$

(d) $P(1 \text{ def.}) + P(2 \text{ def.}) = 0.1437$

5. $S = \{(T_1, T_2, T_3), \cdots, (T_{48}, T_{49}, T_{50})\}$; assumption: the sample is chosen at random; P assigns $1/C(50, 3)$ to each sample point in S.

(a) $\dfrac{C(4, 3)}{C(50, 3)}$ (b) $\dfrac{C(4, 1) \cdot C(46, 2)}{C(50, 3)}$ (c) $\dfrac{C(4, 2) \cdot C(46, 1)}{C(50, 3)}$

(d) $\dfrac{C(4, 1) \cdot C(46, 2)}{C(50, 3)} + \dfrac{C(4, 2) \cdot C(46, 1)}{C(50, 3)}$

7. $S = \{(C_1, C_2, C_3, C_4, C_5, C_6, C_7), \cdots, (C_{46}, C_{47}, C_{48}, C_{49}, C_{50}, C_{51}, C_{52})\}$; assumption: the deck is well-shuffled; the hand is dealt at random from the deck; P assigns $1/C(52, 7) = 1/133{,}784{,}560$ to each sample point in S.

(a) $\dfrac{C(4, 3) \cdot C(48, 4)}{C(52, 7)}$ (b) $\dfrac{C(4, 3) \cdot C(4, 2) \cdot C(44, 2)}{C(52, 7)}$

(c) $\dfrac{C(4, 3) \cdot C(48, 4)}{C(52, 7)} + \dfrac{C(4, 2) \cdot C(48, 3)}{C(52, 7)} - \dfrac{C(4, 3) \cdot C(4, 2) \cdot C(44, 2)}{C(52, 7)}$

$= 0.00582 + 0.07679 - 0.00017 = 0.08244$

(d) $\dfrac{C(13, 4) \cdot C(39, 3)}{C(52, 7)}$ (e) $\dfrac{C(13, 4) \cdot C(13, 2) \cdot C(26, 1)}{C(52, 7)}$

(f) $\dfrac{C(13, 4) \cdot C(39, 3)}{C(52, 7)} + \dfrac{C(13, 2) \cdot C(39, 5)}{C(52, 7)} -$

$\dfrac{C(13, 4) \cdot C(13, 2) \cdot C(26, 1)}{C(52, 7)} = 0.04884 + 0.33568 - 0.01084 = 0.3737$

9. S is the collection of events described by all permutations of 2 books that can be made from the 8 available books; assumption: the books are chosen at random and placed on the shelf; P assigns $1/P(8, 2) = 1/56$ to each sample point in S.

(a) 1/8 (b) 1/28 (c) $1 - 1/28 = 27/28$

11. (a) Sample space S is the collection of events expressed by all combinations of 500 fish that can be selected from N fish. There are $C(N, 500)$ sample points in S. Assumming that all fish in the lake have the same likelihood of being caught, define P to be the function that assigns the same value, $1/C(N, 500)$, to each sample point in S. (b) The number that maximizes the probability of catching 2 tagged fish in the batch of 500 that were caught.

(c) $\dfrac{C(300, 2) \cdot C(N - 300, 498)}{C(N, 500)}$, (d) 75,000

13. Fred Bass has an appropriate probability model for the process of catching one fish, but S is not a sample space for the process of catching two fish. The conclusion that the probability of catching two undersized fish is $C(6, 2)/C(9, 2) = \dfrac{5}{12}$ does not follow from the fisherman's probability model for catching one fish. Fred incorrectly formulated a probability model for the process to begin with, and then switched horses in midstream to obtain a conclusion that is valid with respect to some other probability model, but not the one he had set up. In short, a catastrophic disaster.

The probability model with respect to which his conclusions are valid and that should have been set down is the following: $S = \{(f_1, f_2), (f_1, f_3), \cdots, (f_8, f_9)\}$, where (f_1, f_2) is the event that fish f_1 and f_2 are chosen, and so on. P is defined by $P(f_1, f_2) = P(f_1, f_3) = \cdots P(f_8, f_9) = 1/C(9, 2) = \dfrac{1}{36}$.

15. (a) $S = \{(5, 6), (5, 7), (5, 8), (5, 9), (5, 10), (6, 7), (6, 8), (6, 9), (6, 10), (7, 8), (7, 9), (7, 10), (8, 9), (8, 10), (9, 10)\}$; P assigns $1/C(6, 2) = 1/15$ to each sample point in S.

(b) The sample is chosen at random from Q.

(c) $\bar{x}(5, 6) = 5.5$ $\bar{x}(6, 7) = 6.5$ $\bar{x}(7, 9) = 8$
$\bar{x}(5, 7) = 6$ $\bar{x}(6, 8) = 7$ $\bar{x}(7, 10) = 8.5$
$\bar{x}(5, 8) = 6.5$ $\bar{x}(6, 9) = 7.5$ $\bar{x}(8, 9) = 8.5$

$\bar{x}(5, 9) = 7$　　　　　　$\bar{x}(6, 10) = 8$　　　　　　$\bar{x}(8, 10) = 9$
$\bar{x}(5, 10) = 7.5$　　　　　$\bar{x}(7, 8) = 7.5$　　　　　$\bar{x}(9, 10) = 9.5$
$P(5.5) = P(6) = P(9) = P(9.5) = 1/15$
$P(6.5) = P(7.0) = P(8) = P(8.5) = 2/15;$　　$P(7.5) = 3/15$
(d) (i) 2/5　　　　　　(ii) 13/15

Section 8.2 (p. 256)

1. (a) 0.80,　　　**(b)** 1/4,　　　**(c)** 1/5

3. (a) 1/3,　　　**(b)** 1/3,　　　**(c)** 2/3

5. 1/5

Section 8.3 (p. 264)

1. B: The blood test indicates the absence of the disease. A_1: A person with the disease is selected. A_2: A person not having the disease is selected. From the tree graph of Figure 1 we have:

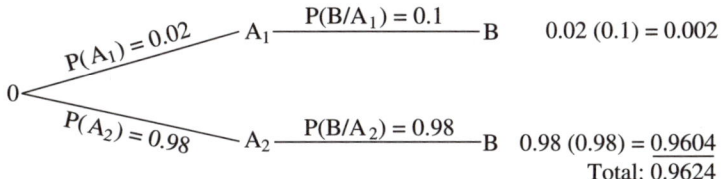

```
P(A₁)=0.02 → A₁ — P(B/A₁)=0.1 → B    0.02 (0.1) = 0.002
P(A₂)=0.98 → A₂ — P(B/A₂)=0.98 → B   0.98 (0.98) = 0.9604
                                      Total: 0.9624
```

Figure 1

$P(A_1 / B) = 0.0020/0.9624 = 0.0021.$

In the long run, approximately 0.21% of the people testing negative will turn out to have the disease.

3. B_1: Mercury content diagnosed as excessive. B_2: Mercury content diagnosed as not excessive. A_1: Mercury content of the fish is excessive. A_2: Mercury content of the fish is not excessive.

(a) $P(A_2 / B_1) = \dfrac{(0.94)(0.04)}{(0.94)(0.04) + (0.06)(0.99)} = 0.388.$

(b) $P(A_1 / B_2) = \dfrac{(0.06)(0.01)}{(0.06)(0.01) + (0.94)(0.96)} = 0.00066.$

(c) In the long run, approximately 38.8% of the fish diagnosed as having excessive mercury content will actually not have it and 0.066% of the fish diagnosed as not having excessive mercury content will have it.

5. B: The ball bearing is defective. A_1: The ball bearing was made by M1. A_2: The ball bearing was made by M2. A_3: The ball bearing was made by M3.

  (a) $P(A_1/B) = 0.4375$; (b) $P(A_2/B) = 0.3125$; (c) $P(A_3/B) = 0.2500$.

7. B: More than 200 units were sold, A_1: The computer is a big seller, A_2: The computer is a good seller, A_3: The computer is a poor seller.

  (a) $P(A_1/B) = 0.726$;   (b) $P(A_2/B) = 0.242$;
  (c) $P(A_3/B) = 0.032$;   (d) $0.726 + 0.242 = 0.968$.

9. (a) 0.957, (b) 0.115

Section 8.4 (p. 276)

1. (a) $\frac{1}{2}$, (b) $\frac{7}{18}$, (c) $\frac{7}{36}$, (d) $\frac{5}{18}$, (e) $\frac{17}{36}$ (2) $\frac{69}{216}$, (3) $\frac{37}{216}$

3. (a) 0.28,   (b) 0.14,   (c) 0.129,   (d) 0.101

5. (a) $T_1 = \begin{bmatrix} \frac{1}{6} & \frac{2}{6} & 0 & \frac{3}{6} \\ \frac{3}{6} & \frac{1}{6} & \frac{2}{6} & 0 \\ 0 & \frac{3}{6} & \frac{1}{6} & \frac{2}{6} \\ \frac{2}{6} & 0 & \frac{3}{6} & \frac{1}{6} \end{bmatrix}$

7. (a)
$$\text{From state} \begin{array}{c} d \\ h \\ r \end{array} \begin{bmatrix} 1 & 0 & 0 \\ \frac{1}{2} & \frac{1}{2} & 0 \\ 0 & 1 & 0 \end{bmatrix} = T_1$$
with column headers "To state" $d\ h\ r$

  (b) (1) $\frac{6}{36}$, (2) $\frac{13}{36}$, (3) $\frac{13}{36}$, (c) (1) $\frac{71}{216}$,   (b) (1) $\frac{1}{4}$, (2) $\frac{1}{8}$, (3) $\frac{1}{16}$

Section 9.1 (p. 281)

1. No; $\frac{1}{3} \neq \frac{1}{2} \cdot \frac{1}{2}$

3. (a) 1/4,   (b) 1/60,   (c) 1/5,   (d) 2/15

Section 9.2 (p. 286)

1. $n = 5$, E = event that a head appears on any one coin toss, $p = 0.5$, $q = 0.5$. Assumptions: Fair coin, probabilities do not change from toss to toss, the outcome of each toss is independent of the outcome of any other toss. A success is

the occurrence of a head on any one trial. X = the number of successes in the 5 trials.

(a) $P(X = 2) = C(5, 2)(0.5)^2(0.5)^3$.

(b) $P(X \geq 3) = P(X = 3) + P(X = 4) + P(X = 5) =$
$C(5, 3)(0.5)^3(0.5)^2 + C(5, 4)(0.5)^4(0.5) + C(5, 5)(0.5)^5(0.5)^0$.

3. $n = 6$, E = event that an even number appears on any one toss, $p = 0.5$ $q = 0.5$. Assumptions: Fair die, probabilities do not change from toss to toss, the outcome of each toss is independent of the outcome of any other toss. A success is the occurrence of a head on any one trial. X = the number of successes in the 5 trials.

$P(X \geq 4) = P(X = 4) + P(X = 5) + P(X = 6) =$
$C(6, 4)(0.5)^4(0.5)^2 + C(6, 5)(0.5)^5(0.5) + C(6, 6)(0.5)^6(0.5)^0$.

5. $n = 1{,}000$, E = event that any one of the light bulbs is defective, $p = 0.01$, $q = 0.99$. Assumption: Whether or not any bulb is defective is independent of whether or not any of the bulbs are. A success is the occurrence of a defective bulb. X = the number of defective bulbs from the 1,000.

$P(X \leq 15) = P(X = 0) + P(X = 1) + \cdots + P(X = 15) =$
$C(1000, 0)(0.01)^0(0.99)^{1{,}000} + C(1000, 1)(0.01)(0.99)^{999} + \cdots$
$+ C(1000, 15)(0.01)^{15}(0.99)^{985}$.

7. $n = 10{,}000$, E = event that an allergic reaction develops, $p = 0.005$, $q = 0.995$. Assumption: Whether or not one person develops an allergic reaction is independent of whether or not another person develops one. X = the number of people out of the 10,000 that develop an allergic reaction.

$P(X < 100) = P(X = 0) + P(X = 1) + \cdots + P(X = 99) =$
$C(10000, 0)(0.005)^0(0.995)^{10{,}000} + C(10000, 1)(0.005)(0.995)^{9{,}999} +$
$\cdots + C(10000, 99)(0.005)^{99}(0.995)^{9{,}901}$.

Section 9.3 (p. 294)

1. 0.4913
3. $0.5000 - 0.4406 = 0.0594$
5. $0.3849 + 0.5000 = 0.8849$
7. $0.4761 - 0.2939 = 0.1822$
9. $0.5000 - 0.3925 = 0.1075$
11. $0.4783 - 0.1700 = 0.3083$
13. $0.3413 - 0.0987 = 0.2426$
15. $0.0793 + 0.1179 = 0.1972$
17. $0.4772 - 0.3413 = 0.1359$
19. $0.2357 + 0.5000 = 0.7357$
21. $0.5000 - 0.1368 = 0.3632$

Section 9.4 (p. 300)

1. $\mu = np$, $\sigma = \sqrt{npq}$; thus $\mu = 40$, $\sigma = 4.9$.

    $z = \dfrac{X - 40}{4.9}$ gives the z value for any X in this problem.

    (a) $P(X = 30) = P(29.5 \leq X \leq 30.5) = P\left(\dfrac{29.5 - 40}{4.9} \leq z \leq \dfrac{30.5 - 40}{4.9}\right)$

    $P(-2.14 \leq z \leq -1.94) = 0.4838 - 0.4738 = 0.0100$.

    (b) $P(X = 25) = P(24.5 \leq X \leq 25.5) = P(-3.16 \leq z \leq -2.96)$
    $= 0.5000 - 0.4985 = 0.0015$.

    (c) $P(X \leq 50.5) = 0.0938$.   (d) $P(X \geq 30.5) = 0.9738$.
    (e) $P(19.5 \leq X < 40.5) = 0.5398$.

3. Bernoulli Model: $n = 10{,}000$; $E =$ occurrence of an allergic reaction; $p = 0.05$; $X =$ number of allergic reactions out of the 10,000 people; Assumption: A reaction to the antibiotic by any one individual does not influence the outcome for anyone else. Since $np = 500 \geq 5$ and $n(1 - p) = 9{,}500 \geq 5$ the normal approximation is applicable with $\mu = 500$ and $\sigma = 21.79$.

    (a) $P(X \geq 551) = P(z \geq 2.32) = 0.0102$.   (b) $P(X \leq 490) = 0.3300$.
    (c) $P(480 < X < 510) = 0.4833$.

5. Bernoulli Model: $n = 100$; $E =$ occurrence of a correct answer; $p = 0.5$; $X =$ number of correct answers out of the 100 questions; Assumption: The outcome of any one toss of the coin does not influence the outcome on any other toss. Since $np = 50 = n(1 - p) = 50 \geq 5$ the normal approximation is applicable with $\mu = 50$ and $\sigma = 5$. Thus $P(X \geq 70) = P(z \geq 3.90) \approx 0$.

7. Bernoulli Model: $n = 500{,}000$; $E =$ a person gets the flu; $p = 0.01$; $X =$ number of people who got the flu from the 500,000; Assumption: the effectiveness of the vaccine for any one person does not influence the outcome for anyone else. Since $np = 5{,}000 \geq 5$ and $n(1 - p) = 495{,}000 \geq 5$ the normal approximation is applicable with $\mu = 5{,}000$ and $\sigma = 70.36$.

    (a) $P(X < 2000) = P(z < -42.64) \approx 0$.   (b) $P(X > 4900) = 0.9207$.
    (c) $P(4{,}950 < X < 5{,}100) = P(-0.70 < z < 1.41) = 0.6787$.

SELF-TESTS FOR CHAPTERS 6–9

SELF-TEST 1 (p. 301)

1. Mathematics did not fail Harry since, if used correctly, it can only guarantee valid conclusions with respect to the background assumed, and Harry's conclu-

sion is indeed valid with respect to his probability model. Mathematics is precise only in the sense of guaranteeing valid conclusions. What went wrong was that Harry's conclusion, while valid, is not true about the die he obtained from his friend's friend, as shown by an even number showing 239 times instead of the predicted 800 times. This can only mean that the probability model presented to Harry is not a realistic description of the die that came with it.

2. (a) $P(A) = 2/10 + 3/10 + 2/10 = 7/10 = 0.70$.

   (b) If Dan's die is tossed a large number of times, an even number will show approximately 70% of the time.

   (c) This evidence is irrelevant to the validity of the conclusion cited in (a). The validity of $P(A) = 0.70$ is established by adding $P(2)$, $P(4)$, and $P(6)$ in the model and obtaining 0.70.

   (d) The evidence cited in (c) tells us that the relative frequency with which an even number has shown for a large number of tosses of Dan's die is 0.40, which is sharply at variance with the relative frequency behavior of the die predicted in (b). This means that the conclusion cited in (a), interpreted in relative frequency terms, is false.

   (e) A: the event that an odd number shows; B: the event that a number less than 4 shows.

   $P(A/B) = P(A \cap B) / P(B) = 0.2 / 0.4 = 0.50$.

   (f) A: the event that a number greater than 2 shows; B: the event that an even number shows.

   $P(A \cap B) = 0.5 \neq P(A) P(B) = (0.7)(0.7) = 0.49$. Therefore, A and B are not independent events.

3. As a subjective probability. It is a numerical measure of A.W.'s degree of belief that he will receive an A in statistics this semester. Whether this belief is based on wishful thinking or exam grades or both, we cannot say.

4. B: The tube selected is defective; A_1: The tube selected was made by $M1$; A_2: The tube selected was made by $M2$; A_3: The tube selected was made by $M3$; Bayes' Theorem (see Figure 1) yields:

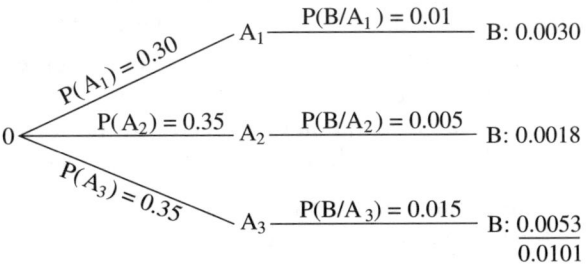

Figure 1

(a) $P(A_1/B) = 0.0030/0.0101 = 0.2970$;
(b) $P(A_2/B) = 0.0018/0.0101 = 0.1782$;
(c) $P(A_3/B) = 0.0053/0.0101 = 0.5248$.

5. $20 \cdot 19 \cdot 18 \cdot 17 = 116,280$ 6. $C(44, 6)$

7. Assuming that the outcome of one drawing (good or defective item is selected) does not influence the outcome of any other drawing, we take as our probability model a Bernoulli trial model with $n = 7, p = 0.05, q = 0.95$ (success here is the event a defective item is drawn). Let X denote the number of defectives drawn. The problem is to find $P(X > 1)$.

$$P(X > 1) = P(X = 2) + P(X = 3) + P(X = 4) + P(X = 5)$$
$$+ P(X = 6) + P(X = 7)$$
$$= C(7,2)(.05)^2(.95)^5 + C(7,3)(.05)^3(.95)^4$$
$$+ C(7,4)(.05)^4(.95)^3 + C(7,5)(.05)^5(.95)^2$$
$$+ C(7,6)(.05)^6(.95) + (.05)^7$$
$$= 0.0406 + 0.0036 + 0.0002 + 0.0000 + 0.0000$$
$$= 0.0444$$

SELF-TEST 2 (p. 302)

1. (a) P(number greater than 3 shows) $= 1/16 + 2/16 + 3/16 = 3/8$.

   (b) If Dan's die is tossed a large number of times, a number greater than 3 will show approximately 37.5% of the time.

   (c) A: A number greater than 3 shows. B: An odd number shows
   $$P(A/B) = \frac{P(A \cap B)}{P(B)} = \frac{2/16}{9/16} = \frac{2}{9}.$$

   (d) A: A number greater than 2 shows. B: An even number shows.
   $P(A \cap B) = 4/16 \neq P(A) \cdot P(B) = (10/16) \cdot (7/16)$; not independent.

   (e) (i) No; the probability was calculated correctly based on the model. The behavior of the die is not relevant to the validity issue.

   (ii) Mathematical reasoning is precise in the sense that it yields valid conclusions with respect to assumptions. What went wrong is that Dan's model was not a realistic description of the behavior of his die. This led to a valid conclusion, but one which was not accurate for Dan's die.

2. (a) Bernoulli trial model defined by $n = 60,000$, E is the event that the vaccine is ineffective on the person given it, $p = 0.001$.

   (b) The effectiveness or non-effectiveness of the vaccine on any person does not influence the outcome on any other person given the vaccine.

(c) Yes; since the effectiveness of the vaccine for one person has to do with his internal biochemistry, which is not relevant to how any other person will react to the vaccine.

(d) $P(X \le 70) = C(60{,}000, 0)(0.001)^0(0.999)^{60{,}000} + \cdots + C(60{,}000, 70)(0.001)^{70}(0.999)^{59{,}930}$.

(e) Yes; since $np = 60 \ge 5$ and $n(1 - p) = 59{,}940 \ge 5$.

(f) $\mu = 60$; $\sigma = \sqrt{(60{,}000)(0.001)(0.999)} = 7.74$,

$$P(X \le 70) \doteq P\left(z \le \frac{70.5 - 60}{7.74}\right) = P(z \le 1.36) = 0.9131.$$

3. (a) $S = \{(2, 6), (2, 10), (2, 14), (2, 16), (6, 10), (6, 14), (6, 16), (10, 14), (10, 16), (14, 16)\}$

   P assigns to each sample point the value $1/C(5, 2) = 0.1$.

   (b) The sample is chosen without bias, at random, from the population.

   (c) $\bar{x}(2, 6) = 4$, $\bar{x}(2, 10) = 6$, $\bar{x}(2, 14) = 8$, $\bar{x}(2, 16) = 9$, $\bar{x}(6, 10) = 8$, $\bar{x}(6, 14) = 10$, $\bar{x}(6, 16) = 11$, $\bar{x}(10, 14) = 12$, $\bar{x}(10, 16) = 13$, $\bar{x}(14, 16) = 15$. This yields $p(x) = P(\bar{x} = x)$ defined as follows: $p(4) = P(6) = p(9) = p(10) = p(11) = p(12) = p(13) = p(15) = 0.1$, $p(8) = 0.2$.

   (d) 0.5

4. (a) Let B denote the event "the plate chosen is diagnosed as containing a flaw," A_1 denote the event "the plate chosen contains a flaw," and A_2 denote the event "the plate chosen does not contain a flaw." The problem is to determine $P(A_1/B)$. The randomness of the selection and the data given lead us to the following assignment of probabilities: $P(A_1) = 0.03$, $P(A_2) = 0.97$, $P(B/A_1) = 0.985$, $P(B/A_2) = 0.03$. Bayes's theorem yields $P(A_1/B) = 0.504$.

   (b) Let B denote the event "the plate chosen is diagnosed as not containing a flaw, A_1 denote the event "the plate chosen contains a flaw," and A_2 denote the event "the plate chosen does not contain a flaw." The problem is to find $P(A_1/B)$. The randomness of the selection and the data given lead us to the following assignment of probabilities: $P(A_1) = 0.03$, $P(A_2) = 0.97$, $P(B/A_1) = 0.015$, $P(B/A_2) = 0.97$. Bayes's theorem yields $P(A_1/B) = 0.000478$.

5. No, because whenever the dice are tossed, both E and F may occur at the same time. This happens when 2 shows, for example.

6. For notational simplicity think of the 52 cards in the deck as numbered C_1, C_2, ..., C_{52}. Let S_1 denote the sample space of events described by all combinations of 6 cards that can be dealt from 52 cards.

   $S_1 = \{(C_1, C_2, C_3, C_4, C_5, C_6), \ldots, (C_{47}, C_{48}, C_{49}, C_{50}, C_{51}, C_{52})\}$,

   where $(C_1, C_2, C_3, C_4, C_5, C_6)$, for example, is the event that cards C_1 through C_6 are dealt in any order; and so on.

Let S_2 denote the sample space of events described by all permutations of 6 cards that can be dealt from 52 cards.

$$S_2 = \{(C_1C_2C_3C_4C_5C_6), (C_2C_1C_3C_4C_5C_6), \cdots, (C_{47}C_{48}C_{49}C_{50}C_{51}C_{52})\},$$

where $(C_2C_1C_3C_4C_5C_6)$, for example, is the event that the cards indicated are dealt in the order C_2 followed by C_1 followed by C_3 followed by C_4 followed by C_5 followed by C_6.

$S_3 = \{P, NP, M\}$, where P is the event that the hand dealt consists entirely of picture cards, NP is the event that the hand dealt consists entirely of nonpicture cards, and M is the event that the hand dealt consists of a mixture of picture and nonpicture cards.

$S_4 = \{0, 1, 2, 3, 4\}$, where 0 is the event that there are no kings in the hand dealt, 1 is the event that there is 1 king in the hand dealt, and so on.

$S_5 = \{H, NH, M\}$, where H is the event that the hand dealt consists entirely of hearts, NH is the event that the hand dealt consists entirely of nonhearts, and M is the event that a mixture of hearts and nonhearts is dealt.

7. (a) $S = \{b_1, b_2, \ldots, b_{10,000}\}$, where b_1 is the event that bulb 1 is chosen, and so on. $P(b_1) = \cdots = P(b_{10,000}) = \frac{1}{10,000}$.

(b) All bulbs have the same likelihood of being chosen; there is no bias in the selection procedure that favors certain bulbs over others. (c) (1) $\frac{60}{10,000}$ (2) $\frac{4950}{10,000}$, (3) $\frac{120}{10,000}$

8. (a) Johnson's conclusion is valid in that it follows as an inescapable consequence of his probability model. This is established by observing that $P(E) = P(2) = 1/3$.

(b) The relative-frequency interpretation of Johnson's conclusion is that if a well-balanced coin is tossed twice in succession a large number of times, then event E—head shows twice in two successive tosses—will occur approximately 33.3% of the time. However, when this process is repeated a large number of times, it is observed that event E occurs in the neighborhood of 25% of the time. This discrepancy between the predicted 33.3% value and the actually obtained value in the neighborhood of 25% establishes that Johnson's conclusion, $P(E) = 1/3$, interpreted in relative-frequency terms, is false; well-balanced coins do not behave in the way predicted by Johnson's model.

What's wrong with Johnson's model? In Johnson's model the same value, 1/3, is assigned to the sample points 0, 1 and 2. 0—no head shows—occurs when the two successive tosses yield TT, tail followed by tail; 1 occurs when the two successive tosses yield TH or HT; 2 occurs when the two successive tosses yield HH. With a well-balanced coin one would expect the sample point 1 to occur roughly twice as often as 0 or 2, and 0 to occur roughly as often as 2 over the long run. This behavior is actually what is observed to occur when such a coin is tossed twice in succession a large number of times, but it is not well-reflected by the probability assignment in Johnson's model. To obtain the probability assignment that best reflects the above behavior, let x denote the probability to be assigned to 0 by a probability function we shall denote by P_1. We have

$$P_1(0) = x \quad P_1(1) = 2x \quad P_1(2) = x$$

Since the sum of the probabilities assigned to all sample points must be 1, we have

$$P_1(0) + P_1(1) + P_1(2) = 4x = 1$$

which yields $x = 1/4$.

Thus the probability model that best reflects the nature of a well-balanced coin in terms of sample space S is with probability function P_1 defined on S by

$$P_1(0) = \frac{1}{4}, \quad P_1(1) = \frac{1}{2}, \quad P_1(2) = \frac{1}{4}$$

9. **(a)** $S = \{s_1, s_2, \ldots, s_{5000}\}$, where s_1 is the event that student 1 is chosen, and so on. $P(s_1) = \cdots = P(s_{5000}) = \frac{1}{5000}$.

   (b) All students have the same likelihood of being chosen; there is no bias in the selection procedure that favors certain students over others. **(c)** (1) $\frac{180}{5000}$, (2) $\frac{1800}{5000}$, (3) $\frac{380}{5000}$

10. Relative-frequency interpretation: if the underlying process is repeated a large number of times, then event E will occur in the neighborhood of 95% of the time. Subjective interpretation: 0.95 is a quantitative expression of an individual's degree of belief in the occurrence of event E in connection with the process. The relative-frequency interpretation presupposes that the underlying process can be repeated a large number of times and has an objective content that is independent of the observer. The assignment of a subjective probability to an event is possible for an event connected with a once-and-only situation and depends very much on the individual making the assignment.

SELF-TEST 3 (p. 305)

1. The probabilistic statement "there is a 75% probability of rain today" should be interpreted in subjective terms. It is a quantitative measure of the degree of belief of weather analysts that rain will fall on the specific day that "today" represents based on an analysis of weather conditions and how often rain fell under similar conditions. The odds in favor of this event are 75 to 25 or 3 to 1.

2. The probability assignment is a quantitative expression of this person's belief that his sister will live to age 70 or longer. It is a subjective probability value

assigned to an event connected with a once-and-only situation based on relative-frequency background (mortality statistics). The odds are 3 to 2.

3. 9/11

4. $P(5, 5) = 120$, $P(5, 2) = 20$, $P(8, 3) = 336$, $C(14, 2) = 91$, $C(20, 3) = 1140$, $C(30, 4) = 27{,}405$

5. The number of games that should be arranged is equal to the number of ways of selecting 2 of 12 teams without regard to order which is $C(12, 2) = 66$. Order is not involved here since only the identities of the two teams playing each other are of interest; A playing B is the same as B playing A.

6. $P(52, 4) = 6{,}497{,}400$,   7. $P(5, 3) = 5 \cdot 4 \cdot 3 = 60$

8. (a) $5 \cdot 4 \cdot 3 \cdot 2 = 120$,   (b) $2 \cdot 4 \cdot 3 \cdot 2 = 48$,   (c) $4 \cdot 3 \cdot 2 \cdot 1 = 24$

9. (a) There are four places to be filled, and each place can be filled with any of the five given digits and thus in 5 ways. Therefore there are $5 \cdot 5 \cdot 5 \cdot 5 = 625$ such 4-digit numbers.

   (b) For the number to be less than 5000, first place must be filled with 1 or 3, and thus can only be filled in two ways. Each of the other places can be filled with any of the five given digits and thus in 5 ways. Therefore there are $2 \cdot 5 \cdot 5 \cdot 5 = 250$ numbers that are less than 5000.

   (c) For the number to be divisible by 5 last place must be filled with 5, and thus can only be filled in one way. Each of the other places can be filled with any of the five given digits and thus in 5 ways. Therefore there are $5 \cdot 5 \cdot 5 \cdot 1 = 125$ numbers that are divisible by 5.

10. $C(12, 4) = 495$

11. The number of ways that the promotions can be granted is equal to the number of ways of choosing 6 assistant professors from the 20 applicants without regard to order, $C(20, 6)$, times the number of ways of choosing 4 associate professors from the 12 applicants without regard to order, $C(12, 4)$, times the number of ways of choosing 2 full professors from the 6 applicants without regard to order, $C(6, 2)$. This yields $C(20, 6) \cdot C(12, 4) \cdot C(6, 2) = 287{,}793{,}000$.

12. $P(30, 2) = 870$

13. $C(6, 3) = 20$,

14. (a) 0.70,   (b) 2/3,   (c) 1/3

15. $S = \{C_1, C_2, C_3, C_4, C_5, C_6), \ldots ,(C_{47}, C_{48}, C_{49}, C_{50}, C_{51}, C_{52})\}$, where $(C_1, C_2, C_3, C_4, C_5, C_6)$ is the event that the hand dealt consists of cards $C_1, C_2, C_3, C_4, C_5,$ and C_6, in any order, and so on. Based on the assumption that the deck is well shuffled, take as the probability function P the function that assigns the same value, $1/C(52, 6) = \frac{1}{61{,}075{,}570}$ to each sample point in S.

(a) $\dfrac{C(4, 3) \cdot C(48, 3)}{C(52, 6)} = 0.0011,$

(b) $\dfrac{C(4, 4) \cdot C(48, 2)}{C(52, 6)} = 0.000018,$

(c) $\dfrac{C(4, 2) \cdot C(4, 3) \cdot C(44,1)}{C(52, 6)} = 0.000017,$

(d) $\dfrac{C(4, 2) \cdot C(48, 4)}{C(52, 6)} + \dfrac{C(4, 3) \cdot C(48, 3)}{C(52, 6)} -$

$\dfrac{C(4, 2) \cdot C(4, 3) \cdot C(44, 1)}{C(52, 6)} = 0.0202$

(e) $\dfrac{C(13, 3) \cdot C(13, 2) \cdot C(26, 1)}{C(52, 6)} = 0.0095$

(f) $\dfrac{C(13, 3) \cdot C(39, 3)}{C(52, 6)} + \dfrac{C(13, 2) \cdot C(39, 4)}{C(52, 6)} -$

$\dfrac{C(13, 3) \cdot C(13, 2) \cdot C(26, 1)}{C(52, 6)} = 0.1383$

16. Yes; since $P(A \cap B) = P(A) \cdot P(B) = \frac{1}{36}$, where A is the event a sum of 7 shows and B is the event green shows 4.

17. (a) $P(A \cap B \cap C) = \dfrac{1}{5} \cdot \dfrac{1}{4} \cdot \dfrac{1}{3} = \dfrac{1}{60}$ (b) $P(A^c \cap B^c \cap C) = \dfrac{4}{5} \cdot \dfrac{3}{4} \cdot \dfrac{1}{3} = \dfrac{12}{60}$

(c) $P(A^c \cap B) = \dfrac{4}{5} \cdot \dfrac{1}{4} = \dfrac{1}{5}$ (d) $P(A \cup B) = \dfrac{1}{5} + \dfrac{1}{4} - \dfrac{1}{5} \cdot \dfrac{1}{4} = \dfrac{2}{5}$

(e) $P(A \cup B^c) = \dfrac{1}{5} + \dfrac{3}{4} - \dfrac{1}{5} \cdot \dfrac{3}{4} = \dfrac{4}{5}$

18. Bernoulli trial model with $n = 6$. Success is the event an even sum shows, $p = \frac{1}{2}$, $q = \frac{1}{2}$. The assumption is that the dice are well balanced. (a) 0.234, (b) 0.344

SELF-TEST 4 (p. 306)

1. (a) $S = \{(f_1, f_2, f_3, f_4, f_5, f_6, f_7, f_8), \ldots, (f_{73}, f_{74}, f_{75}, f_{76}, f_{77}, f_{78}, f_{79}, f_{80})\}$

   P assigns $1/C(80, 8)$ to each sample point.

   (b) The committee is chosen at random from the 80 faculty. There is no bias or favoritism which favors the selection of certain faculty over others.

(c) $\dfrac{C(30, 3)\, C(50, 5)}{C(80, 8)}$

(d) $\dfrac{C(78, 6)}{C(80, 8)}$

(e) $\dfrac{C(30, 3)\, C(25, 2)\, C(25, 3)}{C(80, 8)}$

(f) $\dfrac{C(30, 3)\, C(50, 5)}{C(80, 8)} + \dfrac{C(25, 2)\, C(55, 6)}{C(80, 8)} - \dfrac{C(30, 3)\, C(25, 2)\, C(25, 3)}{C(80, 8)}$

2. (a) Bernoulli trial model defined by $n = 10{,}100$. E is the event a defective item is produced on a trial, $p = 0.01$.

   (b) The production of a defective or good item on any trial has no effect on the outcome of any other trial.

   (c) $P(X \leq 105) = P\left(z \leq \dfrac{105.5 - 101}{10}\right) = P(z \leq 0.45) = 0.6736$.

3. (a) $P(A) = 8/13 = 0.615$.

   (b) If the die in question is tossed a large number of times, an odd number will show approximately 61.5% of the time.

   (c) No. This data is irrelevant to the conclusion's validity. The conclusion's validity follows from $P(1) + P(3) + P(5) = 8/13$.

   (d) Yes; $795/1{,}500 = 0.612$ which is "close" to the long run projection of 61.5% from (b).

   (e) A: an odd number shows; B: a number greater than 3 shows. No; $P(A \cap B) = 0.23 \neq P(A) \cdot P(B) = (8/13)(7/13) = 0.33$.

   (f) A: a number greater than 3 shows; B: an even number shows.
   $$P(A/B) = \dfrac{4/13}{5/13} = 0.8.$$

4. $\mu = 625(0.2) = 125$, $\sigma^2 = (625)(0.2)(0.8) = 100$, $\sigma = 10$, $P(124 \leq X \leq 130) = P(-0.15 \leq z \leq 0.55) = 0.2684$.

5. Subjective terms; it is a quantitative measure of the weather analyst's degree of belief of rain "today" based on an analysis of how often rain fell under similar conditions in the past odds: 75 to 25 or 3 to 1.

6. (a) Let B denote the event "an applicant passes the test," A_1 denote the event "the applicant is competent," and A_2 denote the event "the applicant is incompetent." We have $P(A_1) = 0.80$, $P(A_2) = 0.20$, $P(B/A_1) = 0.9$, $P(B/A_2) = 0.3$. The problem is to find $P(A_1/B)$. Bayes's theorem yields $P(A_1/B) = 0.923$.

   (b) Let B denote the event "the applicant passes the test," A_1 denote the event "the applicant is competent," and A_2 denote the event "the applicant is incompetent." We have $P(A_1) = 0.80$, $P(A_2) = 0.20$, $P(B/A_1) = 0.1$, $P(B/A_2) = 0.7$. The problem is to find $P(A_1/B)$. Bayes's theorem yields $P(A_1/B) = 0.364$.

7. Let B denote the event "the person who entered the store bought a stereo system," A_1 denote the event "the person who entered the store knew of the advertised reductions," and A_2 denote the event "the person who entered the store did not know of the advertised reductions." We have $P(A_1) = 0.70$, $P(A_2) = 0.30$, $P(B/A_1) = 0.20$, $P(B/A_2) = 0.10$. The problem is to find $P(A_1/B)$. Bayes's theorem yields $P(A_1/B) = 0.824$.

SELF-TEST 5 (p. 308)

1. (a) 0.42, (b) 0.94

2. Let B denote the event "the person selected earns more than $25,000," A_1 denote the event "the person selected is in management," A_2 denote the event "the person selected is in marketing," and A_3 denote the event "the person selected is in the miscellaneous category." We have $P(A_1) = 0.05$, $P(A_2) = 0.60$, $P(A_3) = 0.35$, $P(B/A_1) = 0.60$, $P(B/A_2) = 0.20$, $P(B/A_3) = 0.15$. The problem is to find $P(A_2/B)$. Bayes's theorem yields

$$P(A_2/B) = \frac{0.12}{0.2025} = 0.59$$

3. (a) $\frac{65}{144}$, (b) $\frac{33}{144}$, (c) $\frac{419}{1728}$, (d) $\frac{426}{1728}$

4. (a) Bernoulli trial model defined by $n = 50$, E is the event that O.S. lands on target on a jump, $p = 0.40$.
   (b) The outcome of any one jump does not affect the outcome of any other jump.
   (c) $P(X = 22) = C(50, 22)(0.40)^{22}(0.60)^{28}$
   (d) Yes, since both np and nq exceed 5.
   (e) $\mu = np = 20$, $\sigma = \sqrt{np(1-p)} = 3.464$.
   $P(X = 22) = P(0.43 \le z \le 0.72) = 0.0978$.

5. (a) From state

$$T_1 = \begin{array}{c} \\ 1 \\ 2 \\ 3 \\ 4 \\ 5 \end{array} \begin{bmatrix} 0 & 1 & 0 & 0 & 0 \\ \frac{2}{6} & \frac{1}{6} & \frac{3}{6} & 0 & 0 \\ 0 & \frac{2}{6} & \frac{1}{6} & \frac{3}{6} & 0 \\ 0 & 0 & \frac{2}{6} & \frac{1}{6} & \frac{3}{6} \\ 0 & 0 & 0 & 1 & 0 \end{bmatrix}$$

(b) $T_2 = \begin{bmatrix} \frac{12}{36} & \frac{6}{36} & \frac{18}{36} & 0 & 0 \\ \frac{2}{36} & \frac{19}{36} & \frac{6}{36} & \frac{9}{36} & 0 \\ \frac{4}{36} & \frac{4}{36} & \frac{13}{36} & \frac{6}{36} & \frac{9}{36} \\ 0 & \frac{4}{36} & \frac{4}{36} & \frac{25}{36} & \frac{3}{35} \\ 0 & 0 & \frac{12}{36} & \frac{6}{36} & \frac{18}{36} \end{bmatrix}$

(c) (1) $\frac{13}{36}$, (2) $\frac{3}{36}$, (3) $\frac{8}{216}$, (4) $\frac{55}{216}$

6. $\mu = 96$, $\sigma = 7.589$; (a) 0.0094, (b) 0.2358, (c) 0.3369, (d) 0.2776, (e) 0.2285

7. (a) Let x denote the probability value to be assigned to the sample point (i, j) with odd sum ($i + j$ is odd). There are 18 such sample points so that the sum of the probabilities assigned to them is $18x$. The probability value to be assigned to each of the 18 sample points with even sum is $3x$, so that the sum of the probabilities assigned to them is $18(3x) = 54x$. Since the sum of the probabilities assigned to all sample points must be 1, we have

$$18x + 54x = 72x = 1$$

$$x = \frac{1}{72}$$

If we denote the probability function to be defined by P, then P is defined by

$$P(i, j) = \begin{cases} \frac{1}{72}, & \text{if } i + j \text{ is odd} \\ \frac{3}{72}, & \text{if } i + j \text{ is even} \end{cases}$$

(b) $P(i + j = 7) = 6/72$, $P(i + j \geq 8) = 33/72$, $P(i + j < 6) = 18/72$

8. (a) 1000; (b) the second group of squirrels caught is a "close approximation" to being a random sample of the squirrel population of Bell City; (c) Yes, for two reasons: (1) the second group of squirrels caught may not be a close approximation of a random sample; (2) the maximum likelihood estimate is the best estimate in a probability sense, which is still short of "certainty."

9. (a) $S = \{(3, 7), (3, 9), (3, 11), (3, 13), (7, 9), (7, 11), (7, 13), (9, 11), (9, 13), (11, 13)\}$. P assigns $1/10$ to each sample point in S.

(b) We assume that the sample is drawn at random; that is, there is no bias in the sampling which favors certain samples being drawn over others.

(c) $\bar{x}(3, 7) = 5$, $\bar{x}(3, 9) = 6$, $\bar{x}(3, 11) = 7$, $\bar{x}(3, 13) = 8$, $\bar{x}(7, 9) = 8$, $\bar{x}(7, 11) = 9$, $\bar{x}(7, 13) = 10$, $\bar{x}(9, 11) = 10$, $\bar{x}(9, 13) = 11$, $\bar{x}(11, 13) = 12$. This yields $f(x) = P(\bar{x} = x)$ defined as follows; $f(5) = f(6) = f(7) = f(9) = f(11) = f(13) = 0.1$, $f(8) = f(10) = 0.2$.

(d) 0.4

Section 10.1 (p. 316)

1. row 1, column 1, value 1
3. row 1, column 1, value 2

5.

			Column Corporation	
			C_1 (center)	C_2 (off-center)
Row Corp.	(center)	R_1	0.5	0.65
	(off center)	R_2	0.4	0.5

Section 10.1 (p. 324)

7. Play rows 1 and 2 with probabilities $\frac{1}{5}$ and $\frac{4}{5}$, respectively; play columns 1 and 2 with probabilities $\frac{2}{5}$ and $\frac{3}{5}$, respectively; value $\frac{18}{5}$.

9. Play rows 1 and 2 with probabilities $\frac{4}{7}$ and $\frac{3}{7}$, respectively; play columns 1 and 2 with probabilities $\frac{4}{7}$ and $\frac{3}{7}$, respectively; value $\frac{26}{7}$.

Section 10.2 (p. 333)

1. Min $R(x, y) = x + y$
subject to
$x \geq 0, y \geq 0$
$6x + 3y \geq 1$
$2x + 4y \geq 1$
where $p = xm$, $1 - p = ym$,
and $x + y = 1/m$.
Max $C(s, t) = s + t$
subject to
$s \geq 0, t \geq 0$
$6s + 2t \leq 1$
$3s + 4t \leq 1$
where $r = -sn$, $1 - r = -tn$,
and $s + t = -1/n$.

Optimal strategies: play columns C_1 and C_2 with probabilities $\frac{2}{5}$ and $\frac{3}{5}$, respectively; play rows R_1 and R_2 with probabilities $\frac{1}{5}$ and $\frac{4}{5}$, respectively. Value: $\frac{18}{5}$.

3. Min $R(x, y) = x + y$
subject to
$x \geq 0, y \geq 0$
$5x + 2y \geq 1$
$2x + 6y \geq 1$
where $p = xm$, $1 - p = ym$,
and $x + y = 1/m$.
Max $C(s, t) = s + t$
subject to
$s \geq 0, t \geq 0$
$5s + 2t \leq 1$
$2s + 6t \leq 1$
where $r = -sn$, $1 - r = -tn$,
and $s + t = -1/n$.

Optimal strategies: play columns C_1 and C_2 with probabilities $\frac{4}{7}$ and $\frac{3}{7}$, respectively; play rows R_1 and R_2 with probabilities $\frac{4}{7}$ and $\frac{3}{7}$, respectively; value: $\frac{26}{7}$.

5. Since some entries in the payoff matrix are negative, add 2 to each entry to obtain a payoff matrix with positive entries. This modified game has the same optimal strategies; its value is 2 plus the value of the given game.

Min $R(x, y) = x + y$
subject to
$$x \geq 0, y \geq 0$$
$$3x + y \geq 1$$
$$x + 3y \geq 1$$
where $p = xm$, $1 - p = ym$,
and $x + y = 1/m$.
Max $C(s, t) = s + t$
subject to
$$s \geq 0, t \geq 0$$
$$3s + t \leq 1$$
$$s + 3t \leq 1$$
where $r = -sn$, $1 - r = -tn$,
and $s + t = -1/n$.

Optimal strategies: play both columns C_1 and C_2 with probability $\frac{1}{2}$; play both rows R_1 and R_2 with probability $\frac{1}{2}$. The value of the original game is 0.

7. Min $R = x + y + z$
subject to
$$x \geq 0, y \geq 0, z \geq 0$$
$$4x + 2y + 3z \geq 1$$
$$3x + y + 4z \geq 1$$
$$3x + 4y + 2z \geq 1$$
where $p = xm$, $q = ym$, $1 - p - q = zm$, $x + y + z = 1/m$.
Max $C = s + t + v$
subject to
$$s \geq 0, t \geq 0, v \geq 0$$
$$4s + 3t + 2v \leq 1$$
$$2s + t + 4v \leq 1$$
$$3s + 4t + 2v \leq 1$$
where $r = -sn$, $w = -tn$,
$1 - r - w = -vn$, $s + t + v = -1/n$.

Optimal strategies: play columns C_1, C_2, and C_3 with probabilities $\frac{1}{4}$, $\frac{1}{4}$, and $\frac{1}{2}$, respectively; play rows R_1, R_2, and R_3 with probabilities $\frac{1}{8}$, $\frac{3}{8}$, and $\frac{1}{2}$, respectively. Value: $\frac{11}{4}$.

Section 10.3

SELF-TEST 1 (p. 334)

1. (a) row 2, col. 2, value 3

1. (b) row 2, col. 3, value 2

1. (c) For both games play rows 1 and 2 with probabilities 5/11 and 6/11 respectively; play columns 1 and 2 with probabilities 5/11 and 6/11, respectively.

2. (a) Add 4 to each entry in the payoff matrix to obtain a payoff matrix with positive entries.

Min $R(x, y) = x + y$
subject to
$$x \geq 0, y \geq 0$$
$$6x + 3y \geq 1$$
$$x + 6y \geq 1$$
where $p = xm$, $1 - p = ym$,
and $x + y = 1/m$.

Max $C(s, t) = s + t$
subject to
$$s \geq 0, t \geq 0$$
$$6s + t \leq 1$$
$$3s + 6t \leq 1$$
where $r = -sn$, $1 - r = -tn$,
and $s + t = -1/n$.

Optimal strategies: play columns C_1 and C_2 with probabilities $\frac{3}{8}$ and $\frac{5}{8}$, respectively; play rows R_1 and R_2 with probabilities $\frac{5}{8}$ and $\frac{3}{8}$, respectively. Value: $\frac{33}{8}$.

Answers to Odd-Numbered Exercises and Self-Tests

2. (b) First add 2 to all entries of the payoff matrix.

Row player's program
(in terms of modified game)

Minimize $R(x, y) = x + y$

subject to

$x \geq 0, y \geq 0$

$5x + y \geq 1$

$x + 4y \geq 1$

where $p = xm$, $1 - p = ym$,

$x + y = 1/m$.

Column player's program
(in terms of modified game)

Maximize $C(s, t) = s + t$

subject to

$s \geq 0, t \geq 0$

$5s + t \leq 1$

$s + 4t \leq 1$

where $r = -sn$, $1 - r = -tn$,

$s + t = -1/n$.

Optimal strategies: Play columns 1 and 2 with probabilities $\frac{3}{7}$ and $\frac{4}{7}$, respectively; play rows 1 and 2 with probabilities $\frac{3}{7}$ and $\frac{4}{7}$, respectively; value $\frac{5}{7}$.

SELF-TEST 2 (p. 334)

1. (a) Row 2, column 2, minimax = maximin = 2. **(b)** Play rows 1 and 2 with probabilities $\frac{2}{3}$ and $\frac{1}{3}$, respectively; play columns 1 and 2 with probabilities $\frac{2}{3}$ and $\frac{1}{3}$, respectively. Value: $\frac{11}{3}$. **(c)** Play rows 1 and 2 with probabilities $\frac{1}{5}$ and $\frac{4}{5}$, respectively; play columns 1 and 2 with probabilities $\frac{2}{5}$ and $\frac{3}{5}$, respectively. Value: $\frac{13}{5}$. **(d)** Row 1, column 1, minimax = maximin = -2.

2. (a) Row player's program:

Minimize $R(x, y) = x + y$

subject to

$x \geq 0, y \geq 0$

$2x + y \geq 1$

$x + 3y \geq 1$,

where $p = xm$, $1 - p = ym$,

and $x + y = 1/m$.

Column player's program:

Maximize $C(s, t) = s + t$

subject to

$s \geq 0, t \geq 0$

$2s + t \leq 1$

$s + 3t \leq 1$,

where $r = -sn$, $1 - r = -tn$,

and $s + t = -1/n$.

Optimal strategies: Play columns 1 and 2 with probabilities $\frac{2}{3}$ and $\frac{1}{3}$, respectively; play rows 1 and 2 with probabilities $\frac{2}{3}$ and $\frac{1}{3}$, respectively; value $\frac{5}{3}$.

(b) First add 2 to all entries of the payoff matrix to obtain a payoff matrix with positive entries, but with the same optimal strategies. The value of the original game is, however, increased by 2. Optimal strategies: Play columns 1 and 2 with probabilities $\frac{3}{8}$ and $\frac{5}{8}$, respectively; play rows 1 and 2 with probabilities $\frac{3}{8}$ and $\frac{5}{8}$, respectively; value $\frac{7}{8}$.

Section 11.1 (p. 341)

1. Valid; the hypothesis forces Joe Warren into the category of frogs.

3. $C1$ is not valid; $C1$ is not forced by the hypothesis. $C2$ is not valid. $C3$ is not valid. $C4$ is not valid.

5. $C1$ is valid.

7. $C1$ is valid. $C2$ is not valid.

Section 11.2 (p. 346)

1. The problem here is that what was actually proved is not the same as what was required to be proved. The proof given showed that a unique line PE parallel to L is obtained by the argument developed. This does not show that a line M parallel to L obtained from some other construction must be the same as PE. Indeed, other constructions of parallel lines to L at P are known.

Section 11.3 (p. 353)

1. **Model 1.** Zogs: ordered pairs (points) (1, 1), (2, 1), (2, 2), (1, 2); Glob: The set determined by any two of the points given (see Figure 1).

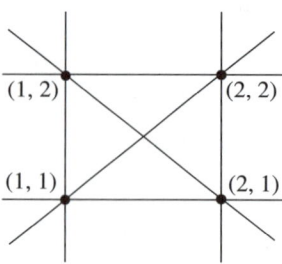

Figure 1

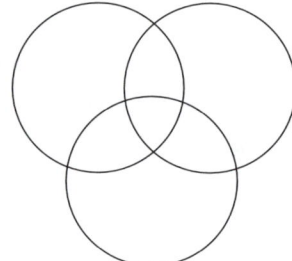

Figure 2

Model 2. Zogs: points inside three overlapping circles that intersect (the globs) as shown in Figure 2.

3. No; the argument does not establish that s and q are distinct.

5. No; that the argument given in 3 is invalid does not by itself exclude the possibility that there is a valid argument establishing 3.

7. (a) **Model 1.** For blobs, consider three distinct points p, q, and r; for neighborhoods consider the sets $P = \{p, q\}$, $R = \{p, r\}$, $Q = \{q, r\}$ (see Figure 3).

  Model 2. For blobs consider three distinct points p, q, and r; for neighborhoods consider the sets $P = \{p, q\}$, $Q = \{p, r\}$, $R = \{p, r\}$, and $S = \{p, q, r, s\}$ (see Figure 4).

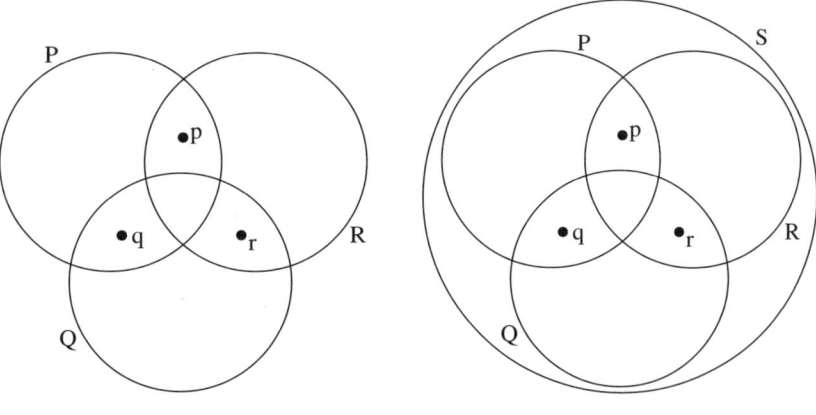

Figure 3   **Figure 4**

(b) C1: "There are at least three neighborhoods," is a theorem of neighborhood theory.

Proof:

Statement	Justification
1. Let p, q and r denote blobs.	1. P1
2. Let R denote a *nbh* containing p but not q, and Q a *nbh* containing q but not p.	2. P3
3. Let P denote a *nbh* containing p and q.	3. P2

4. P, Q, and R are distinct nbhs since they do not contain the same blobs.

(c) C2 is not a theorem. If it were a theorem, then in every model all blobs, as interpreted in the model, could not be contained in any one nbh, as interpreted in the model. That this is not the case is shown by Model 2.

(d) C3 is not a theorem.

9. $P1c$ and $P4c$ clearly satisfy the requirements of $P1$ and $P4$ of Glob Theory. As to $P2c$, let us note that the solutions of (11.1) and (11.2) are given by $x = 3 + 2y - z$, $y = 1 + 3z$, z arbitrary; the solutions of (11.1) and (11.3) are given by $x = 3 + 2y - z$, $y = 3z + 5/2$, z arbitrary; and the solutions of (11.2) and (11.3) are given by x arbitrary, $y = -13/5 + 3x/5$, $z = -7 - 3y + 2x$. Any pair of equations has numerous common solutions and for any two of them there is at least one equation that satisfies them; $P2$ is thus satisfied. Since the system of three equations (11.1), (11.2), and (11.3) has no solution, $P3c$ satisfies the requirements of $P3$.

Since $P1c - P4c$ satisfy the requirements of $P1 - P4$, I_c is a model of Glob Theory.

11. No; $P3e$, for any equation there is a solution (of a pair of equations) which does not satisfy it, is not satisfied since $(3, 1, -1)$ is a common solution of all three equations. Thus $P3e$ does not satisfy the requirements of $P3$ and I_e is not a model of Glob Theory.

Section 11.4 (p. 358)

1. No; as noted at the beginning of this section Mumbo-Jumbo Theory is inconsistent because the theorem "there is a jumbo which is contained by three mumbos" contradicts postulate $P5$: For any jumbo there are exactly two mumbos containing it.

3. No; $C1$, there are at least three neighborhoods in Neighborhood Theory (Sec. 11.3, Ex. 7, p. 354) is a theorem of Neighborhood Theory which contradicts $P4$.

Section 11.5 (p. 364)

1. No; the parallel postulate problem is to show that Euclid's parallel postulate could be deduced from his other postulates or replaced by a simpler equivalent statement that was simple and self-evident.

3. No; Euclidean geometry is not the only possible model for space.

5. Agree.

7. No; a postulate may be true, false or neither true nor false (All x's are y's, for example). It is a statement which with others form the foundation of a system from which valid conclusions are deduced.

Section 11.6 (p. 366)

1. (a) Disagree, (b) Disagree, (c) Agree
   (d) Disagree, (e) Disagree, (f) Disagree
   (g) Disagree, (h) Disagree.

3. The model's assumptions (or postulates).

Section 11.8

SELF-TEST 1 (p. 369)

1. The arguments in all three cases are not valid in that $C1$, $C2$, and $C3$ are not forced on us by the hypothesis (see Figure 5).

2. More than the three cases cited are possible and would have to be considered. Consider, for example, the following possibility: angle 1 + angle 4 > 180° and angle 2 + angle 3 < 180°. This does not contradict the relationship that the sum of the four angles equals 360°.

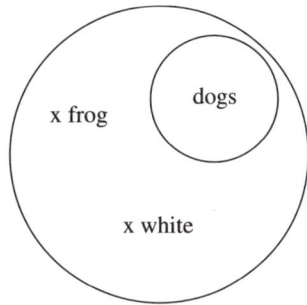

Figure 5

3. No; Jane is in trouble with her first statement which introduces four zogs based on Conjecture 1 that there must be four zogs in Glob Theory. Conjecture 1 has not been proved a theorem and, in fact, is not a theorem since models of Glob Theory with three zogs have been exhibited.

4. (a) Yes; In answer to Exercise 1 of Section 11.3 (p. 353) we have an ideal model of Glob Theory with four zogs.
   (b) No; interpretation I_b of Glob Theory (p. 349) is a model with three zogs.
   (c) Yes

5. (a) Yes; Theorem 2 of Glob Theory says that Glob Theory contains at least three globs so that $P5(a)$ is in agreement with this theorem. Since Glob Theory is consistent, so is Gladys' extension.
   (b) No; it can be proved a theorem from the other postulates.
   (c) Yes; I_b (p. 349) is a model with three globs.

6. No; whether it's contrary to the nature of physical space or not is irrelevant to the question of its consistency. Its consistency depends on whether it itself contains contradictory statements. A model for this system in terms of a sphere is obtained by interpreting point as a point on the surface of the sphere and line as a great circle on the sphere. This model establishes that this geometry (called Riemannian double-elliptic geometry) is as consistent as three dimensional Euclidean geometry.

7. No   8. No   9. No

10. (a) In the sense of being a valid consequence of general relativity.
    (b) It would make clear a limitation of general relativity which might ultimately lead to its modification.

(c) It throws doubt on the validity of the new gyroscope finding in the sense of its credibility.

11. (a) No; the validity of the predictions of quantum theory, outlandish or not, is a matter of whether they follow in a deductive-logical sense from the hypothesis of quantum theory.

(b) The creation of a Bose-Einstein condensate would help settle the question of how realistic are some of the "outlandish" predictions of quantum theory.

SELF-TEST 2 (p. 372)

1. $C1$: valid; $C2$: valid; $C3$: not valid; $C4$: not valid; $C5$: not valid; $C6$: not valid

2. The statement to be proved presupposes that the diagonals $\overline{AC}$ and $\overline{BD}$ meet at a point E. While visually evident, there is nothing in the hypothesis to justify this assumption.

3. (a) **Model 1:** Euclidean plane geometry obtained by interpreting luk and yuk as point and line in the usual way.

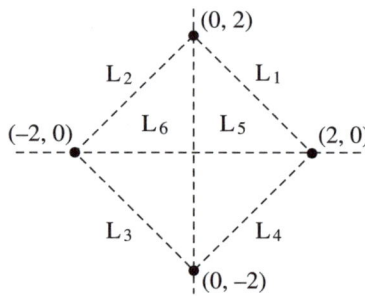

Model 2: luks: the points with coordinates $(2, 0)$, $(0, 2)$, $(-2, 0)$, and $(0, -2)$ (see Figure 6). Yuks: sets of 2 of these points. There are 6 yuks as indicated in Figure 6: $L_1 = \{(2, 0), (0, 2)\}$, $L_2 = \{(0, 2), (-2, 0)\}$, $L_3 = \{(-2, 0), (0, -2)\}$, $L_4 = \{(0, -2), (2, 0)\}$, $L_5 = \{(0, 2), (0, -2)\}$, $L_6 = \{(-2, 0), (2, 0)\}$.

Figure 6

(b) $-(n)$. We run into difficulty when attempting to construct a model for Yuk Theory with fewer than four luks and six yuks. This by itself is not conclusive since it may simply mean that we've not been clever enough and there are such models, but it does suggest that $C2-C6$ are theorems, which they are. The construction of Model 2 with six yuks shows that $C7$ is not a theorem. As to $C1$, it holds in Model 2, but this

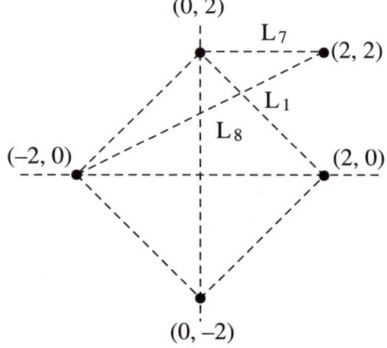

Figure 7

is not conclusive for the general case. We might try to play a bit with Model 2 and come up with a point which is contained by only one set of two points. Consider (2, 2), for example and define $L_7 = \{(0, 2), (2, 2)\}$. (See Fig. 7.) If we add these to Model 2 and leave it at that, the modified system does not satisfy *P5* (For L_1 and (2, 2), there is no L_i containing (2, 2) which is parallel to L_1). We might try to close this gap by introducing $L_8 = \{(-2, 2), (2, 0)\}$; but then we have a violation of *P5* from another point of view (For L_8 and (0, 2) there are two yuks, L_1 and L_5 containing (0, 2) which are parallel to L_8). After a while with this sort of unsuccessful playing around it begins to seem that *C1* might be a theorem. It is.

Theorem 1. Every luk is contained by at least two yuks.

Statement	**Justification**
1. Let p denote any luk in the system. We shall show the p is contained by at least two yuks.	1. *P2*.
2. Let q denote another luk in the system.	2. *P2*.
3. Let L_1 denote a yuk containing p and q.	3. *P3*.
4. Let r denote a luk not contained by L_1.	4. *P4*.
5. Let L_2 denote a yuk containing p and r.	5. *P3*.

In summary, we have that p, a "representative" luk in the system, is contained by distinct yuks L_1 and L_2; L_2 differs from L_1 because it contains r which is not contained by L_1.

4. Model 1 in answer to 3(a) shows that Yuk Theory is as consistent as Euclidean geometry.

5. No; *P6* contradicts *C6*, there are at least six yuks, which is a theorem in the system.

6. (a) No; *P5* is not satisfied. All yuks (great circles) intersect; given a great circle and point not on, there is no great circle passing through the point parallel to the given one. Moreover, *P3* is not satisfied.

   (b) Yes

7. The question reduces to this: Is there a model for postulates *P1–P4* for which *P5* is not satisfied? If so, then the answer is yes, *P5* is independent. Let us go back to Model 2 of Yuk Theory given in answer to 3(a) and add to it the luk (2, −2) and yuks $L_7 = \{(2, 2), (0, 2)\}$, $L_8 = \{(2, 2), (-2, 0)\}$, $L_9 = \{(2, 2), (0, -2)\}$, and $L_{10} = \{2, 2), (2, 0)\}$ (see Figure 3). This system satisfies postulates P1-P4 of Yuk Theory, but not *P5*. For yuk L_8 and luk (0, 2) there are two yuks, L_1 and L_5 containing (0, 2) which are parallel to L_8.

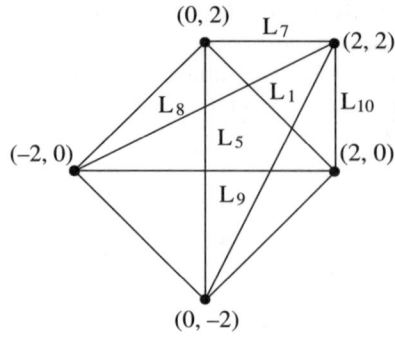

Figure 3

8. The validity of the conclusions of Einstein's general theory of relativity has been established by mathematical reasoning (deductive-logical argumentation) and is therefore beyond dispute. The results obtained by the refined radar technique alluded to, and by any other experimental means, for that matter, pertain to the truth of the conclusions of Einstein's theory. The debate is over the realism of the conclusions of Einstein's theory, not their validity in the logical-deductive sense. The term "validity" as used in this quote is synonymous with realism.

9. The statement quoted means that a valid conclusion of the hypothesis of Einstein's theory of relativity was shown to be true by the experiment and observation means described. Whenever a fundamental valid conclusion of the hypothesis of a theory is shown to be true, this adds support to that hypothesis as a realistic description of the phenomenon in question.

10. (a) No; the term "invalid" here should be replaced by "unrealistic." Data obtained by experimental means is irrelevant to the validity issue in terms of its deductive-logical sense.

    (b) As noted above, it is relevant to the issue of how realistic such theories are.

Index

A

Abelian Group, 358
Abstract postulate system, 339
    interpretation of, 339
Addition of matrices
    definition of, 162
    properties of, 169-170
Addition principle, 232
Advertising-media-selection problem, 112-114
Air-pollution-control problem, 99-100
Assignment problem, 104
Associative property:
    of matrix addition, 169
    of matrix multiplication, 171
Assumption, 114, 208-210, 213-216, 239-241, 244-245, 249-252, 272, 274, 302-304, 306, 308, 310, 337, 403-405, 407-408, 411-412, 418-419, 430
Austin Company, 78, 80-81, 84, 86-87, 89, 95, 118

B

Bank-portfolio-management problem, 100-101
Bass, F., 114
Bayes's theorem:
    application to a marketing situation, 262
    application to a reliability question, 261
    formulation of, 260
Bernoulli, Jacob, 282
Bernoulli trials, 282, 284, 298
Bottom-line Bob, 89
Broaddus, A., 101

C

Cancellation property:
    in connection with matrices, 172
Central limit theorem, 300
Charnes, A., 150
Churchill, N., 181

Closure property
  of matrix addition, 169
  of matrix multiplication, 162, 170-171, 268
Cohen, K. J., 101
Column-selection principle, 141, 150
Combination, 224, 235-236, 274
Commutative property:
  in connection with matrix multiplication, 171
  of matrix addition, 169
Complementary events, 219, 243
Conditional probability, 255-256
Conjecture, 286, 350-351, 353-355, 370-371, 429
Cooper, L., 150
Corner point, 81, 83-87, 97, 120, 122, 130, 139, 142, 154-155, 190
Corner-point interpretation of the simplex method, 153-155
Corner-point method, 91, 130, 190, 388, 399

D

Dantzig, G., 139
Deduction, 339
Deductive reasoning, 337, 339
De Moivre, A., 299
De Moivre-Laplace theorem, 298, 300
Diet problems, 97, 99
Disjoint events, 218-219
Distributive property: of matrices, 172
Dorfman, R., 187
Duality theorem, 136
Dual of a linear program, 133

E

Ecap University, 20, 108, 129, 236, 238-239, 265, 302, 305-306, 366
Ehrenberg, A. S. C., 271-272
Einstein, A., 372, 374-375
Empty set, 196
Engel, J., 113
Equally likely outcomes, 206, 213-214, 239
Equipment purchase problem, 114
Euclid, 358-359
Euclid's Elements, 360
Euclid's parallel postulate, 359-360

Events:
  complementary, 219-221, 243
  disjoint or mutually exclusive, 218
  independent, 279-281
  pairwise disjoint, 218-219, 260
  sample point, 201
  sample space of, 199-201
Expected value:
  of the column player, 322
  of the row player, 317-318
Experiment, 198

F

Feasible point, 82, 118, 130, 136, 150, 189-190, 399-400
Final-demand matrix, 185-186
Finite sample space, 201

G

Game:
  fair, 79, 226, 262-264, 280, 314, 321, 403, 410-411
  matrix, 30, 35, 161-174, 176-183, 185-187, 192-193, 267-271, 273, 275-278, 308-309, 312, 314-317, 321-325, 327, 329, 331-334, 397, 423-425
  maximin strategy for, 313-314
  minimax strategy for, 313-314
  n-person, 311
  probabilistic strategies for, 316-320
  solution by linear programming methods, 325-334
  two-person, 312
  two-person, zero-sum of matrix type, 212
  value of, 314
Gauss, K. F., 299-300, 362
Gensch, D., 113
Glob theory, 348-357, 370-371, 374, 427-429

H

Hadley, G., 150
Hammer, F. S., 101

Hypothesis, 70-72, 83, 337-343, 345, 369, 372-373, 426, 429-430, 432

I

Identity matrices, 171
Income-consolidation problem, 38, 181
Indefinite integral
   Independent events, 280-281
   intuitive view of, 279-280
   probability models based on, 282-286
   properties of, 281
Indicator row, 140-142, 144-145, 147, 150-151, 156-159, 331, 333
Inequalities, 48-51
Inequalities, multivariable: 52-61
Inner product of a row and column, 162
Input-output analysis, 183-186
Integer program:
   definition of, 118
   problems leading to, 102-106, 116-128
Intersection of sets, 197
Invalid argument, 341
Inverse: of a matrix, 162, 164, 170, 172, 185

K

Kantorovich, L. V., 77-78
Kaplan, R. S., 181
Kotler, P., 114

L

Laplace, P. S., 299-300
Least-squares line of best fit, 12, 14
Leontief, W., 183, 187
Line, 1-6
Linear program
   assumptions in the formulation of, 89-93
   corner point of, 82
   definition of, 80
   dual of, 133-134, 191
   existence of solution of, 136
   feasible point of, 136, 189
   objective function of, 80

   solution by the corner-point method, 81-87
   solution by the simplex method, 138-160
   standard form of, 131
Linear programming applications:
   advertising-media-selection problem, 112-114
   air-pollution control, 99
   bank-portfolio management, 100
   diet problem, 97
   game theory, 131, 311-312, 316, 331
   investment, 101, 108, 110, 129, 189
   job assignment, 104
   marketing mix, 111
   production planning, 95
   transportation, 102-103, 122
Linear programming shortfall, 112
Lobachevskian geometry, 360-363
Lobachevsky, N., 362
Lobachevsky's parallel postulate, 361-362
Lonsdale, R., 114

M

Markov chain:
   applied to a problem in genetics, 274
   homogeneous, 267-268
   intuitive background for, 266
   matrices of transition probabilities for, 267-269
   model for a financial situation, 268
   model for consumer brand-choice behavior, 271
Massey, W. F., 272
Math models for a vacation trip, 65
Mathematical model building process, 69, 93
Matrix:
   addition, 123, 162, 169, 172, 232, 237
   approach to solving systems of linear equations, 177-179
   equality, 161
   final-demand, 185-186
   identity, 170-174, 176-177, 244, 345, 358
   input-coefficient, 183, 185-187, 193, 397, 401
   inverse of, 170, 172-177, 191-192, 358
   inversion, 30, 173-179, 181-183, 185, 192
   m by n., 161
   multiplication, 162, 164, 169-172, 192, 229-230, 234-235, 237-238, 250, 268

negative of, 170
nonsingular, 172
of transition probabilities, 267-271, 275-276, 278, 309
output, 77, 95, 183-188, 193, 203, 213-216, 253, 256-257, 260, 265, 287, 302-304, 388, 397, 401, 403-404
payoff, 312-317, 321-325, 327, 329, 331-335, 423-425
properties, 169-171
singular, 172
size of, 161
square, 162
subtraction, 162
zero, 1, 170
Matrix applications
income consolidation, 38, 180-181
input-output analysis, 183-186
Maximum-likelihood estimate, 248-249, 251
McGraw-Edison problem, 113
Minch, R., 181
Morgenstern, O., 311
Morrison, D. G., 272
Multiplication principle, 229-230
Mumbo-Jumbo theory, 354-355, 357, 428
Mutually exclusive events, 221

N

Neighborhood theory, 354, 358, 427-428
Nemchinov, V. S., 78
Normal curve:
  for approximating Bernoulli trial probabilities, 295-299
  standard, 288
  structure of, 287-288
Normal equations, 16
Null set, 196

O

Objective function, 80
One-row, 24, 26-28, 31
Output matrix, 185-186

P

Pairwise disjoint events, 218-219
Parallel postulate, 346, 359-362, 364, 371, 428
Particle in random motion, 266, 276-277, 309
Payoff matrix, 312, 314-317, 321-325, 327, 329, 331-334, 423-425
Permutation, 233, 235-236
Petri, E., 181
Pivoting
  in matrix inversion, 174-177
  in solving systems of linear equations, 22-28
  in the simplex method, 141-151
Pivot-value, 22, 141
Population-estimation problem, 245-249
Postulate
  independent, 222, 266, 279-281, 283, 302-303, 306-307, 309, 357-358, 363-364, 371, 374, 405, 410-411, 413-414, 417, 431
Probability:
  conditional, 255-256
  function, 204
  odds, 226
  posterior, 257, 261
  prior, 257, 261, 366
  relative-frequency interpretation of, 213, 215-216, 252, 262, 264-265, 275, 404, 416
  subjective interpretation of, 221-224
Probability applications:
  consumer brand-choice behavior, 271
  credit-state analysis, 268
  decision-making based on random sampling, 242-243
  estimated gas reserves, 225
  marketing success prediction, 262
  medical test reliability, 261
  Mendelian genetics, 274
  mutation study with *Drosophila*, 287
  particle in random motion, 266, 276-277, 309
  population estimation, 245-249
Probability model:
  assumptions in the formulation of, 206-213
  based on conditional probability, 253-256
  based on equally likely outcomes, 205-206, 239-249
  based on independent events, 282-285
  definition of, 204
  Markov-chain type, 266-276

Problems leading to 0-1 integer programs, 126-128
Problems requiring solutions in integers, 116-124

R

Ramune Company
   dual program of, 133
   linear program of, 96-97
   production planning problem of, 95
   problem solution of, 97, 139-141
Random process, 198
Rasa Company:
   advertising-media-selection problem, 112
   sales volume in terms of advertising expenditures, 12
   Assignment problem, 104
Reflective Ramune., 89
Relative frequency
   definition and properties of, 198, 202

S

Saddle point, 314
Saddle value, 314
Sample point, 201
Sample space, 199-201
Samuelson, P. A., 187
Savage, L. J., 224
Schlaifer, R., 224
Service-charge-allocation problem, 35, 166, 180
Sets, 195-197
Simplex method:
   corner-point interpretation of, 153-155
   procedures for maximum linear programs, 139-151
   procedures for minimum linear programs, 155-159
Slack variable, 140
Slope: of a line, 2
Solow, R. M., 187
Standard form of a linear program, 131
Steinberg, D., 150
Subjective probability, 221-223, 226, 405, 413, 417
Subtraction: of matrices, 162
System of linear equations
   normal equations, 16
   solution by elimination-of-variable method, 6-10
   solution by matrix inversion, 177-179
   solution by tableau method, 20-34, 41-47
System of linear inequalities, 56-57

T

Tableau method, 20-22, 26-30, 34, 139, 141, 174
Television-tube-production problem, 213, 253-255, 257
Traffic-network-flow problem, 44
Transportation problem, 102, 122

U

Union of sets, 196-197
Universe set, 196

V

Valid proof, 337
Validity, 69, 72, 212, 252, 261, 316, 337-341, 343, 345, 347, 349, 351, 353, 355, 357, 359, 361, 363, 365, 367, 369, 371-375, 387-388, 398-399, 401-402, 413-414, 420, 430, 432
Von Neumann, J., 311
Von Neumann's duality theorem, 132, 136-137, 147, 190-191

W

Warshaw, M., 113
Williams, T. H., 181

Y

Yuk theory, 373-374, 430-431

Z

Zero matrix, 170, 193, 397
Zero-one form, 26-28, 31-34, 41-43, 45